Materials Fundamentals of Molecular Beam Epitaxy

Shown on the front cover are key figures taken from Chapters 2-7. Clockwise from upper left they are:

Figure 2.2: Temperature-dependent vapor pressures of monomeric and dimeric Si.

Figure 3.7: Composition and temperature-dependent phase boundaries for GaAs.

Figure 4.3: Experimental and theoretical regular solution parameters for various pseudobinary III/V alloys.

Figure 7.8: Temperature-dependent nonequilibrium impurity partition coefficients during epitaxy.

Figure 6.25: Layer-by-layer oscillations in terrace smoothness during epitaxy.

Figure 5.8: Critical layer thicknesses for given lattice parameter mismatches that separate coherent from semi-coherent films.

Materials Fundamentals of Molecular Beam Epitaxy

Jeffrey Y. Tsao
Sandia National Laboratories
Albuquerque, New Mexico

ACADEMIC PRESS, INC.
Harcourt Brace Jovanovich, Publishers
Boston San Diego New York
London Sydney Tokyo Toronto

This book is printed on acid-free paper. ♾

ACADEMIC PRESS, INC.
1250 Sixth Avenue, San Diego, CA 92101-4311

United Kingdom Edition published by
ACADEMIC PRESS LIMITED
24-28 Oval Road, London NW1 7DX

Tsao, Jeffrey Y.
Materials fundamentals of molecular beam epitaxy / Jeffrey Y. Tsao.
p. cm.
Includes bibliographical references and index.
ISBN 0-12-701625-2
1. Molecular beam epitaxy. 2. Surfaces (Physics) 3. Surfaces (Technology) I. Title.
QC611.6.M64T78 1992
620'.44--dc20 92-33038
CIP

Printed in the United States of Americ

92 93 94 95 96 97 BB 9 8 7 6 5 4 3 2 1

In loving memory
of our extraordinary son
Evan

Contents

Preface

The technology of crystal growth has advanced enormously during the past two decades. Among these advances, the development[1] and refinement of molecular beam epitaxy (MBE) has been among the most important. Crystals grown by MBE are more precisely controlled than those grown by any other method, and today they form the basis for the most advanced device structures in solid-state physics, electronics, and optoelectronics. As an example, Figure 0.1 shows a vertical-cavity surface emitting laser structure grown by MBE.

Broadly stated, MBE is simply crystallization by condensation or reaction of a vapor in ultra-high vacuum (UHV). The *extremeness* of the UHV environment, however, has several important consequences, both for the device physicist interested in the properties of the grown material, and for the materials or surface scientist interested in the growth process itself.

First, MBE surfaces are extremely clean (even cleaner than those commonly studied by surface scientists). Base pressures are typically $\approx 5 \times 10^{-11}$ Torr, near the background-x-ray-induced $\approx 2 \times 10^{-11}$ Torr detection limit of conventional ionization gauges. Even then, the residual gas is composed mainly of H_2, which is largely benign. Partial pressures of hydrocarbons, which are less benign, are typically less than the $\approx 1 \times 10^{-14}$ Torr detection limit of common quadrupole mass spectrometers, particularly when augmented by the now-standard liquid-nitrogen-cooled cryoshrouds. As a result, the device physicist can grow very-high-purity crystals of controlled composition, and the materials/surface scientist can study intrinsic crystal growth apart from extrinsic effects due to impurities or contamination.

Second, crystal growth occurs via the reaction and condensation of molecules that arrive at the surface via molecular, rather than viscous or diffusive, flow. In other words, molecules don't collide with one another enroute to the substrate, and molecules that miss or leave the substrate

[1] A.Y. Cho and J.R. Arthur, "Molecular beam epitaxy," *Prog. Solid State Chem.* **10**, 157 (1975).

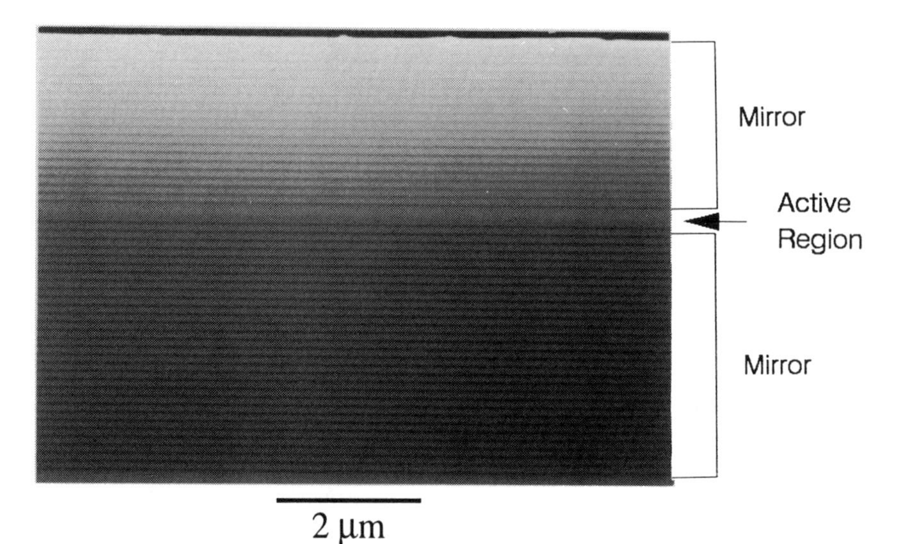

Figure 0.1: Vertical-cavity surface emitting laser structure grown by MBE. The structure is composed of a set of electrically pumped light-emitting GaAs/$Al_xGa_{1-x}As$ quantum wells sandwiched between two sets of layered AlAs/$Al_xGa_{1-x}As$ superlattice interference mirrors.[a]

[a]Photo courtesy of T.M. Brennan and B.E. Hammons, Sandia National Laboratories, Albuquerque, NM 87185-5800.

are pumped away nearly immediately. As a result, the device physicist can grow multilayered structures with extremely abrupt interfaces, and the materials/surface scientist can study microscopic processes occuring on the surface apart from the diffusion-controlled mass and/or heat transport to and from the surface that complicate other forms of crystal growth.

Third, the growing surface is accessible to observation using powerful real-time surface-science diagnostics which require high vacuum. For example, reflection high-energy electron diffraction (RHEED) is routinely used to monitor the structure and microstructure of growing surfaces,[2] reflection mass spectrometry (REMS) and modulated beam mass spectrometry (MBMS) can be used to monitor the chemistry of growing surfaces, and reflectance difference spectroscopy (RDS) can be used to monitor the composition and optical properties of growing surfaces. As a result, the device physicist can control and reproduce the state of the surface (and the subse-

[2]By structure we mean the crystallography of defect-free surfaces; by microstructure we mean the distribution of point and line defects that interrupt that otherwise perfect crystallography.

quent quality of the grown crystal) very precisely, and the materials/surface scientist can study directly the real-time evolution of surface structure, microstructure, and composition.

For all these reasons, MBE is interesting both to device physicists as well as to materials and surface scientists. We do not exaggerate when we note, however, that historically it has been *more* interesting to device physicists. Device physicists have provided, and continue to provide, most of the impetus for research in MBE. As a result, today's MBE practitioners have an enormous number of device "customers" to satisfy. Often, they are hard-pressed to keep abreast of the latest advances in technology, much less explore more fundamental aspects of their craft. Indeed, the technology of MBE is itself rapidly evolving, as is evident from recent monographs reviewing its current state of the art.[3] Figure 0.2 shows, for example, a schematic of a modern commercial MBE system with all of the advanced hardware required for operation in a production environment.

This book is intended to begin to bridge the gap separating MBE practitioners from the more fundamental aspects of their craft by gathering together in a coherent manner the basic materials science principles that apply to MBE. It has two aims. First, it aims to show how the various aspects of MBE "fit" into the perspective of materials science. For this reason, this book may be a useful supplement to intermediate or advanced courses in materials science. Second, it aims to treat the most important aspects of MBE in such depth as to benefit both advanced graduate students as well as professional researchers. It does not aim to discuss superficially all aspects of MBE, but rather comprehensively the most basic materials science aspects of MBE, and particularly those aspects that add richness and insight into other methods of crystal growth. Because MBE is the simplest and most basic method of crystal growth, an appreciation of MBE adds richness and insight into virtually all other methods. For this reason, this book may also be a useful supplement to intermediate or advanced courses in crystal growth.

The book lays heavy emphasis on the statistical and thermodynamic aspects of MBE, although some kinetic aspects are also treated. This choice of emphasis has been unavoidably colored by my own personal preferences and interests, and circumscribed by limits to my own knowledge and technical competence. For example, I have deliberately steered clear of many important topics related to the microscopic physics and chemistry of sur-

[3]E.H.C. Parker, ed., *The Technology and Physics of Molecular Beam Epitaxy* (Plenum Press, New York, 1985); E. Kasper and J.C. Bean, Eds., *Silicon Molecular Beam Epitaxy*, Vols. I and II (CRC Press, Boca Raton, 1988); and M.A. Herman and H. Sitter, *Molecular Beam Epitaxy: Fundamentals and Current Status* (Springer-Verlag, Berlin, 1989).

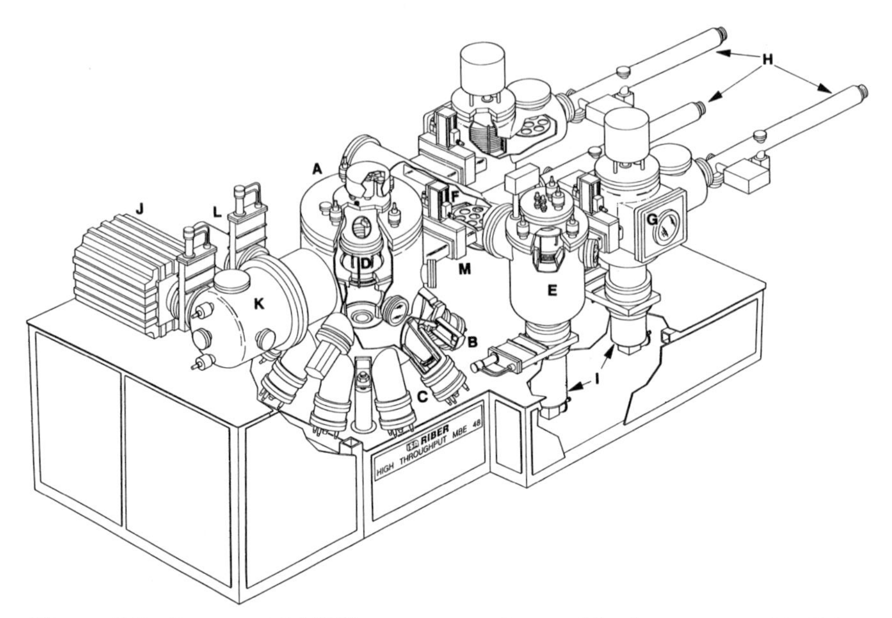

Figure 0.2: Commercial MBE system manufactured by Instruments SA, Riber Division.[a] The system is intended to perform epitaxial growth simultaneously on multiple wafers via evaporation of elemental sources in ultra-high vacuum (UHV). Some of the major parts of the system are labeled: (A) a UHV growth chamber, (B) a shutter for an evaporation source, (C) an evaporation source, (D) a rotating platten station for mounting multiple wafers, (E) a preparation station for wafer degassing, (F) a shuttle mechanism for transporting wafer plattens, (G) a load-locking chamber for transferring plattens between atmosphomeric pressure and UHV, (H) rack-and-pinion mechanisms for transfer of plattens, (I) cryopumps for maintaining UHV, (J) an ion pump, also for maintaining UHV, (K) a sublimation well, (L) another cryopump, and (M) a rectangular gate valve.

[a] Photo courtesy of Riber Division, Instruments SA, 6 Olsen Avenue, Edison, NJ 08820-2419.

faces, such as atomic bonding configurations and reconstructions or atomic mechanisms for adatom migration. However, I treat in great detail the macroscopic materials science principles governing surface morphology and its evolution during MBE.

The book assumes an understanding of solid-state physics and materials science at the introductory graduate level. It is by no means "easy reading," and will need to be supplemented, at times, by intermediate or advanced textbooks in materials science such as those by Haasen[4] or Christian.[5] This unfortunately appears so despite my efforts to make liberal use of clarifying footnotes, exercises, and case studies of technologically important materials systems. Most readers will find the barrier to entry into this book somewhat high, but hopefully will exit from it very nearly at the forefront of current research.

Unlike many other books, this one attempts to cite important original articles. I have two reasons for doing so. First, such citations are the scientist's only formal way of respecting intellectual debts. Hence, I apologize in advance for oversights. They are not intentional, but are caused either by ignorance or by the "obliteration by incorporation" phenomenon,[6] in which the origin of a piece of knowledge becomes obscured as it is incorporated into the existing body of common knowledge. Second, such citations make it easier for students to enter new research areas. Through the use of citation indices,[7] novices in a particular research area can usually bring themselves up to date fairly quickly by searching for current articles which cite in common a few important original works. My experience is that such bibliographic searching, which goes forward in time, is an important complement to the usual form of bibliographic searching, which goes backward in time. In a sense, original articles *define* current research areas, do so better than current articles, and become outdated much less quickly.

Throughout, I have imposed on the book my own peculiar organizational structure. The book is not arranged according to the type of crystal that is being grown (e.g., according to whether the crystal is IV-IV, III-V, II-VI, or metallic), the way an advanced "users" manual might be. Rather, it is arranged according to whether the materials science concepts involve mainly bulk phase equilibria, thin film structure and microstructure, or surface morphology and composition.

Part I contains most of the thermodynamic foundations of MBE, although thermodynamic arguments and ideas will also be sprinkled liberally throughout the rest of the book. Crystal growth is, of course, simply a phase transformation (albeit delicately controlled), and hence the thermodynamics of MBE is in large part the thermodynamics of bulk phase equilibria. Chapter 1 provides an introduction to basic thermodynamic concepts; Chapter 2 discusses equilibria between elemental phases and Chapter 3

---

[4] P. Haasen, *Physical Metallurgy* (Cambridge Univ. Press, Cambridge, 1978).

[5] J.W. Christian, *The Theory of Transformations in Metals and Alloys*, 2nd ed. (Pergamon Press, Oxford, 1975).

[6] R.K. Merton, *Social Theory and Social Structure* (The Free Press, New York, 1968), pp. 26-28, 35-38.

[7] Citation indices, now quite common and easy to use, were pioneered by E. Garfield; see, e.g., E. Garfield, *Citation Indexing — Its Theory and Application in Science, Technology and Humanities* (John Wiley & Sons, New York, 1979).

discusses equilibria between alloy phases.

Part II gives elementary descriptions of thin films grown either homoepitaxially, on substrates of the same material, or heteroepitaxially, on substrates of a different material.[8] In particular, it describes how the structure and microstructure of such films depend on substrate, through what are called "epitaxial constraints." Chapter 4 discusses short- and long-range ordering and phase decomposition in epitaxially constrained thin film alloys, and Chapter 5 discusses the coherency/semicoherency transition during lattice-mismatched epitaxy.

Part III deals with surfaces. Chapter 6 discusses the morphology of surfaces, both equilibrium (in the absence of growth) and nonequilibrium (in the presence of growth). Chapter 7 discusses the composition of surfaces, again both equilibrium and nonequilibrium.

On a more technical note, this book was generated on an IBM-compatible personal computer using mostly the following software: Epsilon[9] for programming and text editing, LaTeX[10] for formatting and typesetting, and Genplot[11] for numerical calculations and graphics. I highly recommend all three. They have enabled me to "program" virtually every aspect of this book, and to make both slight and massive changes easily and quickly.

Finally, it gives me great pleasure to acknowledge: colleagues and friends too numerous to list who have either indirectly, through what they have taught me, or directly, through thoughtful comments, contributed so much to this book; colleagues who read and criticized early chapter drafts, Harry Atwater, Dave Biegelsen, Scott Chalmers, Ben Freund, Kevin Horn, Tom Klitsner, Jim Mayer, Leo Schowalter, Brian Swartzentruber, and Professor David Turnbull; an understanding and supportive management at Sandia National Laboratories, Paul Peercy, Tom Picraux, Harry Weaver, and Del Owyoung; my loving parents, Ching and Matilda Tsao; Frances Koenig, who taught me to value; and my wife Sylvia and son Emil, who taught me value.

Jeffrey Y. Tsao
Albuquerque, New Mexico
June, 1992

---

[8]See, e.g., E.G. Bauer, B.W. Dodson, D.J. Ehrlich, L.C. Feldman, C.P. Flynn, M.W. Geis, J.P. Harbison, R.J. Matyi, P.S. Peercy, P.M. Petroff, J.M. Phillips, G.B. Stringfellow and A. Zangwill, "Fundamental issues in heteroepitaxy – a Department of Energy, Council on Materials Science Panel Report," *J. Mater. Res.* **5**, 852 (1990).

[9]Lugaru Software Ltd., 5740 Darlington Road, Pittsburgh, PA 15217.

[10]L. Lamport, LaTeX*: A Document Preparation System* (Addison-Wesley, Reading, MA, 1986); and D.E. Knuth, *The TeXBook* (Addison-Wesley, Reading, MA, 1984).

[11]Computer Graphic Service, 221 Asbury Road, Ithaca, NY 14850.

List of Figures

List of Tables

Part I

Bulk Phase Equilibria

In this first part, we describe the thermodynamic properties of bulk condensed and vapor phases. Those properties determine the pressures and temperatures at which various phases or phase mixtures are more or less stable with respect to each other, and ultimately define the window in growth conditions within which MBE is preferred over condensation of unwanted phases. The windows are often sharp and unforgiving, and are nearly always the primary consideration in choosing growth conditions. During GaAs MBE, e.g., As_4 or As_2 overpressures must be higher than a critical, temperature-dependent value, because otherwise the surface readily decomposes into Ga liquid and As_2 vapor.

We begin, in Chapter 1, with a concise, general description of free energies. Free energies are the metrics that determine the relative stability of phases when these phases are "open" with respect to exchange with their environment of extensive quantities such as heat, volume or mass. Then, in Chapters 2 and 3, we show how to calculate free energies for elemental and alloy bulk and vapor phases. For elemental phases, discussed in Chapter 2, the free energies determine directly the relative stabilities of the phases. For alloy phases, discussed in Chapter 3, the free energies determine the relative stabilities both of phases and of phase mixtures through what is known as the common tangent construction.

Chapter 1

Free Energies and Open Systems

From a thermodynamics point of view, MBE is ultimately just an exceedingly precise, controlled phase transformation from a vapor to a crystalline solid. As with other phase transformations, there are two basic questions we would like to answer.

The first question is: *why* does the transformation occur — why is the crystalline solid phase favored over the vapor phase as well as over other possible competing solid or liquid phases? In other words, what quantities are maximized or minimized for the system to be in equilibrium? We shall find that the answer depends on whether the system is closed or open to its external environment: in a closed system, the *energy* of the system is minimized, while in an open system, one of a number of *free* energies is minimized.

The second question is: by *how much* does the transformation want to occur — what is the "driving force" for the transformation? In other words, if the system is not in equilibrium, by how much is it not in equilibrium? This question is especially important for transformations, such as MBE, that occur very far from equilibrium. We shall find, again, that the answer depends on whether the system is closed or open with respect to its environment. If a closed system is not in equilibrium because its *energy* can be decreased, then its deviation from equilibrium is the amount by which its energy can be decreased. If an open system is not in equilibrium because its *free energy* can be decreased, then its deviation from equilibrium is the amount by which its free energy can be decreased.

In nearly all situations of interest to MBE, the system we are concerned

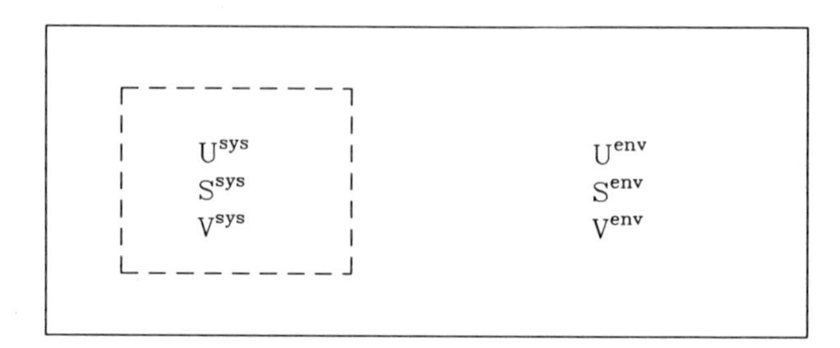

Figure 1.1: Thermodynamic system embedded in a large, closed environment.

with will be open. Therefore, the metric that will be appropriate to quantifying relative thermodynamic stability will be a free energy, usually what is known as the Gibbs free energy. In this first chapter, we will introduce that metric. We begin, in Sections 1.1 and 1.2, with descriptions of the equilibrium thermodynamics of closed and open systems. Then, in Section 1.3, we illustrate the distinction between the two kinds of systems with a concrete example: an electrical capacitor. When disconnected from a battery, the capacitor represents a closed system with respect to exchange of electrical charge; when connected to a battery, it represents an open system with respect to exchange of electrical charge. Finally, in Section 1.4, we discuss the nonequilibrium thermodynamics of closed and open systems.

1.1 Closed Systems

Consider, as illustrated in Figure 1.1, a system embedded in a large environment. The system and its environment are each characterized by internal energies (U^{sys} and U^{env}), entropies (S^{sys} and S^{env}) and volumes (V^{sys} and V^{env}). Let us first treat, in this section, closed systems. Closed systems are simpler to understand, although they are not as appropriate in describing MBE as are open systems.

If a system is closed, i.e., isolated in all respects from its external environment, then, with one exception, its "extensive" quantities are conserved. By extensive quantities, we mean those that scale with the size of the system, such as internal energy (U^{sys}) or volume (V^{sys}). The one extensive quantity not conserved is the entropy (S^{sys}). Indeed, from the second law of thermodynamics, we know that a closed system (in which energy is conserved) will evolve in such a way as to *increase* and eventually maximize its entropy.

Entropy, however, is not as physically intuitive a concept as is energy, and so it is often useful to rephrase mathematically this entropy *maximization* principle in terms of an energy *minimization* principle. That principle is: if entropy *were* conserved, the system would evolve in such a way as to decrease and eventually minimize its energy.[1] It is important to keep in mind, however, that this energy minimization at constant entropy principle is really only a mathematical trick for telling us the correct direction in which the system will evolve, and that it has no physical basis. After all, energy *is* conserved, and hence cannot be minimized.

To clarify this idea, consider the closed system illustrated at the left of Figure 1.2, in which a rock sits at the top of a slide. The system has two kinds of internal energy — the potential energy of the rock and the thermal energy stored in the rock and the slide. Suppose the rock has mass m and the slide has a vertical height h, so that the potential energy is mgh, where g is the gravitational acceleration. Furthermore, suppose the heat capacity (C_V) of the rock and the slide is constant above a certain temperature T_o, so that the thermal energy increases linearly above that temperature according to $Q_o + C_V(T - T_o)$, where Q_o is the thermal energy at T_o. Then, as illustrated in the middle panel of Figure 1.2, if the rock is at the position marked A at the top of the slide, the temperature-dependent internal energy is $U^{sys} = mgh + Q_o + C_V(T - T_o)$, while if the rock is at the position marked B at the bottom of the slide, the temperature-dependent internal energy is only $U^{sys} = Q_o + C_V(T - T_o)$.

Note that although the internal energy depends on whether the rock is at the top or bottom of the slide, the entropy does not. The entropy is purely thermal in origin, and increases with temperature according to $TdS^{sys} = C_V dT$. Then, as illustrated in the right panel of Figure 1.2, the temperature-dependent entropy is $S^{sys} = S_o + C_V \ln(T/T_o)$, where S_o is the entropy at T_o.

Now, suppose we allow the rock to slide down. The potential energy decreases, so on the energy-temperature diagram we drop from point A to point B_1, i.e., U^{sys} *decreases.* Moreover, suppose that we could somehow prevent the energy that was released from becoming thermal energy. Then, since the thermal energy (and hence temperature) would be constant, points A and B_1 on the entropy-temperature diagram would overlay each other. Therefore, if entropy could somehow be kept fixed, the system would evolve in such a way as to decrease and eventually minimize its internal energy.

Note, though, that we really *cannot* prevent the energy that was released from becoming thermal energy, since energy must be conserved. In fact, on

[1] H.B. Callen, *Thermodynamics and an Introduction to Thermostatistics*, 2nd Ed. (Wiley & Sons, New York, 1985).

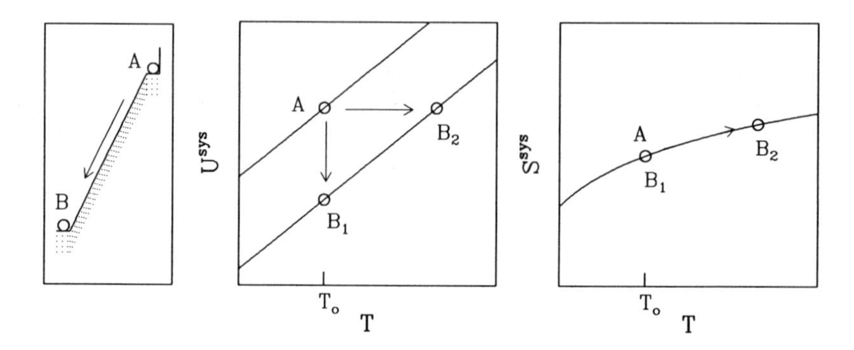

Figure 1.2: Temperature dependence of the internal energies and entropies of a closed thermodynamic system composed of a rock on a slide. In the middle panel, the upper internal energy curve corresponds to the rock at the top of the slide; the lower internal energy curve corresponds to the rock at the bottom of the slide. Point A corresponds to the rock on top of the slide, at a particular temperature. Point B_1 corresponds to the rock having slid to the bottom of the slide, assuming conservation of entropy. Point B_2 corresponds to the rock having slid to the bottom of the slide, assuming conservation of energy.

the energy-temperature diagram, the system will really evolve *laterally* from point A to point B_2, i.e., energy is constant but the *temperature* increases. At the same time, on the entropy-temperature diagram, the entropy rises from point A to point B_2. Therefore, at constant internal energy, the system will actually evolve in such a way as to increase and eventually maximize its entropy.

Nevertheless, in telling us the *direction* in which the system would like to evolve, the two principles are the same. Entropy maximization at constant energy and energy minimization at constant entropy, though physically *inequivalent*, are mathematically *equivalent*.

1.2 Open Systems

In Section 1.1, we treated closed systems whose extensive quantities, such as U^{sys} or V^{sys}, are conserved. In this section, we treat open systems. Such systems are not fully isolated from their external environments, and so their extensive quantities need *not* be conserved. If the environment is a reservoir of heat at a fixed temperature, T, then entropy will be exchanged so as to keep the system at the same fixed temperature. If the external environment is a reservoir of "volume" at a fixed pressure, p, then volume will be exchanged so as to keep the system at the same fixed pressure.

In such an open system, it is well known, but perhaps not physically intuitive, that the equilibrium state of the system is that for which one of a number of "free energies" is minimal. For example, at constant temperature the equilibrium state of the system has a minimimal Helmholtz free energy,

$$F^{\rm sys} = U^{\rm sys} - TS^{\rm sys}. \tag{1.1}$$

At constant pressure it has a minimimal enthalpy,

$$H^{\rm sys} = U^{\rm sys} + pV^{\rm sys}. \tag{1.2}$$

At constant temperature *and* pressure it has a minimimal Gibbs free energy,

$$G^{\rm sys} = U^{\rm sys} - TS^{\rm sys} + pV^{\rm sys}. \tag{1.3}$$

Note that the free energies differ from the internal energy by the addition or subtraction of terms involving extensive quantities, which the system is free to exchange with its environment, and their conjugate intensive parameters. Those terms arise because the equilibrium state of an open system is that which minimizes the internal energy of the system *combined* with that of its environment; they account for possible changes in the internal energy of that environment. Because changes in the internal energy of the external environment are due solely to exchange of extensive quantities with the system, they can be conveniently expressed in terms of those extensive quantities.

Suppose, for example, that a system can freely exchange entropy and volume with its environment, and hence is at constant temperature and pressure. If, due to some transformation, the entropy or volume of the system changes, then there must be equal and opposite changes in the surrounding environment — heat must flow or volume must be exchanged. But by energy conservation, the internal energy of the environment must then also change, by

$$dU^{\rm env} = TdS^{\rm env} - pdV^{\rm env} = -TdS^{\rm sys} + pdV^{\rm sys}. \tag{1.4}$$

Therefore, the change in the combined energies of the system and its environment is

$$d(U^{\rm sys} + U^{\rm env}) = dU^{\rm sys} - TdS^{\rm sys} + pdV^{\rm sys}, \tag{1.5}$$

or, since temperature and pressure are constant,

$$d(U^{\rm sys} + U^{\rm env}) = d(U^{\rm sys} - TS^{\rm sys} + pV^{\rm sys}). \tag{1.6}$$

In other words, minimizing the internal energy of the system *combined* with its environment, $U^{\rm sys} + U^{\rm env}$, is the same as minimizing the Gibbs free energy given by Equation 1.3, of the system alone. *The extra terms in the free energies are merely a convenient way of accounting for thermodynamic changes in the environment, via thermodynamic variables that have only to do with the system itself.*

1.3 Thermodynamics of a Capacitor

In Sections 1.1 and 1.2, we gave general, but abstract, treatments of the equilibrium thermodynamics of closed and open systems. In this section, we concretize the treatments by applying them to the two simple systems shown in Figure 1.3. Both systems consist of a capacitor whose capacitance is C, and a dielectric whose incremental motion toward the capacitor incrementally increases the capacitance by dC. Given a charge, $Q = CV$, that is initially placed on the capacitor, we ask whether or not the dielectric will be attracted to the capacitor.

The system on the left in Figure 1.3 is "closed": charge, which is the relevant extensive quantity in this system, is not free to enter or leave the capacitor. Then, if entropy were constant, the energy of the system, $Q^2/2C$, would *decrease* as the capacitance increases according to

$$\frac{\partial}{\partial C}\left[\frac{1}{2}\frac{Q^2}{C}\right]_Q = -\frac{1}{2}\frac{Q^2}{C^2} = -\frac{1}{2}V^2. \tag{1.7}$$

In other words, the dielectric is *attracted* to the capacitor.

The system on the right in Figure 1.3 is "open": charge *is* free to leave the capacitor, so as to maintain a constant voltage, V, which is the relevant intensive parameter in this system. Then, if entropy were constant, the energy of the system would *increase* as the capacitance increases according to

$$\frac{\partial}{\partial C}\left[\frac{1}{2}CV^2\right]_V = +\frac{1}{2}V^2. \tag{1.8}$$

At first glance the dielectric would seem *not* to be attracted to the capacitor.

However, we have neglected the flow of charge from the battery to the capacitor, which, at constant voltage, has decreased the energy stored in the external battery by $VdQ = d(QV) = d(CV^2)$. If we include that energy, then the energy of the system combined with its environment *does* decrease as the capacitance increases:

$$\frac{\partial}{\partial C}\left[\frac{1}{2}CV^2 - CV^2\right]_V = \frac{1}{2}V^2 - V^2 = -\frac{1}{2}V^2. \tag{1.9}$$

In both systems, then, the dielectric is attracted to the capacitor. In the closed system, the metric that gives us that answer is the internal energy

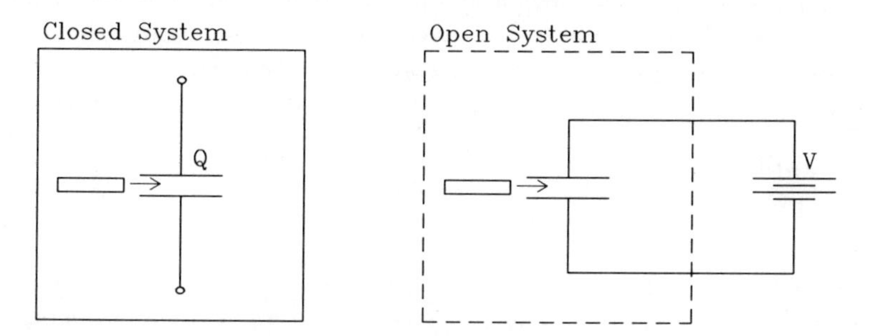

Figure 1.3: Thermodynamic systems consisting of capacitors and movable dielectrics. The system on the left is "closed," in that charge on the capacitor is conserved. The system on the right is "open," in that charge can be exchanged with the external battery in order to maintain a constant voltage V on the capacitor.

of the system, $U^{\rm sys}$. In the open system, however, the metric that gives us that answer is the *free* energy of the system, $U^{\rm sys} - Q^{\rm sys}V$.

1.4 Driving Forces for Transformations

Thus far, in discussing the direction in which systems evolve, we have elaborated on the *mathematical* rule that systems tend to decrease and eventually minimize their energy at constant entropy. In doing so, we have implicitly couched our arguments in the language of so-called "reversible," or equilibrium, thermodynamics, for which entropy is conserved. In all real systems, including the capacitor example just considered in Section 1.3, transformations are to some extent irreversible, in that entropy is not conserved, but increases. Therefore, to quantify not just the direction in which the system will evolve, but the "driving force" for that evolution, we return, in this section, to the *physical* rule that a system will evolve in such a way as to increase and eventually maximize its entropy at constant energy.

Since transformations will generally have some degree of irreversibility *and* reversibility, it is convenient to write changes in the entropy of the system as $dS^{\rm sys} = dS^{\rm sys}_{\rm irr} + dS^{\rm sys}_{\rm rev}$. The irreversible part of the entropy change ($dS^{\rm sys}_{\rm irr}$) must be positive, and is not compensated by any corresponding entropy decrease in the environment. The reversible part ($dS^{\rm sys}_{\rm rev}$) may be positive or negative, but must be compensated exactly by a corresponding change in the entropy of the surrounding environment, i.e., $dS^{\rm env} = -dS^{\rm sys}_{\rm rev}$.

In this language of irreversible thermodynamics, the most basic definition of the stability of a system is that there should *not* exist some transformation for which $dS_{\rm irr}^{\rm sys} > 0$, since otherwise it would be possible to increase the combined entropy of the system and its environment. Hence, $dS_{\rm irr}^{\rm sys}$ is a measure of the degree to which the system deviates from equilibrium.

For example, in a closed system, for which $dS^{\rm env} = -dS_{\rm rev}^{\rm sys} = 0$ and $dS^{\rm sys} = dU^{\rm sys}/T$,

$$dS_{\rm irr}^{\rm sys} = -\frac{dU^{\rm sys}}{T}. \tag{1.10}$$

If the internal energy of a closed system would have decreased in a hypothetical system for which entropy is conserved, then the irreversible entropy is what would actually increase in a real physical system for which energy must be conserved. Hence, *in a closed system the deviation from equilibrium is exactly the amount by which the internal energy would have decreased were entropy to have been conserved.* Thus, the deviation from equilibrium can be evaluated in terms of the physically more intuitive concept of energy, rather than in terms of the physically less intuitive concept of entropy.

In an open system, for which $dS^{\rm env} = -dS_{\rm rev}^{\rm sys} \neq 0$,

$$\begin{aligned} dS_{\rm irr}^{\rm sys} &= (dS_{\rm irr}^{\rm sys} + dS_{\rm rev}^{\rm sys}) - (dS_{\rm rev}^{\rm sys}) \\ &= (dS^{\rm sys}) - \left(\frac{1}{T}dU^{\rm sys} - \frac{p}{T}dV^{\rm sys}\right) \\ &= -\frac{1}{T}dG^{\rm sys}. \end{aligned} \tag{1.11}$$

If the Gibbs free energy of an open system decreases, then the irreversible entropy of the system increases. Note that, unlike the internal energy, the Gibbs free energy need *not* be conserved, because it includes the entropy of the system, which may increase irreversibly. Hence, *in an open system the deviation from equilibrium is exactly the amount by which the Gibbs free energy can decrease.*

In an open system, therefore, Gibbs free energies measure the relative stability of different phases, by telling us how much entropy would be created if the transformation were to occur. They tell us how far from equilibrium a given transformation is, and its degree of reversibility. Near-equilibrium transformations are those for which $dG^{\rm sys}$ is approximately zero. Consequently, $dS_{\rm irr}^{\rm sys}$ is also approximately zero, entropy is conserved, and the transformations are reversible. Far-from-equilibrium transformations are those for which $dG^{\rm sys}$ differs substantially from zero. Then, $dS_{\rm irr}^{\rm sys}$ also differs substantially from zero, entropy is not conserved, and the transformations are irreversible.

Suggested Reading

1. H.B. Callen, *Thermodynamics and an Introduction to Thermostatistics*, 2nd Ed. (Wiley & Sons, New York, 1985).

2. L.D. Landau and E.M. Lifshitz, *Statistical Physics* (Pergamon Press, Oxford, 1969).

Exercises

1. Suppose that, instead of the single slide shown in Figure 1.2, there are many (N) slides all arranged in a circle facing outwards. They each have their own distinct upper levels, but they all share a common lower level. Assuming that there is still only a single rock, the entropy of the system is greater if the rock is on an upper level than if it is on the lower level, by a configurational entropy term, $S_{\text{conf}}^{\text{sys}} = k \ln(N)$. For what temperatures will the equilibrium state of the system be that for which the rock is on one of the upper levels, rather than on the lower level?

2. Suppose that the closed single-rock single-slide system shown in Figure 1.2 were *open*, in that it exchanges heat with its environment so as to maintain constant temperature T. How do the entropies and internal energies of the system and its environment change as the rock slides down? How does the Gibbs free energy of the system change as the rock slides down?

3. Imagine a system divided into two identical halves, each of whose internal energies are purely thermal in origin. Each half has the same heat capacity, C_V, hence their internal energies $U_{1,2} = U_o + C_V T$ and entropies $S_{1,2} = S_o + C_V \ln(T/T_o)$ depend on temperature in the same way. Suppose one half is at temperature T_1, but the other half is at temperature T_2. What is the total entropy of the system? Suppose heat flows between the two halves, until each half has the same temperature. What is the new temperature? What is the new total entropy of the system? Show that the entropy has increased, and calculate by how much. Illustrate the changes graphically on an S vs. T plot.

Chapter 2

Elemental Phases

In Chapter 1, we showed that the metrics that govern the relative stability of phases contained in open systems are free energies of various kinds. For MBE and most other forms of crystal growth, both volume and heat are exchanged with the external environment so as to keep pressure and temperature constant throughout the system and its environment. Hence, the metric that measures the relative stabilities of different phases is the Gibbs free energy defined by Equation 1.3. Describing how to calculate the Gibbs free energies of different kinds of solid, liquid, and vapor phases is therefore the essence of this and the next chapter. In this chapter, we consider elemental phases; in the next chapter we consider alloy phases.

We begin, in Section 2.1, by introducing a standard nomenclature. In our experience, an inconsistent and complex nomenclature can make thermodynamics unnecessarily difficult, and so we have tried to make ours as simple and self-explanatory as possible. In particular, we focus attention on *molar* quantities, such as molar Gibbs free energies, molar entropies and enthalpies, and molar heat capacities.

Then, in Sections 2.2 and 2.3, we describe how to calculate the molar Gibbs free energies of various elemental phases. In doing so, it will be useful to distinguish between condensed and vapor phases. In Section 2.2, we treat condensed phases for which practical calculations are usually based on thermodynamic heat capacities. In Section 2.3, we treat vapor phases, for which calculations are usually based on first-principles statistical mechanical partition functions.

Finally, in Section 2.4, we present detailed case studies of the phases of two simple but important elements: Si and Ge. These case studies illustrate nearly all the essential features of the calculation of molar Gibbs free energies of elemental phases. In particular, we will calculate the molar Gibbs free energies of the condensed crystalline, amorphous and liquid phases, as well as of the vapor phase, which is composed of a mixture of monomers and dimers. In a sense, Figure 2.3 at the end of Section 2.4 is the main result of this chapter.

2.1 Nomenclature and Preliminaries

Let us first introduce, in this section, a standard nomenclature. Consider a phase, α, composed exclusively of N_a moles (or atoms) of a single component, a. The Gibbs free energy of such a phase is the difference between two terms:

$$G^\alpha(p,T,N_\mathrm{a}) = H^\alpha(p,T,N_\mathrm{a}) - TS^\alpha(p,T,N_\mathrm{a}). \tag{2.1}$$

The first term, the enthalpy,

$$H^\alpha(p,T,N_\mathrm{a}) = U^\alpha(p,T,N_\mathrm{a}) + pV^\alpha(p,T,N_\mathrm{a}), \tag{2.2}$$

is itself actually the free energy at constant pressure. The second term, as discussed in Chapter 1, is a constant-temperature correction due both to reversible heat flow into and out of the system and to irreversible entropy creation. The Gibbs free energy can, in a sense, be thought of as the free enthalpy at constant temperature.

From the Gibbs free energy of α is derived the *molar* Gibbs free energy (the Gibbs free energy per mole or per atom) of α,

$$g^\alpha(p,T) = \frac{G^\alpha(p,T,N_\mathrm{a})}{N_\mathrm{a}}, \tag{2.3}$$

and the chemical potential of a in α,

$$\mu_\mathrm{a}^\alpha(p,T) = \frac{\partial G^\alpha(p,T,N_\mathrm{a})}{\partial N_\mathrm{a}}. \tag{2.4}$$

Both $g^\alpha(p,T)$ and $\mu_\mathrm{a}^\alpha(p,T)$ have the same units, namely, energy per mole (or per atom). However, they have very different physical interpretations: $g^\alpha(p,T)$ is a property of a *phase* and is used to compare relative stabilities of different phases, while $\mu_\mathrm{a}^\alpha(p,T)$ is a property of a *component* in a phase, and is used to compare relative propensities of that component to incorporate into different phases. Later on in this book, when we consider two-component phases, the distinction between the two will be important. However, for the single-component phases discussed in this chapter, they have the same magnitudes, and the distinction is not important. Nevertheless, even in this chapter we will for consistency use molar Gibbs free energies in comparing relative stabilities of phases, even though we could just as well use chemical potentials.

Throughout this book we will adopt the chemist's language, in which g^α is referred to as a *molar* quantity and μ_a^α is referred to as a *partial* molar quantity. However, we will *not* attach to these two quantities either the chemist's preferred units of kcal/mole or the materials scientist's preferred units of kJ/mole. Instead, we will attach to them the physicist's preferred units of eV/atom. The reason is that ultimately we would like to understand the energetics of large aggregrates of atoms in terms of the energetics of *individual* atoms. Nevertheless, the units are interchangeable; the conversion factors are 23.061 (kcal/mole)/(eV/atom) and 96.487 (kJ/mole)/(eV/atom).

We will also adhere strictly to the following nomenclature. Extensive quantities, such as the Gibbs free energy, will be denoted by uppercase symbols, such as G. Extensive quantities that have been normalized by some number of moles (or atoms), such as the molar Gibbs free energy, will be denoted by the equivalent lowercase symbols, such as g.[1]

Using this nomenclature, the molar entropies and molar enthalpies of α are then written as

$$s^\alpha(p,T) = \frac{S^\alpha(p,T,N_\mathrm{a})}{N_\mathrm{a}} \quad (2.5)$$

$$h^\alpha(p,T) = \frac{H^\alpha(p,T,N_\mathrm{a})}{N_\mathrm{a}}, \quad (2.6)$$

in terms of which the molar Gibbs free energy is

$$g^\alpha(p,T) = h^\alpha(p,T) - Ts^\alpha(p,T). \quad (2.7)$$

To deduce the molar Gibbs free energy of α, then, we first need to calculate the molar entropies and enthalpies of α. There are two general approaches to that calculation: a semi-empirical "thermodynamic" approach usually applied to complex phases such as condensed phases, and a first-principles "statistical mechanical" approach usually applied to simple phases such as monoatomic or diatomic vapor phases. An approach intermediate between the two may be used for phases of intermediate complexity, such as polyatomic vapor phases.

[1] To keep our language from becoming cumbersome, though, we will sometimes only refer to a quantity as a molar quantity if an ambiguity not resolved by context might otherwise arise. For example, we will occasionally refer to molar Gibbs free energies simply as Gibbs free energies.

2.1.1 Heat Capacities and Thermodynamics

First, let us consider the semi-empirical thermodynamics approach to calculating the molar entropies and enthalpies of α. In this approach a semi-empirical *heat capacity* plays the central role; its integrals define the molar entropy and enthalpy. To see how, suppose an amount of heat, Tds^α, is added (per mole or per atom) to α. Since the constant-pressure molar heat capacity is *defined* as the amount of heat that produces unit temperature rise, $c_p^\alpha \equiv Tds^\alpha/dT$, the molar entropy is

$$s^\alpha(p,T) = s^\alpha(p,T_o) + \int_{T_o}^{T} \frac{c_p^\alpha(p,T')}{T'} dT'. \tag{2.8}$$

As heat is added to the system, the molar enthalpy of the system must of course also increase, by $dh^\alpha = Tds^\alpha$, so that the constant-pressure molar heat capacity could equally well have been defined as $c_p^\alpha \equiv dh^\alpha/dT$. Therefore, the molar enthalpy is

$$h^\alpha(p,T) = h^\alpha(p,T_o) + \int_{T_o}^{T} c_p^\alpha(p,T')dT'. \tag{2.9}$$

Note that although $dh^\alpha/dT = Tds^\alpha/dT$, and at constant temperature $dg^\alpha = dh^\alpha - Tds^\alpha$, the molar Gibbs free energy need not be constant. The reason is that the temperature is *not* constant, but is increasing as heat is added. Indeed, we can also write

$$dg^\alpha = d(h^\alpha - Ts^\alpha) = dh^\alpha - Tds^\alpha - s^\alpha dT = -s^\alpha dT, \tag{2.10}$$

the constant-pressure form of the Gibbs-Duhem relation. It tells us that for positive entropy, the molar Gibbs free energy decreases with increasing temperature. Since entropies are always positive, *all phases are stabilized at higher temperatures, although higher entropy phases are stabilized more than are lower entropy phases.*

Equations 2.7, 2.8 and 2.9 determine the molar enthalpies, entropies and Gibbs free energies of a phase completely in terms of a temperature-dependent heat capacity per atom, and a pair of molar enthalpy and entropy "offsets" at a particular reference temperature T_o. Hence, in the thermodynamic approach, the main task is to calculate molar heat capacities and enthalpy and entropy offsets for particular phases.

2.1.2 Partition Functions and Statistical Mechanics

Second, let us consider the first-principles statistical mechanical approach to calculating the molar entropies and enthalpies of α. In this approach the *partition function* plays the central role; its derivatives define the entropies and enthalpies. To see how, consider an ideal gas composed of N particles contained in a volume V. Denote the quantum mechanical energy levels of the gas as a whole as $\varepsilon_i(N, V)$, and the degeneracies of those levels as $\omega_i(N)$. The partition function, Q, is the sum over the occupation probabilities of those levels, weighted by their degeneracies:

$$Q = \sum_{i=0}^{\infty} \omega_i(N) e^{-\varepsilon_i(N,V)/kT}. \tag{2.11}$$

The "thermodynamic" internal energy is the ensemble average of the level energies, and can be written as a first derivative of the partition function,

$$u \equiv \frac{\sum_{i=0}^{\infty} \omega_i \varepsilon_i e^{-\varepsilon_i/kT}}{N \sum_{i=0}^{\infty} \omega_i e^{-\varepsilon_i/kT}} = kT^2 \left[\frac{\partial \ln Q}{N \partial T}\right]_{N,V}. \tag{2.12}$$

Likewise, the pressure of the system is the ensemble average of the change in the level energies with volume, and can also be written as a first derivative of the partition function,

$$p \equiv \frac{\sum_{i=0}^{\infty} \omega_i [\partial \varepsilon_i / \partial V] e^{-\varepsilon_i/kT}}{\sum_{i=0}^{\infty} \omega_i e^{-\varepsilon_i/kT}} = kT \left[\frac{\partial \ln Q}{N \partial v}\right]_{N,T}. \tag{2.13}$$

Hence, for an ideal gas obeying $v = kT/p = T[\partial v/\partial T]_p$, the molar enthalpy can be written as

$$\begin{aligned} h &\equiv u + pv \\ &= kT^2 \left[\frac{\partial \ln Q}{N \partial T}\right]_{N,V} \left[\frac{\partial T}{\partial T}\right]_p + kT^2 \left[\frac{\partial \ln Q}{N \partial v}\right]_{N,T} \left[\frac{\partial v}{\partial T}\right]_p \\ &= kT^2 \left[\frac{\partial \ln Q}{N \partial T}\right]_{N,p}. \end{aligned} \tag{2.14}$$

Note that this formulation for the molar enthalpy is strictly true only for an ideal gas, and that it requires that the partition function be considered a function of N, p, and T rather than of N, V, and T.

The constant-pressure molar heat capacity is the derivative of the molar enthalpy,

$$c_p \equiv \left[\frac{\partial h}{\partial T}\right]_p = kT^2 \left[\frac{\partial^2 \ln Q}{N \partial T^2}\right]_{N,p} + 2kT \left[\frac{\partial \ln Q}{N \partial T}\right]_{N,p}, \tag{2.15}$$

in terms of which the molar entropy can be deduced to be

$$s \equiv \int_0^T \frac{c_p}{T'} dT' = kT \left[\frac{\partial \ln Q}{N \partial T}\right]_p + k \frac{\ln Q}{N}, \tag{2.16}$$

and the molar Gibbs free energy can be deduced to be

$$g = h - Ts = -kT\frac{\ln Q}{N}. \tag{2.17}$$

Note that the expression for g is much simpler than those for s, h, or c_p. Therefore, it is often easier in practice to first calculate g, and then numerically differentiate to get s, h, and c_p:

$$h = \frac{d}{d(1/T)}\left[\frac{g(T)}{T}\right] \tag{2.18}$$

$$s = \frac{-d}{dT}\left[g(T)\right] \tag{2.19}$$

$$c_p = -T\frac{d^2}{dT^2}\left[g(T)\right]. \tag{2.20}$$

Equations 2.18, 2.19 and 2.17 determine the molar enthalpy, entropy, and Gibbs free energy of a phase completely in terms of a temperature-dependent partition function. Hence, in the statistical mechanical approach, the main task is to calculate the partition function for particular phases.

2.2 Condensed Phases

In Subsection 2.1.1 we showed how, *given* a semi-empirical molar heat capacity, the other thermodynamic quantities of interest (molar entropies, enthalpies and Gibbs free energies) could be calculated. In this section, we describe how to estimate such semi-empirical molar heat capacities for condensed phases. In general, molar heat capacities of condensed phases depend negligibly on pressure at the subatmospheric pressures usually associated with MBE. Therefore, although in Equations 2.8 and 2.9 we explicitly allowed for the possibility of a pressure-*dependent* heat capacity, for condensed phases at subatmospheric pressure we can assume that the heat capacity is pressure-*independent.*

There are a number of well-established theories for this pressure-independent heat capacity. At constant volume, e.g., the contribution from lattice vibrations often obeys quite closely the Debye theory. In that theory, the excitation of a spectrum of harmonic lattice vibrations is calculated as a function of temperature. The heat capacity is near-zero at temperatures so low that nearly all lattice vibrational modes are quantum-mechanically "frozen out." Then, it increases rapidly with temperature as successively higher frequency lattice vibrational modes become excited. Finally, it saturates at temperatures beyond a characteristic Debye temperature Θ_D at which the highest frequency lattice vibrational mode becomes significantly excited. At those temperatures, energy becomes "equipartitioned" in units of $kT/2$ into each of the six (three potential and three kinetic) degrees of vibrational freedom per atom, and the constant-volume heat capacity per atom approaches the classical Dulong-Petit value of $\partial(6kT/2)/\partial T = 3k$.

In practice, however, experimental constant-volume heat capacities do not always agree perfectly with the Debye theory. The theory is only approximate and does not treat, e.g., anharmonicities in lattice vibrations at high temperature, or electronic contributions at low temperature. Furthermore, the heat capacity at constant pressure, which is of greatest interest to us, differs in a temperature-dependent way from the heat capacity at constant volume:

$$c_p = c_v(1 + \gamma\alpha_v T). \tag{2.21}$$

In this equation, a variation of what is known as the Nernst-Lindemann equation, the isobaric volume expansion coefficient is $\alpha_v = [d(\ln v)/dT]_p$ and the isothermal compressibility is $\kappa_T = -[d(\ln v)/dp]_T$. The Gruneisen constant, $\gamma = \alpha_v v/\kappa_T c_v$, is a nearly temperature-independent dimensionless constant typically between one and two.

Because of these deviations, in numerical calculations involving heat capacities it is common to use semi-empirical formulas fit to experimental data in a particular temperature range. Usually, these are algebraic polynomials of the form[2]

$$c_p = a + bT + c/T^2. \tag{2.22}$$

The constant a is positive and usually nearly equal to the Dulong-Petit value expected for the heat capacity at high temperatures. The constant b is also positive, as from Equation 2.21 there is a slight tendency for constant-pressure heat capacities to increase at high temperatures. The constant c is usually negative, because at low temperatures lattice vibrations are quantum mechanically frozen out and heat capacities decrease.

For thermodynamic calculations in a restricted range of medium-to-high temperatures, such polynomials are usually sufficiently accurate. However, for calculations over a wider range of temperatures, and particularly at the low to medium temperatures at which MBE often occurs, some inaccuracy is introduced. Our experience has been that for many solids, the following semi-empirical formula[3] fits experimental data over a significantly wider

[2] C.G. Maier and K.K. Kelley, "An equation for the representation of high-temperature heat content data," *Amer. Chem. Soc.* **54**, 3243 (1932).

[3] J.Y. Tsao, "Two semi-empirical expressions for condensed-phase heat capacities," *J. Appl. Phys.* **68**, 1928 (1990).

temperature range:

$$c_p = \left(\frac{T^2}{T^2 + \Theta_T^2} \right) (c_0 + c_1 T + \ldots) . \tag{2.23}$$

The first, "trigonometric" part of the formula describes the low-temperature region near a semi-empirical critical temperature Θ_T within which the heat capacity is rising sharply. The second, "polynomial" part of the formula describes the high-temperature region in which the heat capacity is approximately constant. In practice it is usually sufficient to truncate the polynomial to linear order, so that only c_0 and c_1 are nonzero.

Importantly, each of the parameters in the formula can be estimated through simple physical arguments. Such estimates are particularly useful if the heat capacity of the phase of interest has not been measured (or is difficult to measure) in the temperature range of interest. Indeed, as we shall see, the art of calculating phase equilibria is to a large extent the art of estimating thermodynamic quantities in temperature ranges over which they are inacessible to measurement.

For example, the semi-empirical critical temperature, Θ_T, is the temperature at which the heat capacity rises to half its saturation value, and is thus related to the Debye temperature. In fact, it is numerically equal to approximately one-fourth the Debye temperature. The reason it is such a small fraction of the Debye temperature is that at the Debye temperature, which corresponds to the energy of the *highest* frequency vibrational mode, all lower frequency vibrational modes are excited, and so the heat capacity has already nearly saturated at the Dulong-Petit value.

The constant c_0 is the approximate saturation value of the heat capacity, and is thus closely related to the Dulong-Petit value for the heat capacity at constant volume, $3k = 0.258\,\text{meV}/(\text{atom} \cdot \text{K})$. The constant c_1 determines the high-temperature increase in the heat capacity, and is thus related to the constant-pressure correction to the heat capacity given in Equation 2.21, $\alpha_v^2 v/\kappa_T$. However, it may also contain contributions due to lattice vibrational anharmonicities. In order of magnitude, the ratio c_1/c_0 typically varies from $10^{-5}\,\text{K}^{-1}$ to $10^{-3}\,\text{K}^{-1}$.

As a semi-empirical expression, Equation 2.23 is a significant improvement over Equation 2.22, but it is far from perfect. In particular, it implies that the very-low-temperature heat capacity approaches zero as T^2, rather than as T^3 as the Debye theory predicts and as is generally observed. However, such low temperatures are usually outside the region of interest even for MBE, and in any case the very-low-temperature region contributes (nearly) negligibly to the integrals in Equations 2.8 and 2.9 when evaluated at higher temperatures.

Finally, an especially useful feature of Equation 2.23 is that, like Equation 2.22, it defines a heat capacity for which both c_p and c_p/T are analytically integrable. Hence, temperature-dependent molar entropies and enthalpies can be conveniently calculated via Equations 2.8 and 2.9:

$$s = s(T_o) + \frac{c_0}{2}\left[\ln(T'^2 + \Theta_T^2)\right]_{T_o}^{T} + c_1\left[T' - \Theta_T \arctan\left(\frac{T'}{\Theta_T}\right)\right]_{T_o}^{T} \tag{2.24}$$

and

$$h = h(T_0) + c_0\left[T' - \Theta_T \arctan\left(\frac{T'}{\Theta_T}\right)\right]_{T_o}^{T} + \frac{c_1}{2}\left[T'^2 - \Theta_T^2 \ln(T'^2 + \Theta_T^2)\right]_{T_o}^{T}. \tag{2.25}$$

Throughout this book, then, condensed-phase heat capacities will be approximated semi-empirically by Equation 2.23, and entropies and enthaplies will be approximated semi-empirically by Equations 2.24 and 2.25.

2.3 Vapor Phases

In Subsection 2.1.2, we showed how, *given* a statistical mechanical partition function, the other thermodynamic quantities of interest (molar entropies, enthalpies, and Gibbs free energies) could be calculated. In this section, we show how to estimate such partition functions for vapor phases. For simplicity, we restrict ourselves to low-density vapors at the subatmospheric pressures associated with MBE, and hence which behave as ideal gases.

Consider, then, an ideal gas composed of N identical, non-interacting molecules occupying a system of volume V. Each molecule considered separately will have its own spectrum of quantum-mechanical energy levels, and hence its own partition function q. The energies of those levels will be all the possible *sums* of the energies of its translational, rotational, vibrational, and electronic quantum-mechanical energy levels. The degeneracies of those levels will be the corresponding *products* of the degeneracies of the translational, rotational, vibrational, and electronic energy levels. From Equation 2.11, we see that the partition function of an individual molecule can therefore be written, conveniently, as the product of the translational, rotational, vibrational and electronic partition functions, $q = q_{\text{tra}}q_{\text{rot}}q_{\text{vib}}q_{\text{ele}}$.

Now, the total partition function for all N atoms is *itself* the product of the partition functions for each molecule (less overcounting of permutations of identical molecules): $Q = q^N/N!$. Therefore, $\ln Q \approx (N \ln q) - (N \ln N)$, and we can write

$$\frac{\ln Q}{N} \approx \ln(q_{\text{tra}}/N) + \ln(q_{\text{rot}}) + \ln(q_{\text{vib}}) + \ln(q_{\text{ele}}). \tag{2.26}$$

In other words, the logarithm of the partition function of the vapor as a whole is essentially a simple sum of the logarithms of the translational, rotational, vibrational, and electronic partition functions of the individual molecules themselves. The only deviation is the extra factor of N, which, as we shall see, converts the volume dependence of the translational partition function into a pressure dependence.

Note that since only $(1/N)\ln Q$ enters into Equations 2.16, 2.14 and 2.15, the molar entropy, enthalpy, and heat capacity may themselves be taken to be a simple sum of translational, rotational, vibrational, and electronic contributions:

$$c_p = c_{p,\text{tra}}(p,T) + c_{p,\text{rot}}(T) + c_{p,\text{vib}}(T) + c_{p,\text{ele}}(T) \quad (2.27)$$

$$s = s_{\text{tra}}(p,T) + s_{\text{rot}}(T) + s_{\text{vib}}(T) + s_{\text{ele}}(T) \quad (2.28)$$

$$h = h(T_o) + [h_{\text{tra}}(T') + h_{\text{rot}}(T') + h_{\text{vib}}(T') + h_{\text{ele}}(T')]_{T_o}^{T} . \quad (2.29)$$

Here, we have assumed that, according to the third law of thermodynamics, the overall entropy offset at zero temperature is zero. In other words, nonzero entropies at zero temperature arise exclusively from ground-state degeneracies in the translational, rotational, vibrational, or electronic levels, and hence are already accounted for. The overall molar enthalpy offset, however, need *not* be zero, and can only be determined *relative* to the molar enthalpy offsets of other phases.

To concretize this discussion, let us consider in the following two subsections two kinds of vapor: one composed purely of monomers and one composed purely of dimers.

2.3.1 Monomeric Vapors

As a first example, consider the simplest vapor, composed of single atoms. Such a vapor, having no internal nuclear degrees of freedom, will have only translational and electronic contributions to its thermodynamic functions.

For the translational contribution, the partition function, in the classical limit, is known[4] to be $q_{\text{tra}} = (2\pi mkT/h^2)^{3/2}V$, or, normalized by N, $q_{\text{tra}}/N = (2\pi mkT/h^2)^{3/2}kT/p$. Using Equation 2.16, the translational contribution to the molar entropy can then be shown to give what is known as the Sackur-Tetrode equation,[5] $s_{\text{SacTet}} = (5k/4)\ln(T^2/\Theta_{\text{T,tra}}^2)$, where the critical temperature, $\Theta_{\text{T,tra}}$, is determined by the pressure, p, and the

[4]The partition function is the sum over (nondegenerate) translational quantum levels whose energies are $\epsilon_{n_x,n_y,n_z} = (h^2/8mV^{2/3})(n_x^2+n_y^2+n_y^2)$ where $n_x, n_y, n_z \in \{1, 2, \ldots\}$. In the classical limit, the discrete sum is approximated by a continuous integral.

[5]O. Sackur, *Ann. Physik* **36**, 598 (1911) and H. Tetrode, *Ann. Physik* **38**, 434 (1912).

atomic mass, m:

$$\Theta_{\mathrm{T,tra}}(p) = \frac{p^{2/5}}{ek}\left(\frac{h^2}{2\pi m}\right)^{3/5} = 1.593\,\mathrm{K}\left(\frac{p}{760\,\mathrm{Torr}}\right)^{2/5}\left(\frac{1\,\mathrm{amu}}{m}\right)^{3/5}. \tag{2.30}$$

Therefore, the heat capacity, again in the classical limit, is constant and given by $c_{p,\mathrm{SacTet}} \equiv T\partial s_{\mathrm{SacTet}}/\partial T = 5k/2$, which is what is known as the Dulong-Petit value for an ideal gas. Physically, it arises because energy is equipartitioned in units of $kT/2$ into each of the three translational directions. Therefore, the constant-volume heat capacity is $c_{v,\mathrm{SacTet}} = 3k/2$. Since $pv = kT$ for an ideal gas, the constant-pressure heat capacity can then be deduced from Equation 2.21 to be $c_{p,\mathrm{SacTet}} = c_{v,\mathrm{SacTet}} + k = 5k/2$.

Note that the Sackur-Tetrode equation is in nearly the same logarithmic form as Equation 2.24, with $c_{1,\mathrm{tra}} = 0$ and $c_{o,\mathrm{tra}} = 5k/2$. It can be brought into *exactly* the same form by making two simple assumptions. The first assumption is that the heat capacity is *not* constant, but decreases to zero at (very) low temperatures according to Equation 2.23:

$$c_{p,\mathrm{tra}}(p,T) = c_{o,\mathrm{tra}}\left(\frac{T^2}{T^2 + \Theta^2_{\mathrm{T,tra}}}\right). \tag{2.31}$$

In fact, heat capacities *must* decrease to zero at zero temperature, lest the entropy integrand in Equation 2.8 become infinite at zero temperature. The Sackur-Tetrode equation does not hold at temperatures less than $\Theta_{\mathrm{T,tra}}$ (where it predicts negative entropies, in violation of the third law of thermodynamics) because it was derived classically, not quantum mechanically. Quantum mechanically, even translational degrees of freedom must ultimately be frozen out at temperatures that are low compared to the spacing of translational energy levels.

The second assumption is that the third law of thermodynamics holds, viz., $s_{\mathrm{tra}}(T_o = 0) = 0$, so that, from Equation 2.24, the translational contribution to the molar entropy is

$$s_{\mathrm{tra}}(T) = \frac{c_{o,\mathrm{tra}}}{2}\ln\left(\frac{T^2 + \Theta^2_{\mathrm{T,tra}}}{\Theta^2_{\mathrm{T,tra}}}\right). \tag{2.32}$$

The translational contribution to the molar enthalpy, in turn, is

$$h_{\mathrm{tra}}(T) = c_{o,\mathrm{tra}}\left[T - \Theta_{\mathrm{T,tra}}\arctan\left(\frac{T}{\Theta_{\mathrm{T,tra}}}\right)\right], \tag{2.33}$$

where the arbitrary integrating constant has been omitted because it may be incorporated into that already present in Equation 2.29.

In practice, the semi-empirical Equation 2.32 is nearly indistinguishable from the Sackur-Tetrode equation, due to difficulties in measuring the thermodynamic properties of vapors at extremely low temperatures. However, Equation 2.32 is physically more satisfying, because it obeys the third law of thermodynamics, and is mathematically more satisfying, because it is defined at all positive temperatures. Therefore, we will use Equations 2.31, 2.32 and 2.33 to describe the translational contributions to the molar heat capacities, entropies and enthalpies of vapor phases just as we will use Equations 2.23, 2.24 and 2.25 to describe those of condensed phases. The only difference is that for the vapor phase $\Theta_{\mathrm{T,tra}}$ is *not* semi-empirical and can be calculated from first principles according to Equation 2.30, while for condensed phases Θ_{T} *is* semi-empirical and is independent of pressure.

For the electronic contribution, the partition function cannot in general be summed analytically. However, in practice usually only a few excited electronic levels have low enough energies to contribute significantly to the heat content. Then, the partition function can be summed over a finite number of levels,

$$q_{\mathrm{ele}} = \sum \omega_{i,\mathrm{ele}} e^{-\varepsilon_{i,\mathrm{ele}}/kT}, \tag{2.34}$$

and the molar enthalpies, entropies and heat capacities calculated numerically from Equations 2.26, 2.17, 2.18, 2.19 and 2.20. Note that if the ground electronic level is degenerate, then, according to Equation 2.16, the zero-temperature molar entropy does not vanish, but rather is $k \ln(\omega_{0,\mathrm{ele}})$.

2.3.2 Dimeric Vapors

As a second example, consider the next simplest vapor, composed of dimer pairs of atoms. For these vapors, the translational and electronic contributions to the thermodynamic functions can be described in the same way as those for the monomeric vapor, with two differences. First, the dimer mass, rather than the monomer mass, must be used in Equation 2.30 for calculating $\Theta_{\mathrm{T,tra}}$. Second, all thermodynamic quantities must be *halved* in order for their units to be per *atom* rather than per dimer.

In addition, the thermodynamic functions for dimer atoms contain contributions from rotational and vibrational motion. The rotational contribution can be treated in nearly exactly the same way as the translational contribution was treated. In the rigid-rotor approximation, the rotational partition function can, in the classical limit, readily be evaluated.[6]

[6] The partition function is the sum over rotational quantum levels whose degeneracies are $(2J+1)/2$ and whose energies are $J(J+1)k\Theta_{\mathrm{T},rot}$, where $J \in \{0, 1, 2, \ldots\}$. In the classical limit, the discrete sum is approximated by a continuous integral.

Then, Equation 2.16 gives the molar entropy. Finally, to "fix" the zero-temperature catastrophe, the molar entropy is recast into the semi-empirical form of Equation 2.24, giving

$$c_{p,\mathrm{rot}}(T) = c_{o,\mathrm{rot}} \left(\frac{T^2}{T^2 + \Theta^2_{\mathrm{T,rot}}} \right) \tag{2.35}$$

$$s_{\mathrm{rot}}(T) = \frac{c_{o,\mathrm{rot}}}{2} \ln \left(\frac{T^2 + \Theta^2_{\mathrm{T,rot}}}{\Theta^2_{\mathrm{T,rot}}} \right) \tag{2.36}$$

$$h_{\mathrm{rot}}(T) = c_{o,\mathrm{rot}} \left[T - \Theta_{\mathrm{T,rot}} \arctan \left(\frac{T}{\Theta_{\mathrm{T,rot}}} \right) \right], \tag{2.37}$$

where the arbitrary integrating constant in the molar enthalpy has again been omitted, as it may be incorporated into that already present in Equation 2.29. In these equations,

$$\Theta_{\mathrm{T,rot}} = \frac{\hbar^2}{2Ik} = 23.93\,\mathrm{K} \left(\frac{1\,\mathrm{amu}}{\mu} \right) \left(\frac{1\,\text{Å}}{r} \right)^2 \tag{2.38}$$

is the critical temperature below which rotational motion freezes out, and $c_{o,\mathrm{rot}} = k/2$. The critical temperature decreases with increasing rotational inertia of the molecule, $I = \mu r^2$, because the spacing of rotational energy levels decreases. Here, μ is what is known as the "reduced" mass of the dimer (half the mass of each atom of the dimer, one-fourth the mass of the dimer itself), and r is the dimer bond length.

The vibrational contribution can in principle also be fit by a similar semi-empirical form. However, in the simple harmonic oscillator approximation, the thermodynamic functions are straightforward to describe exactly, not just semi-empirically[7]:

$$c_{p,\mathrm{vib}}(T) = c_{o,\mathrm{vib}} \left(\frac{\Theta_{\mathrm{T,vib}}}{T} \right)^2 \frac{e^{\Theta_{\mathrm{T,vib}}/T}}{\left(e^{\Theta_{\mathrm{T,vib}}/T} - 1 \right)^2} \tag{2.39}$$

$$s_{\mathrm{vib}}(T) = c_{o,\mathrm{vib}} \left[\frac{\Theta_{\mathrm{T,vib}}/T}{e^{\Theta_{\mathrm{T,vib}}/T} - 1} - \ln \left(1 - e^{-\Theta_{\mathrm{T,vib}}/T} \right) \right] \tag{2.40}$$

$$h_{\mathrm{vib}}(T) = c_{o,\mathrm{vib}} \left[\frac{\Theta_{\mathrm{T,vib}}}{2} + \frac{\Theta_{\mathrm{T,vib}}}{e^{\Theta_{\mathrm{T,vib}}/T} - 1} \right]. \tag{2.41}$$

[7]The partition function is the sum over (nondegenerate) quantum levels whose energies are $(n+\frac{1}{2})k\Theta_{\mathrm{T,vib}}/2$, where $n \in \{0, 1, 2, \ldots\}$. The partition function can be summed analytically, without approximation by a continuous integral. Therefore, the result is fully quantum mechanical, not classical, and Equation 2.39 correctly describes the heat capacity's approach to zero at low temperature. It essentially reproduces Einstein's calculation of the heat capacity of solids.

In these equations,

$$\Theta_{\mathrm{T,vib}} = h\nu/k \tag{2.42}$$

is the temperature associated with a vibration at frequency ν and $c_{o,\mathrm{vib}} = k/2$. The leading term in the equation for h_{vib} is due to the zero-point $h\nu/2$ vibration of the oscillator.

Note that $\Theta_{\mathrm{T,rot}}$ and $\Theta_{\mathrm{T,vib}}$, unlike $\Theta_{\mathrm{T,tra}}$, are independent of pressure, having only to do with *internal* degrees of freedom of the molecule. Also note that at typical MBE pressures, $\Theta_{\mathrm{T,vib}}$ is much higher than $\Theta_{\mathrm{T,rot}}$, which in turn is much higher than $\Theta_{\mathrm{T,tra}}$. In fact, both $\Theta_{\mathrm{T,rot}}$ and $\Theta_{\mathrm{T,tra}}$ are usually so low as to be experimentally unobservable. At all *normal* temperatures and pressures, the translational and rotational contributions to the heat capacity (per atom, not per dimer) are constant and equal to $5k/4$ and $k/2$, respectively, giving a total of $7k/4$.

2.4 Two Simple Elements: Si and Ge

In Sections 2.2 and 2.3, we presented general analytic expressions for the molar heat capacities, entropies, and enthalpies of condensed and vapor phases. These expressions are characterized by either semi-empirical or first-principles parameters of various kinds, each of which must ultimately be estimated or calculated for a particular phase.

In this section, we illustrate such estimations and calculations by considering as case studies the various phases of Si and Ge. For both of these elements, four phases occur at normal (subatmospheric) pressures: two solid (amorphous and crystalline), one liquid, and one vapor. Following Kubaschewski's notation,[8] we denote the liquid phases by braces ($\{\mathrm{Si}\}$ and $\{\mathrm{Ge}\}$) and the vapor phases by parentheses ((Si) or $(\frac{1}{2}\mathrm{Si}_2)$ and (Ge) or $(\frac{1}{2}\mathrm{Ge}_2)$). We denote solid phases by angled brackets, and by subscripts outside the brackets if multiple solid phases need to be distinguished ($\langle\mathrm{Si}\rangle_\mathrm{a}$, $\langle\mathrm{Si}\rangle_\mathrm{c}$, $\langle\mathrm{Ge}\rangle_\mathrm{a}$, and $\langle\mathrm{Ge}\rangle_\mathrm{c}$). The components themselves, Si and Ge, are *not* enclosed by brackets, braces, or parentheses of any sort.

We divide the discussion into three subsections. In Subsection 2.4.1, we estimate the molar heat capacities of the various condensed and vapor phases. These molar heat capacities determine the temperature dependences of the molar entropies and enthalpies, up to an "integrating constant," or offset. Then, in Subsection 2.4.2, we estimate those molar entropy and enthalpy offsets. Finally, in Subsection 2.4.3, we gather together all of these estimates and use them to calculate the molar Gibbs free energies as functions of pressure and temperature.

[8] O. Kubaschewski and C.B. Alcock, *Metallurgical Thermochemistry*, 5th Ed. (Pergamon Press, Oxford, 1979).

2.4.1 Molar Heat Capacities

Let us begin, in this subsection, by estimating the molar heat capacities of all of the phases listed above as functions of temperature and, for the vapor phases, also of pressure. Throughout, we will make liberal use of estimates based on physical arguments. The reason is that the molar heat capacity of a phase can often only be measured if that phase is the equilibrium one, since otherwise the phase will have a tendency to transform to a new one during the measurement. Therefore, it is important to develop methods for *estimating* heat capacities in temperature regimes in which they have not been measured.

Figure 2.1 shows estimated temperature dependences of the heat capacities of the various phases of Si and Ge. For the condensed phases, heat capacities were calculated using the semi-empirical Equation 2.23. For the monomer and dimer vapors, heat capacities were calculated using Equation 2.27. The numerical values for the heat-capacity parameters are given in Tables 2.1, 2.2 and 2.3, and were estimated in the following ways.

Condensed Phases

For $\langle \mathrm{Si} \rangle_\mathrm{c}$ and $\langle \mathrm{Ge} \rangle_\mathrm{c}$, the heat capacity parameters were deduced by nonlinear least-squares fits to experimental data. It can be seen from Figure 2.1 that the experimental data are fit by the semi-empirical forms exceedingly well. In fact, only at very low temperatures (for $\langle \mathrm{Si} \rangle_\mathrm{c}$ less than 50 K and for $\langle \mathrm{Ge} \rangle_\mathrm{c}$ less than 30 K) does the percentage deviation become significant, and even then the absolute deviation is, for our purposes, negligible.

As expected, the semi-empirical critical temperatures for $\langle \mathrm{Si} \rangle_\mathrm{c}$ and $\langle \mathrm{Ge} \rangle_\mathrm{c}$ are nearly equal to one-fourth their respective Debye temperatures, $\Theta_\mathrm{D}^{\langle \mathrm{Si} \rangle_\mathrm{c}} \approx 640\,\mathrm{K}$ and $\Theta_\mathrm{D}^{\langle \mathrm{Ge} \rangle_\mathrm{c}} \approx 374\,\mathrm{K}$. For both $\langle \mathrm{Si} \rangle_\mathrm{c}$ and $\langle \mathrm{Ge} \rangle_\mathrm{c}$, the c_0 parameters are essentially equal to the Dulong-Petit value of $3k = 0.258\,\mathrm{meV/(atomK)}$. The c_1 parameters, however, are about one order of magnitude greater than those ($\approx 2.4 \times 10^{-6}\,\mathrm{eV/(atomK^2)}$) that would be calculated from Equation 2.21, probably due to vibrational anharmonicities in the diamond-cubic lattice.[9]

For $\langle \mathrm{Si} \rangle_\mathrm{a}$ and $\langle \mathrm{Ge} \rangle_\mathrm{a}$, the experimental heat capacities are not accurately known, largely because of the difficulty of forming large thermal masses of very pure $\langle \mathrm{Si} \rangle_\mathrm{a}$ and $\langle \mathrm{Ge} \rangle_\mathrm{a}$. High-purity thin films of amorphous material may be prepared either by MBE at very low temperatures or by ion-implantation, but usually only on thick substrates whose thermal mass

[9]P.C. Trivedi, H.O. Sharma and L.S. Kothari, "Lattice anharmonicity of diamond-structure crystals," *J. Phys. C: Solid State Phys.* **10**, 3487 (1977).

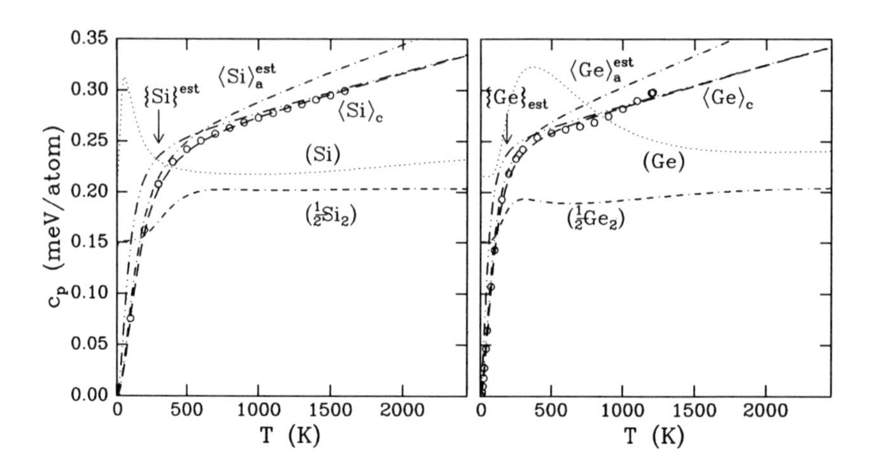

Figure 2.1: Measured and estimated temperature dependences of the heat capacities of the various phases of Si (left) and Ge (right). Experimental data (open circles) for $\langle \mathrm{Si} \rangle_c$[a] and $\langle \mathrm{Ge} \rangle_c$[b] are also shown.

[a]Data below 90 K are from H.-Matsuo Kagaya and T. Soma, *Properties of Silicon*, EMIS Datareviews Series No. 4 (INSPEC, 1988); data above 90 K are from D.R. Stull and H. Prophet, *JANAF Thermochemical Tables*, 2nd Ed., **NSRDS–NBS 37** (U.S. National Bureau of Standards, June 1971).

[b]R. Hultgren, P.D. Desai, D.T. Hawkins, M. Gleiser, K.K. Kelley, and D.D. Wagman, *Selected Values of the Thermodynamic Properties of the Elements* (American Society for Metals, Metals Park, Ohio, 1973), pp. 204-209.

would dominate the measurement. Therefore, estimates of various kinds must be made.

To first order, we expect the heat capacities of the amorphous phases to be fairly similar to those of the crystalline phases. For example, we expect the c_0 parameters for the amorphous phases to be very nearly the Dulong-Petit values, just as they were for the crystalline phases. Here, we estimate that they are in fact the same as those for the crystalline phases.

To second order, though, we expect differences. Although the amorphous phases retain the overall tetrahedral coordination and sp^3 bonding of the crystalline phases, their bond lengths and angles nevertheless deviate *locally* from those of perfect tetrahedra. Therefore, we expect their vibrational properties to be somewhat different. Indeed, both low-temperature calorimetry[10] and room-temperature sound velocity measurements[11] indi-

[10]M. Mertig, G. Pompe and E. Hegenbarth, "Specific heat of amorphous silicon at low temperatures," *Solid State Communications* **49**, 369 (1984).

[11]S.I. Tan, B.S. Berry and B.L. Crowder, "Elastic and anelastic behavior of ion-

Phase	Θ_T (K)	c_0 (meV/(atomK))	c_1 (10^{-5} meV/(atomK^2))
$\langle \mathrm{Si} \rangle_c$	149	0.241	3.85
$\langle \mathrm{Si} \rangle_a$	134	0.241	5.25
$\{\mathrm{Si}\}$	82	0.241	3.85
$\langle \mathrm{Ge} \rangle_c$	87	0.246	3.95
$\langle \mathrm{Ge} \rangle_a$	78	0.246	5.91
$\{\mathrm{Ge}\}$	48	0.246	3.95

Table 2.1: Heat capacity parameters for the condensed phases of Si and Ge.

cate Debye temperatures that are somewhat reduced from those of the crystalline phases. The reduction factor is only $\approx 10\%$ for the clean and dense films prepared by ion implantation, while it exceeds $\approx 30\%$ for sputter-deposited films. Since we expect the cleanest and densest films to be most representative of the fully relaxed amorphous phases, we estimate the semi-empirical critical temperatures of the amorphous phases to be reduced from those of the crystalline phases by only 10%.

Perhaps the most difficult parameters to estimate are the c_1 parameters. At medium to high temperatures, however, limited measurements[12] indicate that the heat capacity of $\langle \mathrm{Ge} \rangle_a$ exceeds that of $\langle \mathrm{Ge} \rangle_c$ by an amount that depends linearly on temperature, $\approx 3.92 \times 10^{-5} T$ meV/(atomK). Those measurements, however, were made on films deposited from the vapor under non-ultra-high-vacuum conditions. For clean, dense films we would expect the excess to be less. Indeed, by analogy to the reductions in sound velocities discussed above, we might expect them to be less by one-half or more. Here, we estimate them to be less by one-half, and hence estimate that the c_1 parameters for the amorphous phases are greater than those for the crystalline phases by the temperature coefficients 1.96×10^{-5} meV/(atomK) for Ge and, through scaling with the melting temperatures, by 1.40×10^{-5} meV/(atomK) for Si.

For $\{\mathrm{Si}\}$ and $\{\mathrm{Ge}\}$, experimental heat capacities are only known above the melting temperature, because it is experimentally very difficult to achieve significant supercoolings of those two liquids below their freezing temperature. For both liquids, though, the heats capacities just above the melting temperature are simple extrapolations of those of the crystalline solids just

implanted silicon," *Appl. Phys. Lett.* **20**, 88 (1972). For a recent review, see I.R. Cox-Smith, H.C. Liang and R.O. Dillon, "Sound velocity in amorphous films of germanium and silicon," *J. Vac. Sci. Technol.* **A3**, 674 (1985).

[12] H.S. Chen and D. Turnbull, "Specific heat and heat of crystallization of amorphous germanium," *J. Appl. Phys.* **40**, 4214 (1969).

Phase	$\Theta_{T,tra}$ (K)	$c_{o,tra}$	$\Theta_{T,rot}$ (K)	$c_{o,rot}$	$\Theta_{T,vib}$ (K)	$c_{o,vib}$
(Si)	$0.215\,(p/760\,\text{Torr})^{0.4}$	$5k/2$	–	0	–	0
$(\frac{1}{2}Si_2)$	$0.142\,(p/760\,\text{Torr})^{0.4}$	$5k/4$	0.338	$k/2$	735	$k/2$
(Ge)	$0.122\,(p/760\,\text{Torr})^{0.4}$	$5k/2$	–	0	–	0
$(\frac{1}{2}Ge_2)$	$0.080\,(p/760\,\text{Torr})^{0.4}$	$5k/4$	0.110	$k/2$	532	$k/2$

Table 2.2: Heat capacity parameters for the vapor phases of Si and Ge.

below the melting temperature. Hence, in both cases, it is reasonable to use the same values for the c_0 and c_1 parameters as those for the crystalline solids.

However, the vibrational properties of the liquid phases differ substantially from those of the crystalline phases, and so we expect their semi-empirical critical temperatures to be different. To first order, we can guess that the semi-empirical critical temperatures, just as the Debye temperatures, will scale as the velocity of sound in the phase. Since sound velocities in the liquid are approximately 0.55 of the velocities in the crystals,[13] we estimate that the semi-empirical critical temperatures of the liquids are approximately 0.55 of the critical temperatures of the crystals.

Vapor Phases

For the vapor phases (Si), (Ge), $(\frac{1}{2}Si_2)$ and $(\frac{1}{2}Ge_2)$, the parameters that enter into the translational, rotational and vibrational contributions to the heat capacities are given in Table 2.2. The translational critical temperatures follow from Equation 2.30 using masses of 28.08 amu and 56.16 amu for Si and Si_2, and masses of 72.6 amu and 145.2 amu for Ge and Ge_2. The rotational and vibrational temperatures follow from Equations 2.38 and 2.42 using bond lengths and vibrational frequencies of 2.246 Å and 510.98 cm^{-1} for Si_2,[14] and of 2.44 Å and 370 cm^{-1} for Ge_2.[15]

The parameters that enter into the electronic contributions to the heat capacities, viz., the energies and degeneracies of the lowest lying electronic levels, are listed in Table 2.3. For the Si and Ge monomers, the values were

[13]V.V. Baidov and M.B. Gitis, "Velocity of sound in and compressibility of molten germanium and silicon," *Sov. Phys. Semicond.* **4**, 825 (1970).

[14]From measurements by R.D. Verma and P.A. Warsop, "The absorption spectrum of the Si_2 molecule," *Can. J. Phys.* **41**, 152 (1963).

[15]From *ab initio* pseudopotential self-consistent-field calculations by G. Pacchioni, "Theoretical investigation of the electronic structure and of the potential energy curves for the lowest lying states of Ge_2," *Mol. Phys.* **49**, 727 (1983).

Molecule	Level ($^{(2S+1)}L_J$)	Degeneracy ($\omega_{i,\text{ele}}$)	Relative energy (eV)
Si	$^3P_0$	1	0
	$^3P_1$	3	0.00956
	$^3P_2$	5	0.02769
Ge	$^3P_0$	1	0
	$^3P_1$	3	0.069
	$^3P_2$	5	0.175
Si_2	$^3\Sigma_g^-$	3	0
	$^3\Pi_u$	6	0.13
	$^1\Delta_g$	2	0.48
	$^1\Pi_u$	2	0.66
	$^1\Sigma_g^+$	1	0.85
Ge_2	$^3\Sigma_g^-$	3	0
	$^3\Pi_u$	6	0.06
	$^1\Delta_g$	2	0.48
	$^1\Pi_u$	2	0.64
	$^1\Sigma_g^+$	1	0.57

Table 2.3: Energies and degeneracies for the lowest lying electronic levels of Si, Ge, Si_2 and Ge_2.

taken from the standard experimental tables of Moore.[16] For the Si_2 and Ge_2 dimers, the values were taken from theoretical calculations.[17] In all cases, the usual notation $^{(2S+1)}L_J$, where S, L and J are the spin, orbital and total angular momentum quantum numbers, has been used to denote the individual electronic levels. Note that the lowest lying electronic levels of (Si) and (Ge) have unusually low (fractions of an eV) energies. As a consequence, their low-temperature heat capacities are unusually high, as can be seen from Figure 2.1.

[16] C.E. Moore, *Atomic Energy Levels Vols. I-III*, Circular of the National Bureau of Standards **467** (June 15, 1949).

[17] The calculation for Si_2 is by A.D. Mclean, B. Liu and G.S. Chandler, "Second row homopolar diatomic molecules. Potential curves, spectroscopic constants, and dissociation energies at the basis set limit for SCF and limited CI wave functions," *J. Chem. Phys.* **80**, 5130 (1984). The calculation for Ge_2 is by G. Pacchioni, "Theoretical investigation of the electronic structure and of the potential energy curves for the lowest lying states of Ge_2," *Mol. Phys.* **49**, 727 (1983).

2.4.2 Molar Entropy and Enthalpy Offsets

In Subsection 2.4.1, we estimated the molar heat capacities of the various phases of Si and Ge. These estimates are summarized in Tables 2.1, 2.2 and 2.3, and allow one to deduce the *relative* temperature dependences of the molar entropies and enthalpies of the various phases of Si and Ge through Equations 2.24 and 2.25. However, to deduce the *absolute* temperature dependences, we also need to know molar entropy and enthalpy offsets for each phase at particular temperatures. In this subsection, we estimate these offsets. They are summarized in Table 2.4, and were deduced in the following ways. We will first consider the condensed phases and then the vapor phases.

Condensed Phases

For the crystalline phases, the third law of thermodynamics states that the entropy at absolute zero of a perfect crystalline substance is zero. Therefore, we can use as the entropy offset the entropy at 0 K, namely zero. The enthalpy scale, however, is not fixed by an equivalent law, so, as is customary, we arbitrarily fix the enthalpy of the equilibrium phases of Si or Ge at standard temperature and pressure (298 K and 760 Torr) to be zero. Since both $\langle Si \rangle_c$ and $\langle Ge \rangle_c$ are the equilibrium phases of Si or Ge at standard temperature and pressure, we set their enthalpies to zero at 298 K.

For the amorphous solid phases, we can estimate fairly well the molar enthalpy offsets *relative* to the molar enthalpies of the crystalline phases. In particular, the molar enthalpies of the amorphous phases relative to the molar enthalpies of the crystal phases are fixed by the latent heats of crystallization, which have been measured[18] to be 0.139 eV/atom at $\approx$ 950 K for Si and to be 0.119 eV/atom at $\approx$ 720 K for Ge. The molar entropy offsets, however, can only be tentatively estimated. Based on model-building studies, Spaepen[19] has calculated that the configurational entropy due to lattice disorder in the amorphous phases is approximately $0.2k$. Since at zero temperature that should be the only contribution to the entropy (the vibrational contribution should be frozen out), we estimate that the zero-temperature molar entropy for the amorphous phases is $0.2k$.

For the liquid phases, we can deduce experimentally *both* molar entropy and enthalpy offsets relative to the molar entropies and enthalpies of the crystalline phases. In particular, the molar enthalpies of the liquid phases

[18]E.P. Donovan, F. Spaepen, D. Turnbull, J.M. Poate and D.C. Jacobson, "Calorimetric studies of crystallization and relaxation of amorphous Si and Ge prepared by ion implantation," *J. Appl. Phys.* **57**, 1795 (1985).

[19]F. Spaepen, "On the configurational entropy of amorphous Si and Ge," *Phil. Mag.* **30**, 417 (1974).

Phase	T_o (K)	$h(T_o)$ (eV/atom)	T_o (K)	$s(T_o)$ (eV/(atomK))
$\langle Si\rangle_c$	298	0	0	0
$\langle Si\rangle_a$	950	$h^{\langle Si\rangle_c} + 0.139$	0	$0.2k$
$\{Si\}$	1685	$h^{\langle Si\rangle_c} + 0.525$	1685	$s^{\langle Si\rangle_c} + 0.525/1685$
(Si)	0	$h^{\langle Si\rangle_c} + 4.62$	0	0
$(\frac{1}{2}Si_2)$	0	$h^{\langle Si\rangle_c} + 3.07$	0	0
$\langle Ge\rangle_c$	298	0	0	0
$\langle Ge\rangle_a$	720	$h^{\langle Ge\rangle_c} + 0.119$	0	$0.2k$
$\{Ge\}$	1213	$h^{\langle Ge\rangle_c} + 0.382$	1213	$s^{\langle Ge\rangle_c} + 0.382/1213$
(Ge)	0	$h^{\langle Ge\rangle_c} + 3.81$	0	0
$(\frac{1}{2}Ge_2)$	0	$h^{\langle Ge\rangle_c} + 2.43$	0	0

Table 2.4: s and h offsets for the various phases of Si and Ge.

relative to the molar enthalpies of the crystal phases are fixed by experimental measurements of the latent heats of fusion at the freezing temperatures. As summarized in Table 2.4, the latent heats of fusion and freezing temperatures for Si are 0.525 eV and 1685 K and for Ge are 0.382 eV and 1213 K.

Note that exactly at the melting temperature, the crystal and liquid phases coexist, and so their molar Gibbs free energies must be the same. Hence, if the crystal and liquid phases have different molar enthalpies, they must also have different molar entropies, in exactly offsetting amounts. Therefore, *exactly at the melting temperature*, $T_o = T^{\langle\rangle_c\{\}}$, the entropies of fusion can be deduced from the molar enthalpies of fusion via

$$s^{\{\}}(T_o) - s^{\langle\rangle_c}(T_o) \equiv \frac{h^{\{\}}(T_o) - h^{\langle\rangle_c}(T_o)}{T_o}. \tag{2.43}$$

It is important to keep in mind that we have now fixed the liquid molar entropy and enthalpy offsets relative to the crystal molar entropies and enthalpies at the melting temperature. However, the crystal molar entropies and enthalpies were themselves fixed at *different* temperatures, namely, 0 K for the entropy and 298 K for the enthalpy. Therefore, to deduce the *absolute* molar entropies and enthalpies of the liquid we must first use Equations 2.8 and 2.9 (along with the heat capacity of the crystal) to deduce the change in the molar entropy of the crystal between 0 K and the melting temperature and the change in the molar enthalpy of the crystal between 298 K and the melting temperature. Because of this, our estimates of the absolute molar entropies and enthalpies of the liquid depend both on the measured molar entropies and enthalpies of fusion as well as on the assumed molar heat capacity of the crystal.

Vapor Phases

For the vapor phases, the entropy offsets are fixed at zero at 0 K by the third law of thermodynamics.[20] The molar enthalpy offsets can be fixed relative to the enthalpies of either the crystal or liquid phases. For example, the molar enthalpies of the vapor phases relative to the molar enthalpies of the crystal can be fixed by experimental measurements of the latent heats of sublimation. Equivalently, they can be fixed relative to the molar enthalpies of the liquid by experimental measurements of the latent heats of evaporation.

The complication, though, is that experimental determinations of latent heats of sublimation or evaporation are usually not based directly on calorimetry, but rather indirectly on measurements of the equilibrium vapor pressure or of its temperature dependence. To see how, note that if, at a particular temperature, either the crystal or the liquid is in equilibrium with a certain pressure of vapor, then the molar Gibbs free energies of the crystal or liquid must equal the molar Gibbs free energy of vapor at that pressure. For example, if we consider sublimation of the crystal, we can write

$$g^{\langle\rangle_c} = g^{()} \tag{2.44}$$

$$h^{\langle\rangle_c} - Ts^{\langle\rangle_c} = h^{()} - Ts^{()}. \tag{2.45}$$

For both the monomer and dimer vapors, we see from Equations 2.32 and 2.30 that the translational contributions to the entropies separate into the sum of (nearly) pressure-independent and temperature-independent parts:

$$s^{()}(p,T) = s^{()}(p_o,T) - \frac{2}{5}c_{o,\mathrm{tra}} \ln\left(\frac{p}{p_o}\right), \tag{2.46}$$

where the reference pressure, p_o, is arbitrary and can be taken, e.g., to be 760 Torr, and the reference temperature, T_o, has been taken to be 0. Therefore, the pressure dependence can be separated from the molar Gibbs free energy of the vapor:

$$g^{()}(p,T) = g^{()}(p_o,T) + \frac{2}{5}c_{o,\mathrm{tra}} T \ln\left(\frac{p}{p_o}\right). \tag{2.47}$$

[20]Except, as noted above, for a possible degeneracy in the ground electronic level.

In other words, the equality between the molar Gibbs free energies of the crystal and the vapor becomes

$$\ln\left(\frac{p}{p_o}\right) = \frac{g^{\langle\rangle_c}(T) - g^{()}(p_o, T)}{2c_{o,\mathrm{tra}}T/5}. \tag{2.48}$$

This equation, when expanded into its entropy and enthalpy components, is the basis for what is known as the Rankin-Kirchoff semi-empirical equation,

$$\ln(p) = \alpha + \delta \ln T - \frac{\beta}{T} + \gamma T, \tag{2.49}$$

often used to describe vapor pressures.[21]

Since from the heat capacities we know the *relative* temperature dependences of the molar entropies and enthalpies of both the vapor and the crystal, Equation 2.48 defines the equilibrium vapor pressures in terms of the molar entropy and enthalpy offsets between the vapor and crystal. Therefore, those relative molar entropy and enthalpy offsets (essentially the latent entropies and heats of sublimation) can be deduced from a comparison between Equation 2.48 and experimental measurements.

In making that comparison, two methods are commonly used. In the first method, known as the "third-law" method, one assumes that the third law of thermodynamics holds. Then, the 0 K entropies of the crystal and vapor are taken to be zero, except for a possible degeneracy of the ground electronic state.

The difference between the molar enthalpies of the crystal and the vapor at a reference temperature T_o is then used to fit the absolute magnitude of the equilibrium vapor pressure at a particular temperature. Such fits are illustrated in Figure 2.2, which shows experimental measurements of the equilibrium vapor pressures of the crystals (below the freezing temperatures) and of the liquids (above the freezing temperatures), along with monomer and dimer vapor pressures calculated using Equation 2.48 and the molar enthalpy offsets listed in Table 2.4. The fits are extremely sensitive to those molar enthalpy offsets: the higher the latent heat of vaporization, the lower the equilibrium vapor pressure. Note that the dimer vapor pressures for both Si and Ge are only a fraction of the monomer vapor pressures, and hence for practical purposes may be neglected.

It should be noted that the offsets listed in Table 2.4 are the heats of sublimation at 0 K. At any other temperature those heats will be slightly different, due to heat capacity differences between the crystal and vapor, or

[21] An extensive tabulation of empirical values for the coefficients in this equation for a number of semiconductors may be found in O. Kubaschewski and C.B. Alcock, *Metallurgical Thermochemistry*, 5th Ed. (Pergamon Press, Oxford, 1979).

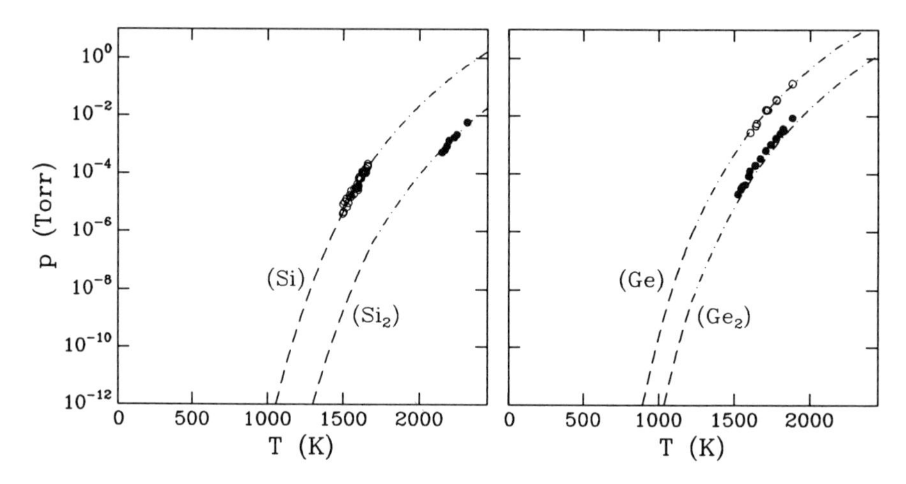

Figure 2.2: Equilibrium vapor pressures over the crystals (dashed lines) and liquids (dot-dashed lines). Experimental measurements are shown as open circles for the monomers Si[a] and Ge[b] and as filled circles for the dimers Si_2[c] and Ge_2.[d]

[a] J.L. Souchiere and V.T. Binh, "The evaporation rate of silicon," *Surf. Sci.* **168**, 52 (1986); and R.L. Batdorf and F.M. Smits, "Diffusion of impurities into evaporating silicon," *J. Appl. Phys.* **30**, 259 (1959).

[b] A.W. Searcy and R.D. Freeman, "Measurement of the molecular weights of vapors at high temperature. II. The vapor pressure of germanium and the molecular weight of germanium vapor," *J. Chem. Phys.* **23**, 88 (1955).

[c] J. Drowart, G. De Maria, and M.G. Inghram, "Thermodynamic study of SiC utilizing a mass spectrometer," *J. Chem. Phys.* **29**, 1015 (1958).

[d] J. Drowart, G. De Maria, A.J.H. Boerboom, and M.G. Inghram, "Mass spectrometric study of inter-group IVB molecules," *J. Chem. Phys.* **30**, 308 (1959).

between the liquid and vapor. For example, the temperature dependence of the heat of sublimation is given by

$$\Delta h^{()c()}(T) = \Delta h^{()c()}(T_o) + \int_{To}^{T} \Delta c_p^{()c()}(T')dT'. \tag{2.50}$$

In the second method, known as the "second-law" method, the latent heat of sublimation is deduced from the *variation* with temperature of the equilibrium vapor pressure, rather than from the absolute magnitude of the equilibrium vapor pressure. Indeed, using Equations 2.18 and 2.48, the temperature variation of the equilibrium vapor pressure is seen to be related to the latent heat of sublimation,

$$\frac{\partial}{\partial(1/T)}\left[k \ln\left(\frac{p}{p_o}\right)\right] = h^{()c}(T) - h^{()}(p_o, T). \tag{2.51}$$

As illustrated in Figure 2.2, Equation 2.48, using the molar enthalpy offsets listed above, reproduces reasonably well both the absolute magnitudes as well as the temperature variations of the equilibrium vapor pressures. Therefore, the two methods agree, and that agreement gives confidence in the experimental measurements themselves.

2.4.3 Molar Gibbs Free Energies

In Subsection 2.4.1, we estimated the molar heat capacities summarized in Tables 2.1 and 2.2; in Subsection 2.4.2, we estimated the molar entropy and enthalpy offsets summarized in Table 2.4. In this subsection, we combine the two to deduce the *absolute* temperature dependences of the molar entropies, enthalpies, and Gibbs free energies of the various phases of Si and Ge. For the condensed phases we use Equations 2.7, 2.24 and 2.25, and for the vapor phases we use Equations 2.7, 2.27, 2.28 and 2.29.

The top four panels of Figure 2.3 show the molar entropies and enthalpies. Note that the molar entropies and enthalpies behave quite differently as functions of temperature. The molar enthalpies are initially flat at the lower temperatures where the molar heat capacity is low, then increase linearly at the higher temperatures where the molar heat capacity is high and approximately constant. The molar entropies, in contrast, rise more quickly at the lower temperatures, then saturate at the higher temperatures due to the inverse temperature dependence of the integrand in Equation 2.8. However, this saturation of the entropies at high temperatures does *not* imply that entropy becomes increasingly unimportant at high temperatures. Rather, entropy enters into the molar Gibbs free energy multiplied by T, and so continues to be of major importance in determining phase stability.

As one might expect, the magnitudes of the molar enthalpies and entropies of the solid phases are quite similar, those of the liquid phases are somewhat higher, and those of the vapor phases are very much higher. In other words, the difference between solid phases is less than that between liquid and solid phases, which in turn is much less than that between vapor and liquid phases.

Note that the zero-temperature molar entropies of the liquid phases do not vanish. Instead, they are approximately 3K. Such residual entropies arise from the *configurational* disorder present even in the "frozen" liquid. That disorder is analogous to, though much larger than, that of the amorphous solids, whose zero-temperature molar entropies were estimated to be only $\approx 0.2\,\mathrm{K}$.

The bottom two panels of Figure 2.3 show the molar Gibbs free energies. At any temperature, the stablest phase is that whose molar Gibbs free energy is lowest, and the least stable phase is that whose molar Gibbs free energy is highest. Therefore, in these examples, the crystalline phases are the stablest ones at low temperature and the vapor phases are the stablest ones at high temperature. In between, depending on the pressure of the vapor, there may be a temperature window in which the liquid is stablest.

The reason for this trend is that, as mentioned earlier, at any particular temperature the slope of a molar Gibbs free energy curve is the negative of the molar entropy, and its zero-temperature intercept is the molar enthalpy. Therefore, since all phases have positive entropy, molar Gibbs free energies must decrease with increasing temperature, i.e., all phases are stabilized at higher temperature. However, the molar Gibbs free energies of higher entropy phases decrease faster than those of lower entropy phases, and so higher entropy phases ultimately become stabler than lower entropy phases at high temperature.

Note that the amorphous phases are *never* the stablest phases. At low temperatures the crystalline phases are more stable; at high temperatures the liquid phases are more stable. Despite this, the amorphous phases *may* be observed and studied, because at room temperature they transform into the crystalline phases exceedingly sluggishly. It has even been possible, through pulsed-laser-annealing experiments, to deduce that the melting temperature of $\langle \mathrm{Si} \rangle_a$ is *lower* than that of $\langle \mathrm{Si} \rangle_c$ by $\approx 225\,\mathrm{K}$.[22] From Figure 2.3, such a reduction is expected: the molar Gibbs free energies of the amorphous phases lie above those of the crystalline phases, and must intersect those of the liquid phases at lower temperatures.

Note that the temperature at which the molar Gibbs free energies of the vapor intersect those of the various condensed phases depends on pressure. That dependence is what determines the vapor pressure of the condensed phase. Note that above the melting temperature, the vapor and liquid molar Gibbs free energy curves intersect at a higher temperature than the vapor and crystal molar Gibbs free energy curves intersect. Hence, to be in equilibrium with a given vapor pressure, the liquid must be hotter than the crystal. Equivalently, for a given temperature above the melting temperature, the liquid has a lower vapor pressure than the (superheated) crystal. Note that such a lowering of the vapor pressure upon melting is difficult to observe experimentally, because crystals usually melt at only slight overheatings above their melting temperature.

Finally, once we know the molar Gibbs free energies of the various phases of Si and Ge, we can calculate the "driving force" for MBE. As can be seen from Figure 2.3, the driving force can be quite large. For example, a typical

[22]M.O. Thompson, G.J. Galvin, J.W. Mayer, P.S. Peercy, J.M. Poate, D.C. Jacobson, A.G. Cullis, and N.G. Chew, "Melting temperature and explosive crystallization of amorphous silicon during pulsed laser irradiation," *Phys. Rev. Lett.* **52**, 2360 (1984).

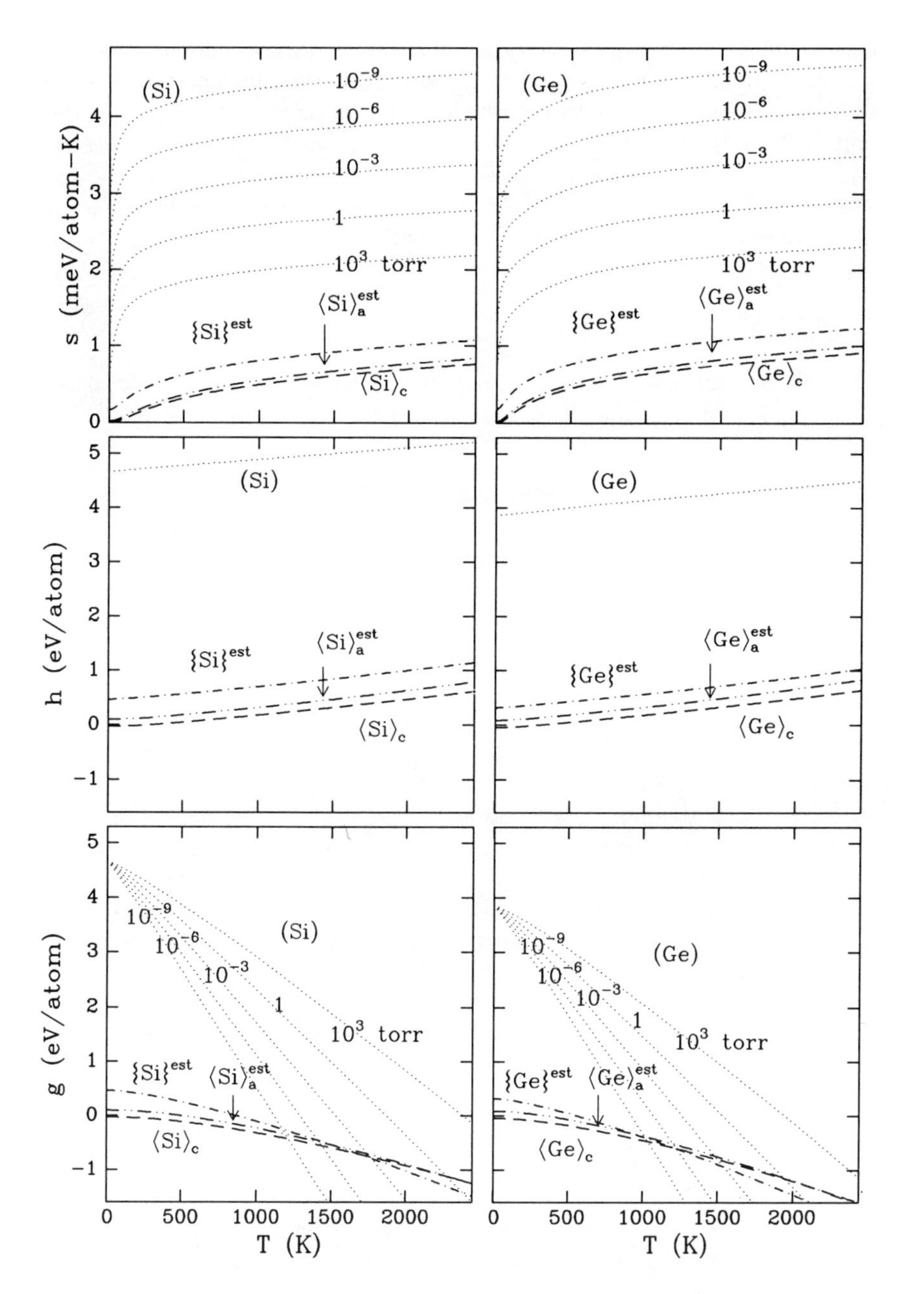

Figure 2.3: s, h, and g for the various phases of Si (left) and Ge (right).

growth temperature for Si MBE is 800 K, and a typical effective pressure for the impinging atoms is 10^{-6} Torr. Under those conditions the difference between the molar Gibbs free energies of the vapor and crystal is approximately 2.5 eV/atom. At that same temperature, the difference between the molar Gibbs free energies of the liquid and crystal is only approximately 0.25 eV/atom, about one order of magnitude less. The vapor, in a sense, is much more supercooled than the liquid with respect to the crystal.

Suggested Reading

1. O. Kubaschewski and C.B. Alcock, *Metallurgical Thermochemistry*, 5th Ed. (Pergamon Press, Oxford, 1979).

2. D.A. McQuarrie, *Statistical Mechanics* (Harper & Row, New York, 1976).

3. R.A. Swalin, *Thermodynamics of Solids*, 2nd Ed. (Wiley-Interscience, New York, 1972).

Exercises

1. Using Equation 2.21, show, for an ideal gas that obeys $pv = kT$, that $c_p = c_v + k$. Explain, *physically*, why c_p is greater than c_v.

2. We calculated, using Equation 2.43, molar entropy offsets in terms of measured molar enthalpy offsets for $\{\mathrm{Si}\}$ and $\{\mathrm{Ge}\}$ relative to the crystalline phases $\langle\mathrm{Si}\rangle_\mathrm{c}$ and $\langle\mathrm{Ge}\rangle_\mathrm{c}$. Could we have calculated, in a similar way, the molar entropy offsets for the amorphous phases $\langle\mathrm{Si}\rangle_\mathrm{a}$ and $\langle\mathrm{Ge}\rangle_\mathrm{a}$ relative to the crystalline phases $\langle\mathrm{Si}\rangle_\mathrm{c}$ and $\langle\mathrm{Ge}\rangle_\mathrm{c}$?

3. Show, using Equations 2.8 and 2.9, that dg/dT is $-s$, and hence that Equation 2.19 is correct. Show that $[d/d(1/T)][g/T]$ is h, and hence that Equation 2.18 is correct.

4. Expand the right side of Equation 2.48 into molar enthalpy and entropy components, and deduce explicit forms for the coefficients in Equation 2.49.

5. Assuming Dulong-Petit heat capacities (neglecting electronic contributions), and given the latent heats of sublimation listed in Table 2.4, what are the *zero-temperature* molar enthalpies of formation of $(\frac{1}{2}\mathrm{Si}_2)$ and $(\frac{1}{2}\mathrm{Ge}_2)$ from (Si) and (Ge), respectively?

6. Although, at 800 K, the difference between the molar Gibbs free energies of (Si) at 10^{-6} Torr and $\langle \mathrm{Si} \rangle_c$ is large ($\approx 2.5\,\mathrm{eV/atom}$), the difference between the molar enthalpies of (Si) and $\langle \mathrm{Si} \rangle_c$ is larger still ($\approx 5\,\mathrm{eV/atom}$). That enthalpy is "liberated" when a Si atom impinges on and sticks to a Si substrate and ultimately becomes thermal energy. How does that liberated thermal energy compare to the thermal energy per atom ($\approx 3kT$) of the Si substrate at 800 K?

Chapter 3

Alloy Phases

In Chapter 2, we described how to calculate the molar Gibbs free energies of phases composed of a single component or element. For condensed phases it was sufficient to understand the temperature dependence; for vapor phases it was necessary to understand both the pressure and temperature dependences.

In this chapter, we introduce *alloy* phases, i.e., phases composed of more than one component or element. For these phases, the molar Gibbs free energies depend on yet another parameter: composition. Therefore, we begin, in Section 3.1, by introducing a simple and consistent nomenclature for alloy phases that includes composition. At the same time, we will ask the general question: given a pressure, temperature, *and* composition, what is the equilibrium mix of phases of the system? The answer is determined, again, by minimizing the Gibbs free energy. We shall find that under some conditions, the Gibbs free energy is minimized when only a single phase is present. Under other conditions, though, the Gibbs free energy is minimized when two phases are present in a particular proportion.

Then, in Section 3.2, we describe the commonly used semi-empirical expressions for the composition dependence of the molar Gibbs free energies of various phases. The expressions are classified according to whether they apply to "perfect" solutions (mixtures of ideal gases), "ideal" solutions (condensed-phase mixtures of chemically similar components), or "regular" solutions (condensed-phase mixtures of chemically dissimilar components).

Finally, Sections 3.3 and 3.4 are devoted to detailed case studies of two particular alloy systems, selected because they are both of current interest and importance in MBE, and because they each illustrate different but important aspects of molar Gibbs free energy calculations. In order of increasing complexity, we consider, in Section 3.3, the alloy phases of

$Si_{1-x}Ge_x$, and then, in Section 3.4, the stoichiometric compound and other phases of $Ga_{1-x}As_x$.

3.1 Nomenclature and Preliminaries

Let us start by introducing a standard nomenclature. As in Chapter 2, we have tried to make ours as simple and self-explanatory as possible. We begin, in Subsection 3.1.1, by describing the properties of the phases themselves. Then, given an understanding of those properties, we discuss, in Subsection 3.1.2, under what conditions two (or more) phases can *coexist*.

3.1.1 One Phase, Two Components

Consider a phase, α, this time composed of N_{a}^{α} moles (or atoms) of component a and N_{b}^{α} moles (or atoms) of component b. The total Gibbs free energy of such a phase is denoted $G^{\alpha}(p,T,N_{\mathrm{a}}^{\alpha},N_{\mathrm{b}}^{\alpha})$. From it are derived the *molar* Gibbs free energy of the α phase,

$$g^{\alpha}(p,T,x^{\alpha}) = \frac{G^{\alpha}(p,T,N_{\mathrm{a}}^{\alpha},N_{\mathrm{b}}^{\alpha})}{N_{\mathrm{a}}^{\alpha}+N_{\mathrm{b}}^{\alpha}}, \tag{3.1}$$

and the chemical potentials of the two components in the α phase,

$$\begin{aligned} \mu_{\mathrm{a}}^{\alpha}(p,T,x^{\alpha}) &= \left[\frac{\partial G^{\alpha}(p,T,N_{\mathrm{a}}^{\alpha},N_{\mathrm{b}}^{\alpha})}{\partial N_{\mathrm{a}}^{\alpha}}\right]_{N_{\mathrm{b}}^{\alpha}} \\ \mu_{\mathrm{b}}^{\alpha}(p,T,x^{\alpha}) &= \left[\frac{\partial G^{\alpha}(p,T,N_{\mathrm{a}}^{\alpha},N_{\mathrm{b}}^{\alpha})}{\partial N_{\mathrm{b}}^{\alpha}}\right]_{N_{\mathrm{a}}^{\alpha}}. \end{aligned} \tag{3.2}$$

Notice that the atomic fractions of the two components, $x_{\mathrm{a}}^{\alpha} \equiv N_{\mathrm{a}}^{\alpha}/(N_{\mathrm{a}}^{\alpha}+N_{\mathrm{b}}^{\alpha})$ and $x_{\mathrm{b}}^{\alpha} \equiv N_{\mathrm{b}}^{\alpha}/(N_{\mathrm{a}}^{\alpha}+N_{\mathrm{b}}^{\alpha})$, are not independent of each other, but sum to unity. Therefore, the molar Gibbs free energies and the chemical potentials can be written as functions of only one of the atomic fractions. As is customary, we arbitrarily choose x_{b}^{α}, which we write as x^{α}.

As before, it is important to keep in mind that g^{α} is a property of a *phase* and is used to compare relative stabilities of different phases. In contrast, $\mu_{\mathrm{a}}^{\alpha}$ and $\mu_{\mathrm{b}}^{\alpha}$ are properties of *components* in a phase, and are used to compare relative propensities of those components to incorporate into different phases.

Often, it is convenient to express the two chemical potentials in terms of the *molar* Gibbs free energies, rather than the Gibbs free energies:

$$\mu_{\mathrm{a}}^{\alpha}(p,T,x^{\alpha}) = g^{\alpha}(p,T,x^{\alpha}) - (x^{\alpha})\frac{\partial g^{\alpha}(p,T,x^{\alpha})}{\partial x^{\alpha}} \tag{3.3}$$

$$\mu_{\mathrm{b}}^{\alpha}(p,T,x^{\alpha}) \quad = \quad g^{\alpha}(p,T,x^{\alpha}) + (1-x^{\alpha})\frac{\partial g^{\alpha}(p,T,x^{\alpha})}{\partial x^{\alpha}}. \tag{3.4}$$

These two Equations can be derived from Equations 3.2 (see Exercise 1 at the end of this chapter) through use of the identity

$$G^{\alpha}(p,T,N_{\mathrm{a}}^{\alpha},N_{\mathrm{b}}^{\alpha}) = (N_{\mathrm{a}}^{\alpha}+N_{\mathrm{b}}^{\alpha})g(N_{\mathrm{b}}^{\alpha}/[N_{\mathrm{a}}^{\alpha}+N_{\mathrm{b}}^{\alpha}]). \tag{3.5}$$

Mathematically, they imply that the intercepts of the tangent to the molar Gibbs free energy vs. composition curve with the $x = 0$ and $x = 1$ axes are the chemical potentials themselves. The two Equations thus form the basis for what is known as the tangent construction for *graphically* deducing chemical potentials.

That construction is illustrated in Figure 3.1. Notice that for the hypothetical molar Gibbs free energy curve drawn, as x^{α} approaches 0 (so that α becomes poor in component b), the chemical potential of b in α, $\mu_{\mathrm{b}}^{\alpha}$, becomes very negative. In other words, the less of component b the phase α has, the more that component is attracted to the phase. In real materials at commonly encountered temperatures, this will often be the case; it is due to the entropy, described later, gained by mixing the two components a and b.

Using these equations, the molar Gibbs energy and the slope of the molar Gibbs free energy can in turn be expressed in terms of $\mu_{\mathrm{a}}^{\alpha}$ and $\mu_{\mathrm{b}}^{\alpha}$:

$$g^{\alpha}(p,T,x^{\alpha}) \quad = \quad (1-x^{\alpha})\mu_{\mathrm{a}}^{\alpha}(p,T,x^{\alpha}) + (x^{\alpha})\mu_{\mathrm{b}}^{\alpha}(p,T,x^{\alpha}) \tag{3.6}$$

$$\frac{\partial g^{\alpha}(p,T,x^{\alpha})}{\partial x^{\alpha}} \quad = \quad \mu_{\mathrm{b}}^{\alpha}(p,T,x^{\alpha}) - \mu_{\mathrm{a}}^{\alpha}(p,T,x^{\alpha}). \tag{3.7}$$

Equation 3.6 tells us that the molar Gibbs free energy is the sum of the chemical potentials of each component, weighted according to the atomic fraction of those components. Equation 3.7 tells us that the derivative of the molar Gibbs free energy with respect to composition is the difference between the two chemical potentials.

3.1.2 Two Phases, Two Components

Suppose, now, that there are two phases: α, an a-rich phase, and β, a b-rich phase. Are there conditions under which these two phases can *coexist* in equilibrium? The equilibrium condition, of course, is that the *total* Gibbs free energy,

$$G^{\mathrm{tot}} = G^{\alpha}(p,T,N_{\mathrm{a}}^{\alpha},N_{\mathrm{b}}^{\alpha}) + G^{\beta}(p,T,N_{\mathrm{a}}^{\beta},N_{\mathrm{b}}^{\beta}), \tag{3.8}$$

be minimized with respect to transfer of atoms from one phase to the other. If so, then the change in the total Gibbs free energy should vanish if we

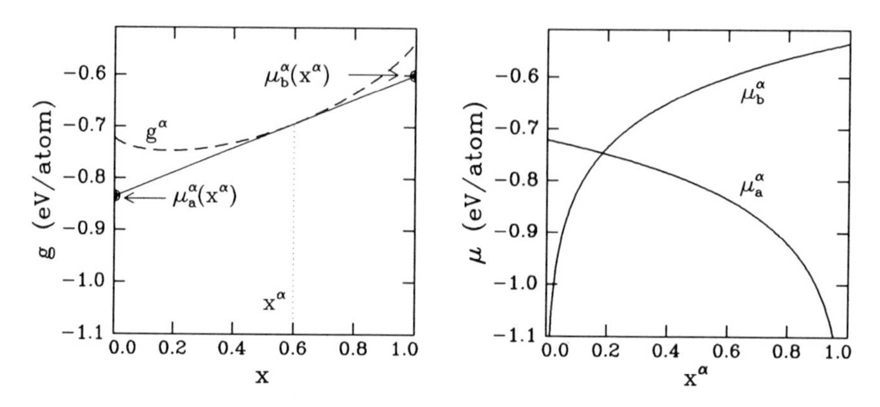

Figure 3.1: The tangent construction for deducing chemical potentials. On the left, a hypothetical molar Gibbs free energy for a phase, g^α, is plotted versus composition, x. The intercepts of the tangent at $x = x^\alpha$ with the $x = 0$ and $x = 1$ axes are the two chemical potentials $\mu_a^\alpha(x^\alpha)$ and $\mu_b^\alpha(x^\alpha)$. As x^α varies, the tangents sweep around the arc of the molar Gibbs free energy curve, and the intercepts of those tangents trace out the chemical potentials at the various x^α. On the right, the two chemical potentials obtained in this way are plotted versus composition, x^α.

increase N_a^α while decreasing N_a^β so that the overall number of a atoms is constant:

$$\frac{\partial G^{\text{tot}}}{\partial N_a^\alpha} = \frac{\partial G^\alpha}{\partial N_a^\alpha} + \frac{\partial G^\beta}{\partial N_a^\alpha} = \frac{\partial G^\alpha}{\partial N_a^\alpha} - \frac{\partial G^\beta}{\partial N_a^\beta} = \mu_a^\alpha - \mu_a^\beta = 0. \tag{3.9}$$

Similarly, the change in the total Gibbs free energy should vanish if we increase N_b^α while decreasing N_b^β so that the overall number of b atoms is constant:

$$\frac{\partial G^{\text{tot}}}{\partial N_b^\alpha} = \frac{\partial G^\alpha}{\partial N_b^\alpha} + \frac{\partial G^\beta}{\partial N_b^\alpha} = \frac{\partial G^\alpha}{\partial N_b^\alpha} - \frac{\partial G^\beta}{\partial N_b^\beta} = \mu_b^\alpha - \mu_b^\beta = 0. \tag{3.10}$$

In other words, the chemical potentials of a in the two phases must be equal, as must the chemical potentials of b in the two phases. To see why, remember that the chemical potentials measure the propensity for atoms to incorporate into a phase. If those propensities were not the same for two phases, then there would be a tendency for atoms to transfer between the two phases, and the two phases could not have been in equilibrium.

Now, we know from Equations 3.3 and 3.4 that the chemical potentials are the intercepts of the tangents to the Gibbs free energy curves. Hence, if the two phases α and β are to be in equilibrium with each other,

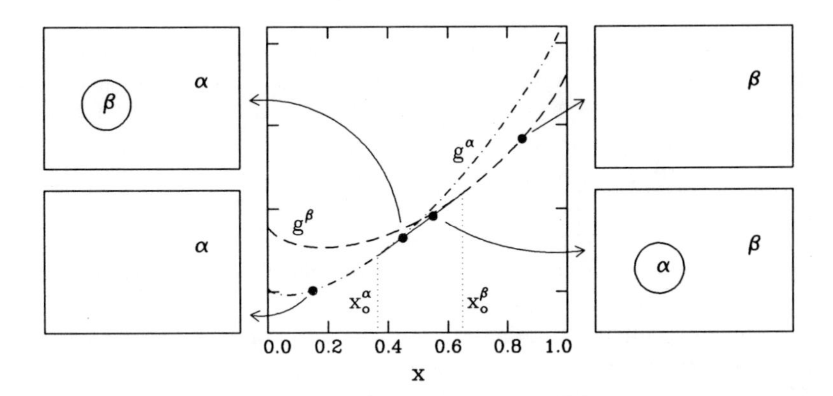

Figure 3.2: Hypothetical molar Gibbs free energies for two phases, α and β, plotted versus composition, x. Their (common) tangent defines the compositions x_o^β and x_o^α at which the two phases are in equilibrium with each other. The intercepts of their (common) tangent with the $x = 0$ and $x = 1$ axes are their two (common) chemical potentials. For overall system compositions and molar Gibbs free energies indicated by the various solid circles, the two phases will be present in the proportions indicated schematically in the various surrounding panels.

their compositions x_o^α and x_o^β must be such that the tangents to g^α and g^β at those compositions have the same intercepts. In other words, as illustrated in Figure 3.2, the two phases must *share a common tangent.*[1] That simple geometric construction can be deduced mathematically by equating Equation 3.3 for μ_{a}^α with the analogous equation for μ_{a}^β, and by equating Equation 3.4 for μ_{b}^α with the analogous equation for μ_{b}^β:

$$
\begin{aligned}
\left[\frac{\partial g^\alpha(p,T,x^\alpha)}{\partial x^\alpha}\right]_{x_o^\alpha} &= \frac{g^\beta(p,T,x_o^\beta) - g^\alpha(p,T,x_o^\alpha)}{x_o^\beta - x_o^\alpha} \\
\left[\frac{\partial g^\beta(p,T,x^\beta)}{\partial x^\beta}\right]_{x_o^\beta} &= \frac{g^\beta(p,T,x_o^\beta) - g^\alpha(p,T,x_o^\alpha)}{x_o^\beta - x_o^\alpha}.
\end{aligned}
\tag{3.11}
$$

Equations 3.11 are the central equations of this section. Their simultaneous solution determines the compositions, x_o^α and x_o^β, at which two phases are in equilibrium with each other, in terms of the composition-dependent molar Gibbs free energies of each phase. If those dependences are simple, then the equations may sometimes be solved analytically, at fixed temperature and pressure, for those equilibrium compositions x_o^α and

[1] J.W. Gibbs, "On the equilibrium of heterogeneous substances," *The Collected Works of J. Willard Gibbs*, Vol. I: Thermodynamics (Yale University Press, New Haven, 1957), pp. 55-353.

x_o^β. More commonly, however, the equations must be solved numerically.[2]

Once we know the equilibrium compositions x_o^α and x_o^β, then we also know, for a given overall *system* composition, x, what fraction $1 - f_o$ of the system is α and what fraction f_o is β. As illustrated in Figure 3.2, if $x < x_o^\alpha$ or $x > x_o^\beta$ (the two extreme solid circles in the center panel of Figure 3.2), then the system will either be pure α or pure β, respectively. In other words, pure α at a composition $x < x_o^\alpha$ has a lower molar Gibbs free energy than any mixture of α at composition x^α and β at composition $x^\beta = 2x - x^\alpha$. Likewise, pure β at a composition $x > x_o^\beta$ has a lower molar Gibbs free energy than any mixture of β at composition x^β and α at composition $x^\alpha = 2x - x^\beta$.

If, however, $x > x_o^\alpha$ and $x < x_o^\beta$ (the two middle solid circles lying on the tangent in the center panel of Figure 3.2), then the equilibrium state of the system will be a *mixture* of the α and β phases, whose fractions are given by what is known as the lever rule,

$$\frac{1 - f_o}{f_o} = \frac{x_o^\beta - x}{x - x_o^\alpha}. \tag{3.12}$$

In other words, the ratio between the β and α fractions of the system is equal to the ratio of the differences between the system composition x and the equilibrium compositions x_o^β and x_o^α. In terms of these fractions, the overall molar Gibbs free energy of the system is then

$$g_o^{\text{tot}} = (1 - f_o)g^\alpha(x_o^\alpha, T) + (f_o)g^\beta(x_o^\beta, T). \tag{3.13}$$

It is important to keep in mind that two phases need not necessarily be in equilibrium with each other. Indeed, they will often *not* be in equilibrium, particularly if the migration rate of various components between phases is slow. Then, the overall molar Gibbs free energy of the system will be

$$g^{\text{tot}} = (1 - f)g^\alpha(x^\alpha, T) + (f)g^\beta(x^\beta, T). \tag{3.14}$$

---

[2]The simplest algorithms are based explicitly on the common tangent construction. For example, in the algorithm implemented by M.O. Thompson and L.R. Doolittle in *PHASE5: A Program for Calculating Binary Phase Diagrams* (Cornell Research Foundation, Computer Graphic Service, 221 Asbury Rd., Ithaca, NY 14850), the composition-dependent molar Gibbs free energies of all the phases of interest are calculated. Then, their minimal envelope is determined. That minimal envelope will be characterized, in general, by various maxima and minima separated by points of inflection. Straddling the leftmost point of inflection must lie the leftmost pair of equilibrium compositions. That pair is found by numerically searching for the common tangent nearest that point of inflection that lies below all other points on the minimal envelope. Then, working rightward, successive pairs of equilibrium compositions are found straddling successive points of inflection.

The actual fraction of the system that is in the β phase, f, need not be that given by the lever rule, and the actual compositions x^α and x^β of those phases need not be those determined by Equations 3.11. When this is so, g^{tot} will be greater than g_o^{tot}, and the amount by which it is greater is a measure of the deviation of the system from equilibrium.

Another way of looking at this is to note that if two phases are not in equilibrium with each other, then Equations 3.9 and 3.10 are, by definition, not satisfied. Then, either the chemical potential of a in α differs from that of a in β, or the chemical potential of b in α differs from that of b in β (or both). Hence, there will be a *driving force* for a or b atoms (or both) to transfer between the α and β phases, and the two phases could *not* have been in equilibrium with each other.

3.2 Models of Solutions

In Section 3.1, we discussed how, *given* dependences of the molar Gibbs free energies of two (or more) phases on composition, various quantities of experimental interest could be calculated. For example, the compositions of phases in equilibrium with each other can be calculated using the common tangent construction, and the degree to which a mixture of phases at a given composition deviates from equilibrium is the difference between Equations 3.13 and 3.14.

In this Section, we discuss how the molar Gibbs free energies of various phases depend on composition. This question has long occupied a prominent place in materials science and solid-state physics. Ultimately, rapid advances in theory and computational hardware may allow it to be answered by first-principles calculations. Indeed, in many cases it is already possible to predict the order of magnitude of alloy heats of formation.[3] Except in a few model systems, however, such calculations are not yet accurate enough for quantitative work. Therefore, it is common to describe experimentally determined composition dependences of the molar Gibbs free energies of alloys by simple semi-empirical expressions. These expressions are usually algebraic, and contain parameters fit either to experimental data or, occasionally, to first-principles calculations.

The various expressions currently in common use are summarized in Table 3.1. In discussing them, we will find it convenient to classify them according to the strength of the interaction between the two components. "Perfect" solutions are those in which the two components do not interact at all with each other, and are discussed in Subsection 3.2.1. "Ideal"

[3]See, e.g., A.R. Miedema, P.F. de Châtel, and F.R. de Boer, "Cohesion in alloys – fundamentals of a semi-empirical model," *Physica* **100B**, 1 (1980).

Type of Solution	h_{mix}	s_{mix}
Perfect	0	0
Ideal	0	$s_{\mathrm{mix,ideal}}$
Strictly Regular	$\Omega x(1-x)$	$s_{\mathrm{mix,ideal}}$
Quasi Regular	$\Omega_{\mathrm{h}}(1-x)$	$s_{\mathrm{mix,ideal}}$ $+\Omega_{\mathrm{s}}x(1-x)$
Sub Regular	$\Omega_{\mathrm{h,sym}}x(1-x)$ $+\Omega_{\mathrm{h,asy}}(x-0.5)x(1-x)$	$s_{\mathrm{mix,ideal}}$ $+\Omega_{\mathrm{s,sym}}x(1-x)$ $+\Omega_{\mathrm{s,asy}}(x-0.5)x(1-x)$

Table 3.1: Molar enthalpies and entropies of mixing for various semi-empirical models of solutions. The ideal entropy of mixing, $s_{\mathrm{mix,ideal}}$, is given by Equation 3.24, and the molar Gibbs free energy of mixing is $g_{\mathrm{mix}} = h_{\mathrm{mix}} - Ts_{\mathrm{mix}}$. Ω, Ω_{h} and Ω_{s} are regular solution parameters, as discussed in the text.

solutions are those in which the two components interact with each other in the same way as they do among themselves, and are discussed in Subsection 3.2.2. "Regular" solutions are those in which the two components interact differently with each other than they do among themselves, and are discussed in Subsection 3.2.3.

3.2.1 "Perfect" Solutions

We start, in this subsection, by considering alloy phases for which the two components do not interact at all with each other. Then, as we might expect, the molar Gibbs free energy of the solution phase is the average of the molar Gibbs free energies of the pure-component phases, weighted according to mole fraction.

For example, for an ideal-gas mixture, $(\mathrm{a}_{1-x}\mathrm{b}_x)$, of noninteracting *monomeric* components a and b, we have

$$\begin{aligned} g^{(\mathrm{a}_{1-x}\mathrm{b}_x)} &= (1-x)g^{(\mathrm{a})}(p^{(\mathrm{a})},T) + (x)g^{(\mathrm{b})}(p^{(\mathrm{b})},T) \\ &= (1-x)g^{(\mathrm{a})}([1-x]p,T) + (x)g^{(\mathrm{b})}(xp,T). \qquad (3.15) \end{aligned}$$

where we have made use of the proportionality between the partial pressures and mole fractions, $p^{(\mathrm{a})} = (1-x)p$ and $p^{(\mathrm{b})} = xp$, where p is the total pressure. If the pressure dependences of the molar Gibbs free energies are factored out through use of Equation 2.48, then we have the explicit form

$$g^{(\mathrm{a}_{1-x}\mathrm{b}_x)} = (1-x)\left\{g^{(\mathrm{a})}(p_o,T) + \frac{2}{5}c_{o,\mathrm{tra}}^{(\mathrm{a})}T\ln\left[\frac{(1-x)p}{p_o}\right]\right\}$$

$$+(x)\left[g^{(\mathrm{b})}(p_o,T)+\frac{2}{5}c_{o,\mathrm{tra}}^{(\mathrm{b})}T\ln\left(\frac{xp}{p_o}\right)\right]. \qquad (3.16)$$

For an ideal-gas mixture, $(\mathrm{a}_{1-x}\mathrm{b}_x)$, of noninteracting *nonmonomeric* components, the proportionality between the partial pressures and mole fractions is somewhat more complicated. For example, suppose component a is monomeric, but component b is dimeric. The total pressure is still the sum of the partial pressures, $p = p^{(\mathrm{a})} + p^{(\mathrm{b}_2)}$, but the mole fractions are

$$x = \frac{2p^{(\mathrm{b}_2)}}{p^{(\mathrm{a})}+2p^{(\mathrm{b}_2)}} \qquad 1-x = \frac{p^{(\mathrm{a})}}{p^{(\mathrm{a})}+2p^{(\mathrm{b}_2)}}, \qquad (3.17)$$

and the partial pressures are

$$p^{(\mathrm{a})} = \frac{1-x}{1-x/2}p \qquad p^{(\mathrm{b}_2)} = \frac{x/2}{1-x/2}p. \qquad (3.18)$$

Hence, the molar Gibbs free energy of the mixture becomes

$$\begin{aligned} g^{(\mathrm{a}_{1-x}\mathrm{b}_x)} &= (1-x)g^{(\mathrm{a})}\left(p^{(\mathrm{a})},T\right)+(x)g^{(\frac{1}{2}\mathrm{b}_2)}\left(p^{(\mathrm{b}_2)},T\right) \\ &= (1-x)g^{(\mathrm{a})}\left(\frac{[1-x]p}{1-x/2},T\right)+(x)g^{(\frac{1}{2}\mathrm{b}_2)}\left(\frac{[x/2]p}{1-x/2},T\right). \end{aligned} \qquad (3.19)$$

If the pressure dependences of the molar Gibbs free energies are factored out through use of Equation 2.48, then, just as in Equation 3.16, we have the explicit form

$$\begin{aligned} g^{(\mathrm{a}_{1-x}\mathrm{b}_x)} &= (1-x)\left\{g^{(\mathrm{a})}(p_o,T)+\frac{2}{5}c_{o,\mathrm{tra}}^{(\mathrm{a})}T\ln\left[\frac{(1-x)p}{p_o}\right]\right\} \\ &\quad + (x)\left\{g^{(\frac{1}{2}\mathrm{b}_2)}(p_o,T)+\frac{2}{5}c_{o,\mathrm{tra}}^{(\frac{1}{2}\mathrm{b}_2)}T\ln\left[\frac{(x/2)p}{p_o}\right]\right\}. \end{aligned} \qquad (3.20)$$

Note that the translational heat capacity parameter for b dimers, $c_{o,\mathrm{tra}}^{(\frac{1}{2}\mathrm{b}_2)}$, is half that for b monomers, $c_{o,\mathrm{tra}}^{(\mathrm{b})}$.

To deduce the chemical potentials of a and b, we must begin with the *total* Gibbs free energy. For example, for monomeric components, we have

$$G^{(\mathrm{a}_{1-x}\mathrm{b}_x)} = N_\mathrm{a}^{(\mathrm{a}_{1-x}\mathrm{b}_x)}g^{(\mathrm{a})}(p^{(\mathrm{a})},T)+N_\mathrm{b}^{(\mathrm{a}_{1-x}\mathrm{b}_x)}g^{(\mathrm{b})}(p^{(\mathrm{b})},T), \qquad (3.21)$$

where the partial pressure of a, $p^{(\mathrm{a})} \equiv (1-x)p$, is independent of the number of b atoms, $N_\mathrm{b}^{(\mathrm{a}_{1-x}\mathrm{b}_x)}$, and the partial pressure of b, $p^{(\mathrm{b})} \equiv xp$, is independent of the number of a atoms, $N_{\mathrm{a}}^{(\mathrm{a}_{1-x}\mathrm{b}_x)}$. From Equation 3.2, the chemical potentials of a and b in $(\mathrm{a}_{1-x}\mathrm{b}_x)$ are then found to be

$$\begin{aligned}\mu_{\mathrm{a}}^{(\mathrm{a}_{1-x}\mathrm{b}_x)} &= g^{(\mathrm{a})}(p^{(\mathrm{a})},T) = g^{(\mathrm{a})}(p_o,T) + \frac{2}{5}c_{o,\mathrm{tra}}^{(\mathrm{a})}\ln\left[\frac{(1-x)p}{p_o}\right] \\ \mu_{\mathrm{b}}^{(\mathrm{a}_{1-x}\mathrm{b}_x)} &= g^{(\mathrm{b})}(p^{(\mathrm{b})},T) = g^{(\mathrm{b})}(p_o,T) + \frac{2}{5}c_{o,\mathrm{tra}}^{(\mathrm{b})}\ln\left[\frac{xp}{p_o}\right] \qquad (3.22)\end{aligned}$$

where $c_{o,\mathrm{tra}} = 5/2$ for monomeric components, and we have identified the molar Gibbs free energies of the pure-component phases $g^{(\mathrm{a})}(p_o,T)$ and $g^{(\mathrm{b})}(p_o,T)$ with $\mu_{\mathrm{a}}^{(\mathrm{a})}(p_o,T)$ and $\mu_{\mathrm{b}}^{(\mathrm{b})}(p_o,T)$, respectively.

It can be seen that the chemical potentials of each component are proportional to simple logarithms of that component's partial pressure, and are independent of the presence or absence of the other component. It is important to note, though, that they can be *indirectly* dependent on the presence or absence of the other component if instead of considering the two partial pressures to be the independent variables, the total pressure and the mole fraction are considered to be the independent variables. Then, for a fixed total pressure, increasing the mole fraction increases the partial pressure of one component, but decreases the partial pressure of the other component, thereby changing *both* chemical potentials.

3.2.2 "Ideal" Solutions

In Subsection 3.2.1, we considered alloy phases for which the two components do not interact at all with each other. In this subsection, we consider alloy phases for which the two components do interact with each other. Then, the molar Gibbs free energies will not be the average of the molar Gibbs free energies of the pure-component phases, weighted according to mole fraction. For example, for a solid-phase solution $\langle \mathrm{a}_{1-x}\mathrm{b}_x\rangle$ of components a and b, we must write

$$g^{\langle \mathrm{a}_{1-x}\mathrm{b}_x\rangle} = (1-x)g^{\langle \mathrm{a}\rangle} + (x)g^{\langle \mathrm{b}\rangle} + g_{\mathrm{mix}}^{\langle \mathrm{a}_{1-x}\mathrm{b}_x\rangle}, \qquad (3.23)$$

where $g_{\mathrm{mix}}^{\langle \mathrm{a}_{1-x}\mathrm{b}_x\rangle}$ is the additional molar Gibbs free energy associated with the mixing.

Such a molar Gibbs free energy of mixing will exist even if the two components interact with each other in the same way that they interact with themselves. At minimum, even if they are chemically identical, each component will, when confined to a condensed phase, exclude the other component from occupying the same atomic volume. Unlike the components of ideal gases, the components of condensed phases cannot, upon mixing, occupy the same volume. *Condensed-phase volumes are additive upon mixing, while vapor-phase volumes are not.* Therefore, component a, previously confined to an initial volume $V^{\langle a \rangle}$, is, upon mixing, free to diffuse into a larger volume $V^{\langle a \rangle} + V^{\langle b \rangle}$. The partition function associated with the mixing is the degeneracy of the system, i.e., the number of ways N_a "a" atoms can position themselves on $N_a + N_b$ lattice sites, less overcounting for permutations among the (indistinguishable) "a" atoms: $(N_a + N_b)!/(N_a!N_b!)$. In the absence of a temperature dependence, the entropy is, from Equation 2.16, proportional to the logarithm of the partition function. Hence, using Stirling's approximation,

$$\begin{aligned} s_{\text{mix,ideal}} &= \frac{k}{N_a + N_b} \ln \left[\frac{(N_a + N_b)!}{N_a! N_b!} \right] \\ &= \frac{k}{N_a + N_b} \left[(N_a + N_b) \ln(N_a + N_b) - N_a \ln N_a - N_b \ln N_b \right] \\ &= -k \left[(1 - x) \ln(1 - x) + x \ln x \right]. \end{aligned} \quad (3.24)$$

Equation 3.24 is the configurational entropy associated with the random mixing of two components into a condensed-phase. For an ideal condensed-phase mixture, then, the molar Gibbs free energy of mixing is given by

$$g_{\text{mix}}^{\langle a_{1-x} b_x \rangle} = -T s_{\text{mix,ideal}}. \quad (3.25)$$

Note that $\ln(x)$ and $\ln(1 - x)$ are negative quantities, so that the entropy of mixing is positive.

From Equations 3.3, 3.4, 3.23 and 3.24, the chemical potentials of a and b in $\langle a_{1-x} b_x \rangle$ are then found, after some algebra, to be

$$\begin{aligned} \mu_a^{\langle a_{1-x} b_x \rangle}(x, T) &= g^{\langle a \rangle}(T) + kT \ln(1 - x) \\ \mu_b^{\langle a_{1-x} b_x \rangle}(x, T) &= g^{\langle b \rangle}(T) + kT \ln(x). \end{aligned} \quad (3.26)$$

Note that Equations 3.26, which apply to components in condensed phases, are quite similar to Equation 2.47, which applies to components in vapor phases. Physically, the reason is that in both cases the logarithmic parts of the chemical potentials arise from entropies of mixing, one due to mixing of a component into a condensed-phase lattice, the other due to mixing of a vapor into the vacuum.

3.2.3 "Regular" Solutions

In Subsection 3.2.2, we considered alloy phases for which the two components interact with each other, but in the same way that they interact with themselves. In this subsection, we consider the more usual case: condensed-phase alloys for which the two components interact with each other substantially and *differently* from the way they interact among themselves. Then, the molar Gibbs free energy of mixing will contain contributions additional to that due to the ideal entropy of mixing. For these alloys, it is common to represent those additional contributions with semi-empirical formulas fit to experimental data. Usually, these are algebraic polynomials of the form[4]

$$g_{\text{mix}}^{\langle \mathrm{a}_{1-x}\mathrm{b}_x \rangle} = -Ts_{\text{mix,ideal}} + x(1-x)\left[b + c(2x-1) + d(2x-1)^2 + \cdots\right]. \quad (3.27)$$

The $x(1-x)$ factor in the second term on the right side of Equation 3.27 guarantees that that term will vanish in the absence of mixing, i.e., for the pure-component phases at $x = 0$ and $x = 1$. The polynomial factor in that term consists of even and odd powers of $2x - 1$ that are, respectively, symmetric and antisymmetric with respect to $x = 0.5$. In its most general form, the b, c, and d coefficients in the expansion may also depend on temperature.

Strictly and Quasi-Regular Solutions

If only the first term (b) is nonzero, then Equation 3.27 reduces to what is known as the regular solution expression,[5]

$$g_{\text{mix,reg}}^{\langle \mathrm{a}_{1-x}\mathrm{b}_x \rangle} = -Ts_{\text{mix,ideal}} + \Omega x(1-x). \quad (3.28)$$

For such a regular solution, the chemical potentials of a and b can then be deduced, again using Equations 3.3 and 3.4 (and again after some algebra), to be

$$\begin{aligned} \mu_{\mathrm{a}}^{\langle \mathrm{a}_{1-x}\mathrm{b}_x \rangle}(x,T) &= g^{\langle \mathrm{a} \rangle}(T) + kT\ln(1-x) + \Omega x^2 \\ \mu_{\mathrm{b}}^{\langle \mathrm{a}_{1-x}\mathrm{b}_x \rangle}(x,T) &= g^{\langle \mathrm{b} \rangle}(T) + kT\ln(x) + \Omega(1-x)^2. \end{aligned} \quad (3.29)$$

If, furthermore, the interaction parameter Ω is independent of temperature, then $\Omega x(1-x)$ may be considered a mixing enthalpy, and the solution is sometimes called "strictly" regular. Physically, strictly regular solutions are those in which the two components are chemically only slightly dissimilar. Then, they will mix very nearly randomly, and the mixing entropy will

[4] O. Redlich and A.T. Kister, "Algebraic representation of thermodynamic properties and the classification of solutions," *Ind. Eng. Chem.* **40**, 345 (1948).

[5] J.H. Hildebrand, J.M. Prasnitz and R.L. Scott, *Regular and Related Solutions: The Solubility of Gases, Liquids, and Solids* (Van Nostrand Reinhold, New York, 1970).

be very nearly the ideal entropy of mixing itself. In fact, for a truly random mixture, it is straightforward to calculate the composition-dependent probability that an atom of one component will be surrounded by various numbers of atoms of the other component. Then, if the atoms interact only with their nearest neighbors, Ω can be shown to be the difference between nearest-neighbor interactions between like ($\varepsilon_{aa} + \varepsilon_{bb}$) and unlike ($\varepsilon_{ab}$) components, $\Omega = \varepsilon_{ab} - \frac{1}{2}(\varepsilon_{aa} + \varepsilon_{bb})$. Note, though, that atoms rarely interact only with their nearest neighbors, and so this simple interpretation of Ω can be at best qualitative. However, Ω is still a useful semi-empirical measure of the chemical "dissimilarity" between two components of an alloy.

If, instead, the interaction parameter depends *linearly* on temperature, $\Omega \equiv \Omega_h - T\Omega_s$, then the solution is sometimes called "quasi-regular," or "simple."[6] The temperature-independent part of the Gibbs free energy of mixing can still be considered an enthalpy of mixing. However, the temperature-dependent part must now be considered an entropy of mixing over and above the ideal entropy of mixing, in part due to non-neglible deviations from randomness.

If the interaction parameter depends *nonlinearly* on temperature, then the enthalpic and entropic contributions to the molar Gibbs free energy of mixing are difficult to distinguish. One way of doing so is to assume that the heat capacity of the solution deviates from the weighted average of the heat capacities of the pure-component phases predicted by what is known as the Neumann-Kopp rule. That deviation defines a "heat capacity of mixing,"

$$c_{p,\mathrm{mix}} \equiv c_p^{\langle \mathrm{a}_{1-x}\mathrm{b}_x \rangle}(x,T) - \left[(1-x)c_p^{\langle \mathrm{a} \rangle}(T) + (x)c_p^{\langle \mathrm{b} \rangle}(T)\right], \tag{3.30}$$

which, in the regular solution approximation, would be written as

$$c_{p,\mathrm{mix}} = \chi(T)x(1-x). \tag{3.31}$$

If $\chi(T)$ were independent of temperature, then the temperature-dependent entropy and enthalpy of mixing would be, by analogy to Equations 2.8 and 2.9,

$$s_{\mathrm{mix}} = [s_{\mathrm{mix}}(T_o) + \chi \ln(T/T_o)]\, x(1-x) \tag{3.32}$$

$$h_{\mathrm{mix}} = [h_{\mathrm{mix}}(T_o) + \chi(T - T_o)]\, x(1-x). \tag{3.33}$$

If, instead, $\chi(T)$ has the more general temperature dependence given by Equation 2.23, then the entropy and enthalpy of mixing take the more complicated forms given by Equations 2.24 and 2.25. Notice that the entropies and enthalpies of mixing may both be negative or positive, depending on the sign of the heat capacity of mixing.

[6] E.A. Guggenheim, *Thermodynamics* (North-Holland, Amsterdam, 1959).

Sub-Regular Solutions

Finally, return to Equation 3.27, if the first two terms (b and c) are nonzero, then that equation reduces to what is known as the sub-regular solution expression,[7]

$$g_{\mathrm{mix,subreg}}^{\langle \mathrm{a}_{1-x}\mathrm{b}_x \rangle} = [(\Omega_{\mathrm{h,sym}} - T\Omega_{\mathrm{s,sym}}) + (x - 0.5)(\Omega_{\mathrm{h,asy}} - T\Omega_{\mathrm{s,asy}})]\, x(1 - x). \tag{3.34}$$

This equation is also sometimes referred to as Margules' equation. It is the simplest expression for which the interaction parameter depends on composition. Physically, such a composition dependence might arise if the two components had different atomic sizes, if that size mismatch led to long-range elastic strains in the mixture, and if those strains were the dominant contribution to the heat of mixing. Then, if the elastic properties of the mixture were a weighted average of the elastic properties of the pure-component phases, the interaction parameter would itself be expected to be a weighted average of the pure-component interaction parameters.[8]

3.3 A Nearly Ideal Solution: SiGe

In Section 3.2, we discussed the various commonly used semi-empirical expressions for the composition dependences of the molar Gibbs free energies of alloy phases. In this and the next Section, we illustrate that discussion through two detailed case studies of the thermodynamic equilibria between the various alloy phases based on Si and Ge and on Ga and As.

We begin, in this Section, with Si and Ge alloys. In particular, we consider the equilibria between the (monomeric) $(\mathrm{Si}_{1-x}\mathrm{Ge}_x)$ vapor phase, the $\{\mathrm{Si}_{1-x}\mathrm{Ge}_x\}$ liquid phase, and the $\langle \mathrm{Si}_{1-x}\mathrm{Ge}_x \rangle$ crystal phase. We will start, in Subsection 3.3.1, by discussing the composition and temperature-dependent molar Gibbs free energy functions of these phases. Then, in Subsection 3.3.2, we discuss the x-T phase diagrams deduced from these molar Gibbs free energy functions using the common tangent construction. Finally, in Subsection 3.3.2, we discuss x-p phase diagrams deduced in a similar way, and use them to understand the kinetic competition between condensation and sublimation of SiGe alloys.

[7] H.K. Hardy, "A 'sub-regular' solution model and its application to some binary alloy systems," *Acta Met.* **1**, 202 (1953).

[8] A.W. Lawson, "On simple binary solid solutions," *J. Chem. Phys.* **15**, 831 (1947).

3.3.1 Free Energies

Let us start by estimating the composition, pressure, and temperature dependences of the molar Gibbs free energies of the $(\mathrm{Si}_{1-x}\mathrm{Ge}_x)$ vapor, $\{\mathrm{Si}_{1-x}\mathrm{Ge}_x\}$ liquid, and $\langle\mathrm{Si}_{1-x}\mathrm{Ge}_x\rangle$ crystal alloy phases of Si and Ge.

The $(\mathrm{Si}_{1-x}\mathrm{Ge}_x)$ vapor phase is just a mixture of two ideal gases, and hence can be considered to be a "perfect" solution. Therefore, its molar Gibbs free energy is the weighted sum of the molar Gibbs free energies of the pure-component vapors:

$$g^{(\mathrm{Si}_{1-x}\mathrm{Ge}_x)}(x,p,T) = (1-x)g^{(\mathrm{Si})}\left((1-x)p,T\right) + (x)g^{(\mathrm{Ge})}\left(xp,T\right). \quad (3.35)$$

In this expression, the molar Gibbs free energies of the pure-component vapor phases can be taken to be those calculated in Chapter 2.

The $\{\mathrm{Si}_{1-x}\mathrm{Ge}_x\}$ liquid and $\langle\mathrm{Si}_{1-x}\mathrm{Ge}_x\rangle$ crystal phases are somewhat more complicated. From the "shapes" of the experimentally measured boundaries between those phases, it has long been known[9] that both phases are nearly, but not quite, ideal. To describe their small degree of nonideality, it is common to use a strictly regular solution model for their molar Gibbs free energies of mixing, so that

$$\begin{aligned} g^{\{\mathrm{Si}_{1-x}\mathrm{Ge}_x\}}(x,T) &= (1-x)g^{\{\mathrm{Si}\}}(T) + (x)g^{\{\mathrm{Ge}\}}(T) - Ts_{\mathrm{mix,ideal}} \\ &\quad + \Omega^{\{\mathrm{Si}_{1-x}\mathrm{Ge}_x\}}x(1-x) \quad (3.36) \\ g^{\langle\mathrm{Si}_{1-x}\mathrm{Ge}_x\rangle}(x,T) &= (1-x)g^{\langle\mathrm{Si}\rangle_c}(T) + (x)g^{\langle\mathrm{Ge}\rangle_c}(T) - Ts_{\mathrm{mix,ideal}} \\ &\quad + \Omega^{\langle\mathrm{Si}_{1-x}\mathrm{Ge}_x\rangle}x(1-x). \quad (3.37) \end{aligned}$$

In fact, experimental data have thus far not been sufficient to determine uniquely *both* of the regular solution interaction parameters. Therefore, here we use the crystal interaction parameter estimated from *ab initio* calculations,[10] $\Omega^{\langle\mathrm{Si}_{1-x}\mathrm{Ge}_x\rangle} = 0.045$ eV/atom. Then, a value for the liquid interaction parameter of $\Omega^{\{\mathrm{Si}_{1-x}\mathrm{Ge}_x\}} = 0.069$ eV/atom is found to fit the experimental liquid–solid phase diagram quite accurately,[11] as illustrated in the left half of Figure 3.3. Note that these interaction parameters are comparable to thermal energies at rather modest ($\approx 300°$C) temperatures, and hence are quite small.

[9] C.D. Thurmond, "Equilibrium thermochemistry of solid and liquid alloys of germanium and of silicon. I. The solubility of Ge and Si in elements of Groups III, IV and V," *J. Phys. Chem.* **57** 827 (1953).

[10] A. Qteish and R. Resta, "Thermodynamic properties of Si–Ge alloys," *Phys. Rev.* **B37**, 6983 (1988). We neglect the slight dependence of the interaction parameter on composition and temperature found in those calculations.

[11] These values are numerically very nearly those ($\Omega^{\{\mathrm{Si}_{1-x}\mathrm{Ge}_x\}} = 0.067$ eV/atom and $\Omega^{\langle\mathrm{Si}_{1-x}\mathrm{Ge}_x\rangle} = 0.037$ eV/atom) found empirically in R.W. Olesinski and G.J. Abbaschian, "The Ge–Si (germanium-silicon) system," *Bull. Alloy Phase Diagrams* **5**, 180 (1984).

3.3.2 Phase Equilibria

In Subsection 3.3.1, we estimated the composition, pressure, and temperature dependences of the molar Gibbs free energies of the various alloy phases of Si and Ge. From these molar Gibbs free energies, we can calculate, using the common tangent rule, which phase or combination of phases minimizes the total Gibbs free energy (and hence represents the equilibrium configuration of the system) for a given *overall* system composition. Usually, if a particular combination of phases minimizes the total Gibbs free energy at one overall system composition, it will also do so within a range of overall system compositions. Of special interest then are the critical compositions that separate the compositions for which one or another combination of phases is the equilibrium configuration of the system. Like the molar Gibbs free energies, those critical compositions change with temperature and pressure, thereby defining "phase boundaries" in x-T-p space. For ease of presentation, it is usually convenient to illustrate those boundaries on x-T phase diagrams at fixed p, or on x-p phase diagrams at fixed T.

In this subsection, we consider x-T phase diagrams at fixed p. Figures 3.3 and 3.4 illustrate such phase diagrams at fixed pressures ranging from 10^0 Torr to 10^{-9} Torr. To make the derivation of the diagrams more concrete, we also show the molar Gibbs free energies of the three phases of interest at 1600 K and 1300 K, and the common tangents at those temperatures.

At the highest pressure, 10^0 Torr, there are three distinct regions in which, at equilibrium, the system contains only one phase (vapor, liquid, or crystal). At high temperatures the vapor is stablest, at intermediate temperatures the liquid is stablest, and at low temperatures the crystal is stablest.

Between these one-phase regions lie two regions in which, at equilibrium, the system contains two phases (liquid plus vapor and crystal plus liquid). In these two-phase regions, the molar Gibbs free energies of none of the phases lie completely below the others. Instead, the molar Gibbs free energies of two of the phases intersect. When this is the case, there will always be a range of compositions around that intersection within which the molar Gibbs free energy is minimized if the two phases coexist. That range is bounded by the critical compositions given by the common tangent construction described algebraically by Equations 3.11. Two of the critical composition boundaries have special names: that dividing the pure liquid from the two-phase liquid plus crystal region is called the *liquidus*, and that dividing the pure crystal from the two-phase liquid plus crystal region is called the *solidus*.

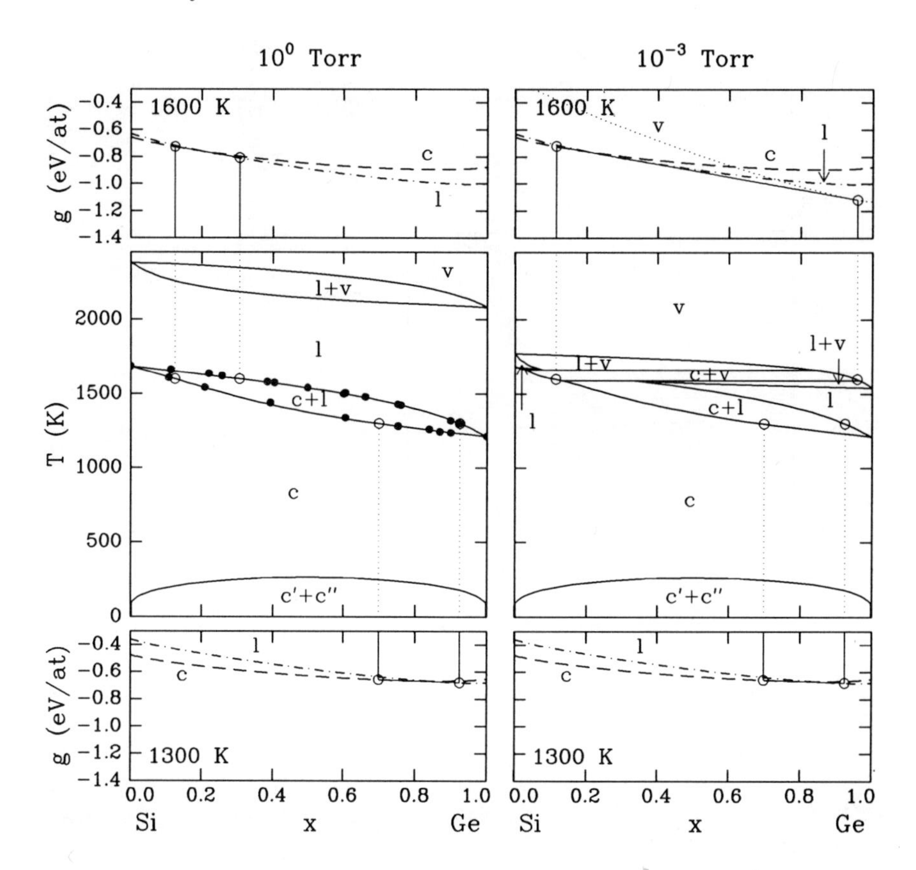

Figure 3.3: x-T phase diagrams for $Si_{1-x}Ge_x$ at pressures of 10^0 Torr (left) and 10^{-3} Torr (right). Above and below each phase diagram are also shown the molar Gibbs free energies of the various phases at 1600 K and 1300 K, their common tangents, and the critical compositions (open circles) determined by those common tangents. The solid circles in the left diagram represent experimental measurements.[a]

[a] H. Storh and W. Klemm, "Über zweistoff systeme mit germanium. I." *Z. Anotr. Chem.* **241**, 305 (1939); and F.X. Hassion, A.J. Goss and F.A. Trumbore, "On the germanium-silicon phase diagram," *J. Phys. Chem.* **59**, 1118 (1955).

At all pressures, there is also a two-phase crystal plus crystal region at very low temperatures. The reason is that at low enough temperatures, the contribution of the entropy of mixing to the molar Gibbs free energy of mixing [Equation 3.28] becomes negligible. Then, a positive interaction parameter causes the molar Gibbs free energy of mixing of the crystalline

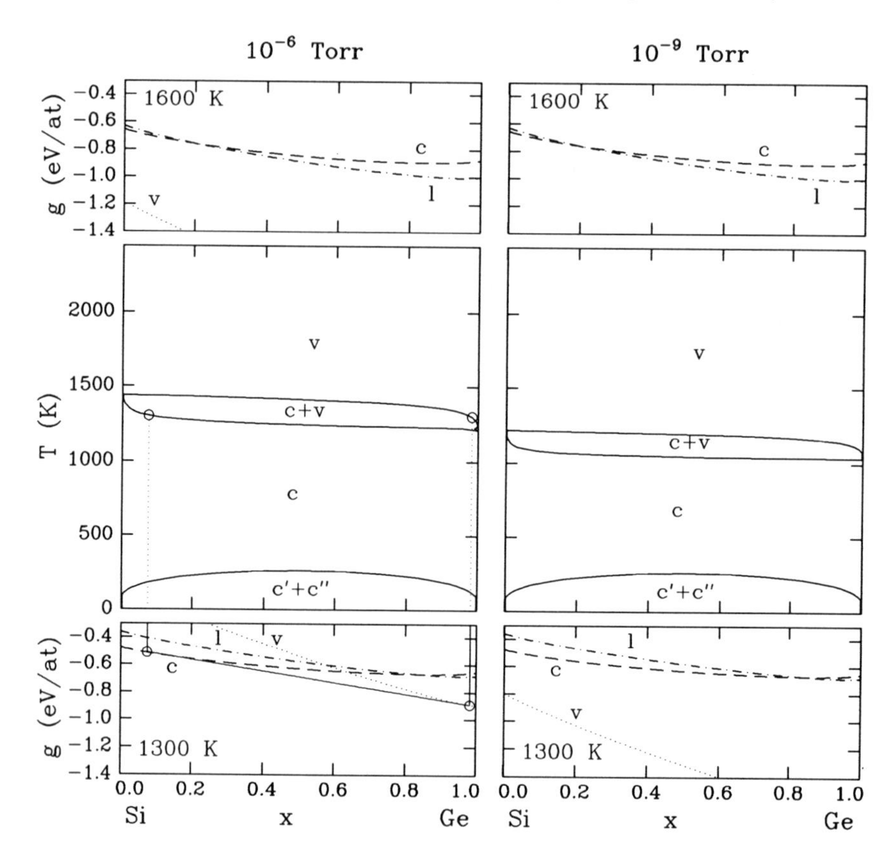

Figure 3.4: x-T phase diagrams for $Si_{1-x}Ge_x$ at pressures of 10^{-6} Torr (left) and 10^{-9} Torr (right). Above and below each phase diagram are also shown the molar Gibbs free energies of the various phases at 1600 K and 1300 K, their common tangents, and the critical compositions (open circles) determined by those common tangents.

solid to bow upward at intermediate compositions. The molar Gibbs free energy will then be minimized if the crystal decomposes into two crystals (c' and c") of different compositions. At higher temperatures, the contribution of the entropy of mixing outweighs that of the mixing enthalpy, causing the molar Gibbs free energy of mixing to bow downward, and this "miscibility" gap vanishes. The critical temperature at which the miscibility gap vanishes therefore occurs when $\partial^2 g_{\text{mix}}/\partial x^2 = 0$, or when

$$T_{\text{misc}} = \frac{\Omega}{2k}. \tag{3.38}$$

For an estimated interaction parameter of $\Omega^{\langle Si_{1-x}Ge_x \rangle} = 0.045$ eV/atom, the miscibility temperature is approximately 261 K. Note, though, that the phase separation would be extremely slow and difficult to observe at those temperatures, due to the sluggishness of solid-state diffusion.

As the pressure decreases, the vapor becomes increasingly stable relative to the liquid and crystal. Consider the top panels of Figures 3.3 and 3.4, which show the molar Gibbs free energies of the various phases at 1600 K. At 10^0 Torr the molar Gibbs free energy of the vapor is so high that it is off the scale of the figure. At 10^{-3} Torr it has moved downward far enough to intersect the molar Gibbs free energies of the crystal; a two-phase vapor plus crystal region then opens up at compositions straddling that intersection. At 10^{-6} Torr it lies well below the molar Gibbs free energies of the crystal and the liquid, so that only the vapor phase is stable. Finally, at 10^{-9} Torr, it has moved so low that it is again off the scale of the figure.

A similar behavior can be seen in the bottom panels of Figures 3.3 and 3.4, which show the molar Gibbs free energies of the various phases at 1300 K. At this temperature, the molar Gibbs free energies of the vapor are all higher than they were at 1600 K, both absolutely, and relative to the molar Gibbs free energies of the crystal and liquid. Only at 10^{-6} Torr has it decreased enough to intersect the molar Gibbs free energy of the crystal, and by 10^{-9} Torr although it lies well below the molar Gibbs free energies of the crystal, it is still visible on the scale of the figure.

The consequence of this increasing stability of the vapor at lower pressures is that the vapor-phase regions in the x-T diagrams move downward in temperature with decreasing pressure, impinging first on the liquid-phase regions, and then on the crystalline-phase regions. In fact, at pressures below 10^{-6} Torr, as shown in Figure 3.4, the liquid-phase regions vanish entirely. Then, as the temperature of the system is raised or lowered, crystal sublimes directly into vapor,[12] and vapor condenses directly into crystal, bypassing the liquid phase.

3.3.3 Condensation and Sublimation

In Subsection 3.3.2, we discussed x-T phase diagrams at fixed p in the SiGe alloy system. There, we found that, at low enough pressures, the liquid phase is absent entirely from the phase diagram: crystal sublimes directly into vapor, and vapor condenses directly into crystal. In this subsection, we

[12] A sampling of the vapor (e.g., through a small orifice in an effusion cell) under such "Knudsen" conditions is a classic method for indirectly measuring the thermodynamic properties of condensed phases. The method depends on the thermodynamic properties of the vapor being well understood, and on the vapor and condensed phases being truly in equilibrium with each other.

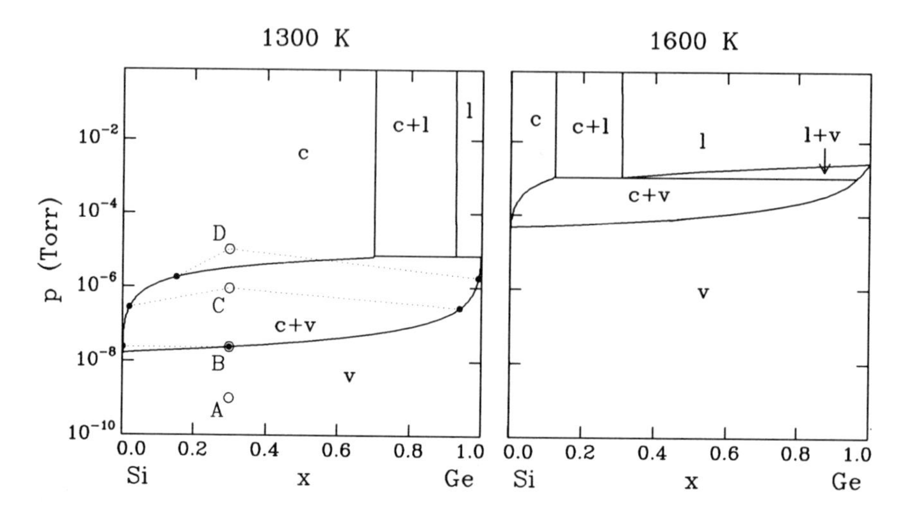

Figure 3.5: x-p phase diagrams for $Si_{1-x}Ge_x$ at temperatures of 1300 K (left) and 1600 K (right). On the left diagram, the open circles correspond to stable (A and B) and unstable (C and D) vapors at various pressures. The filled circles indicate the compositions of the crystals that would condense from those vapors, and the compositions and pressures of the vapors that would then sublime from those crystals.

examine these sublimation and condensation processes in more detail. To do so, let us consider the x-p diagrams at fixed T illustrated in Figure 3.5. In particular, consider a system at a temperature of 1300 K, so that the phase diagram on the left of Figure 3.5 applies.

First suppose the system to be composed of vapor at a composition and pressure corresponding to the point labeled A in that Figure. Since that point lies clearly in the one-phase vapor region, the vapor is stable with respect to the crystal.

Now suppose we increase the pressure of the vapor, so that the vapor has a composition and pressure corresponding to the point labeled B in Figure 3.5. The system is now just on the boundary of the two-phase crystal plus vapor region. The vapor can now *coexist* with crystal. However, if the crystal is to neither grow nor shrink at the expense of the vapor, its composition and pressure would have to correspond to a point on the leftmost boundary of the two-phase region. In other words, the composition of the crystal would not be the same as that of the vapor, but would instead be much more Si-rich.

Now suppose we increase the pressure even further, so that the vapor has a composition and pressure corresponding to the point labeled C in

Figure 3.5. Since the system is now clearly in a two-phase crystal plus vapor region, the pure vapor is unstable with respect to decomposition into a mixture of crystal and vapor, and the system would not be in equilibrium. To understand how this decomposition will actually occur, it is helpful to adopt a *kinetic* point of view, in which the equilibrium between crystal and vapor is considered to be a balance between simultaneous, independent processes of condensation and sublimation. At equilibrium, condensation and sublimation are balanced; but away from equilibrium, they are imbalanced, and net condensation or sublimation occurs.

That view is illustrated in Figure 3.6, which also shows an expanded (and linearized) portion of the x-p diagram around point C of Figure 3.5. In a sense, decomposition of vapor into crystal plus vapor can be thought of as occurring by sequential condensation and sublimation. First, vapor condenses out into crystal. If the sticking coefficients are unity for both species, as they are for Si and Ge, then the composition of the crystal will initially be (nearly) the same as that of the vapor.[13] Then, the crystal sublimes partially back into vapor. Because of the higher vapor pressure of Ge, the vapor that sublimes will be Ge rich, and so the crystal left behind will be Ge poor. The steady-state composition of the *net* condensing crystal is therefore determined by a competition between *congruent* condensation of vapor and *incongruent* sublimation of crystal.

If we denote the compositions and pressures of the incoming (condensing) and outgoing (subliming) vapors as x_{in}^{v}, p_{in}, $x_{\text{out}}^{\text{v}}$, and p_{out}, respectively, and denote the steady-state composition of the growing crystal as x^{c}, then that competition can be written mathematically as

$$x_{\text{in}}^{\text{v}} p_{\text{in}} - x_{\text{out}}^{\text{v}} p_{\text{out}} = x^{\text{c}} (p_{\text{in}} - p_{\text{out}}). \tag{3.39}$$

The left side of the equation is proportional to the net rate at which Ge atoms are transferred from the vapor to the crystal, i.e., the condensation rate minus the sublimation rate. The right side of the equation is proportional to the net rate at which the crystal grows, weighted by the Ge fraction in the crystal. In the limit $x^{\text{c}} \to 0$, so that pure Si is condensing, this equation states that $x_{\text{in}}^{\text{v}} p_{\text{in}} = x_{\text{out}}^{\text{v}} p_{\text{out}}$, i.e., Ge condensation just balances Ge sublimation.

Note that there is only one "unknown" in this equation: the sublimation pressure, p_{out}. Given p_{out}, the phase diagram determines uniquely the compositions, x^{c} and $x_{\text{out}}^{\text{v}}$, of the crystal and vapor that are in equilibrium with each other. Therefore, the problem is to find the sublimation pressure that self-consistently satisfies Equation 3.39, or its slightly more convenient

[13] In this semi-quantitative treatment, we neglect the mass (and hence composition) dependence of the conversion factor between pressure and incident flux.

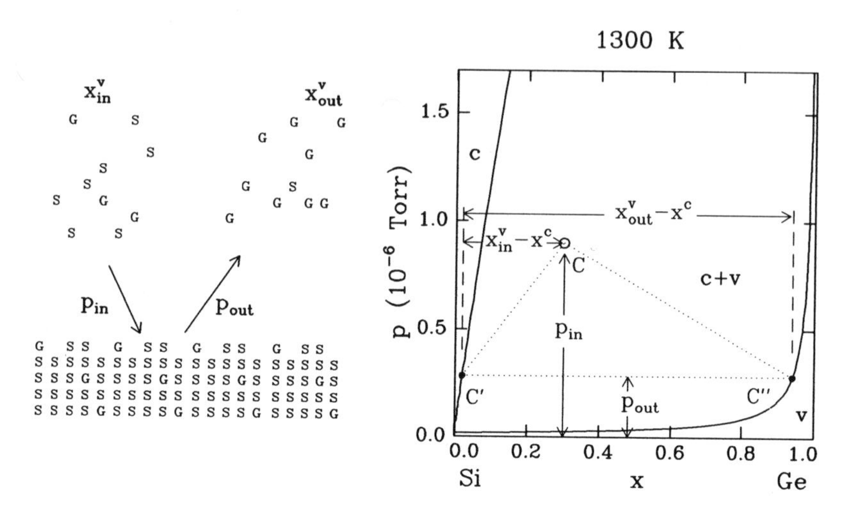

Figure 3.6: Kinetic (and expanded) view of the decomposition, depicted in Figure 3.5, of vapor (point C) into Si-rich crystal (point C') and Ge-rich vapor (point C").

form,

$$\frac{p_{\text{out}}}{p_{\text{in}}} = \frac{x_{\text{in}}^{\text{v}} - x^{\text{c}}}{x_{\text{out}}^{\text{v}} - x^{\text{c}}}. \tag{3.40}$$

An example of such a self-consistent solution is shown graphically on the right side of Figure 3.6. As p_{out} decreases, the left side of Equation 3.40 decreases. At the same time, $x_{\text{out}}^{\text{v}}$, the equilibrium composition of the vapor at p_{out}, approaches x_{in}^{v}, so the right side of Equation 3.40 increases. At the p_{out} shown in Figure 3.6, Equation 3.40 just balances, and the compositions of the growing crystal and the subliming vapor are given by the points C' and C".

For a vapor initially at the composition and pressure corresponding to point C in Figures 3.5 and 3.6, then, the growing crystal is much more Si rich than the condensing vapor itself. However, as the ambient pressure of the vapor increases through the sequence of points B, C, and D on Figure 3.5, the composition of the growing crystal approaches increasingly closely that of the condensing vapor. The reason is that p_{out} becomes increasingly neglible compared to p_{in}, and the left side of Equation 3.40 approaches zero. Then, x^{c} must approach x_{in}^{v}, and the transformation from vapor to crystal becomes increasingly "congruent." In practice, MBE of Si and Ge nearly always occurs under those conditions.

3.4 A Stoichiometric Compound: GaAs

In Section 3.3, we presented a detailed case study of alloy phases based on Si and Ge. In that case, there were only three phases of interest, vapor, liquid and crystal, all of which are continuous solutions throughout the entire composition range.

In this Section, we present a detailed case study of the thermodynamic equilibria between the various alloy phases based on Ga and As. In particular, we consider five phases: the $\{Ga_{1-x}As_x\}$ liquid phase, the $(Ga_{1-x}As_x)$ vapor phase, the $\langle Ga_{1-x}As_x\rangle_\gamma$ nearly pure-Ga orthorhombic crystalline phase, the $\langle Ga_{1-x}As_x\rangle_\alpha$ nearly pure-As rhombohedral crystalline phase, and the $\langle Ga_{1-x}As_x\rangle_c$ nearly stoichiometric zincblende-structure compound phase.[14]

Note that, unlike in the Si–Ge system, in the Ga–As system only the vapor and liquid phases form continuous solutions throughout the entire composition range. For the crystalline phases we must consider not just one, but *three* phases. The reason is that orthorhombic crystalline Ga and rhombohedral crystalline As have different lattice structures and symmetries. Since the two structures are inequivalent, and cannot be transformed one into the other by changing composition, the two phases must be considered distinct. In fact, Ga is only very slightly soluble in crystalline As, and As is only very slightly soluble in crystalline Ga. Instead, when "forced" to mix, Ga and As form a nearly stoichiometric compound crystalline phase. This phase has yet another lattice structure and symmetry, and hence must be considered yet another distinct phase.

We will start, in Subsection 3.4.1, by discussing the composition, pressure and temperature-dependent molar Gibbs free energy functions of these five phases. Then, in Subsection 3.4.2, we discuss the x-T phase diagrams deduced from these free energy functions using the common tangent construction. In Subsection 3.4.3, we use both x-T as well as x-p diagrams to define an "MBE window," the window in temperature and As-overpressure within which the stoichiometric compound coexists solely with the vapor, and not with any other unwanted phase. Then, in Subsection 3.4.4, we use p-T diagrams to understand what is known as the congruent sublimation temperature, below which the stoichiometric GaAs compound sublimes directly into the vapor, bypassing the liquid. Finally, in Subsection 3.4.5, we

[14] Vapor-solid-liquid equilibria in ternary III-III-V or III-V-V alloys, such as AlGaAs or InAsSb, are somewhat more difficult to treat; see, e.g., R. Heckingbottom, "Thermodynamic aspects of molecular beam epitaxy: high temperature growth in the GaAs/GaAlAs System," *J. Vac. Sci. Technol.* **B3**, 572 (1985), and H. Seki and A. Koukitu, "Thermodynamic analysis of molecular beam epitaxy of III-V semiconductors," *J. Cryst. Growth* **78**, 342 (1986).

discuss in more detail the vapor pressures of Ga and As_2 over the various condensed phases and phase mixtures, both above and below the congruent sublimation temperature.

3.4.1 Free Energies

Let us begin, in this subsection, by asking what the thermodynamic properties of these five phases are. Just as we did for the Si–Ge system, here we follow the usual prescription of characterizing the thermodynamic properties of the various phases first at their end point compositions, and then at intermediate compositions.

End Point Compositions

At its end point composition, the zincblende phase is the exactly stoichiometric $\langle Ga_{0.5}As_{0.5}\rangle_c$ compound. At their end point compositions, the vapor, liquid, and rhombohedral and orthorhombic crystalline phases become the three phases of pure Ga [(Ga), {Ga}, and $\langle Ga\rangle_\gamma$] and the three phases of pure As [$(\frac{1}{2}As_2)$, {As}, and $\langle As\rangle_\alpha$].

Note that for the Ga vapor phase, we consider only the most significant species: monomeric Ga. For the As vapor phase, however, we consider only the second most significant species: dimeric As. Although tetrameric As should, at commonly encountered MBE temperatures, be much more abundant than dimeric As *in equilibrium*, in fact As is found experimentally to sublime preferentially as dimeric As from GaAs surfaces under normal MBE conditions.[15] Therefore, dimeric and tetrameric As are *not* in equilibrium with each other during GaAs MBE, and we cannot treat the vapor phase as if it were composed mainly of tetrameric As. Instead, we consider here the opposite extreme, in which tetrameric As is absent entirely.

Note that such "constrained" equilibria can only be an approximate description of GaAs MBE. It will, e.g., describe GaAs MBE in which the As source is tetrameric only to the extent that the temperature dependence of the incorporation rate of incident As_4 into growing GaAs is similar to that of incident As_2. In fact, this approximation is not a bad one. The incorporation rates of both species depend mainly on Ga surface coverage, and their saturated (high Ga surface coverage) incorporation rates differ by at most a factor of two.[16]

---

[15] An upper bound of 1% has been placed on the As_4 to As_2 ratio in vapors sublimed from GaAs; see, e.g., C.T. Foxon, J.A. Harvey and B.A. Joyce, "The evaporation of GaAs under equilibrium and nonequilibrium conditions using a modulated beam technique," *J. Phys. Chem. Solids* **34**, 1693 (1973).

[16] C.T. Foxon and B.A. Joyce, "Interaction kinetics of As_4 and Ga on {100} GaAs surfaces using a modulated molecular beam technique," *Surf. Sci.* **50**, 434 (1975); and

Phase	Θ_T (K)	c_o (meV/(at-K))	c_1 (10^{-5} meV/(atomK^2))
$\langle Ga\rangle_\gamma$	66	0.276	0.00
$\{Ga\}$	66	0.276	0.00
$\langle Ga_{0.5}As_{0.5}\rangle_c$	82.6	0.250	4.22
$\langle As\rangle_\alpha$	84	0.257	4.23
$\{As\}$	84	0.257	4.23

Table 3.2: Heat capacity parameters for the condensed phases of Ga and As at their end point compositions.

Such constrained equilibria will also describe GaAs MBE only in the absence of other condensed phases that might "catalyze" the formation of As_4.[17] Therefore, a general treatment of phase equilibria during GaAs MBE would need to include tetrameric As as a constituent of the vapor phase in the presence of these other condensed phases, but exclude them (as we do here) in their absence.

To estimate the thermodynamic properties of these seven phases, we follow the procedure outlined in Chapter 2. Each phase is characterized both by a temperature-dependent heat capacity, and by enthalpy and entropy offsets at particular temperatures.

For the condensed phases, the heat capacities can be described by our standard form [Equation 2.23], using the parameters listed in Table 3.2. For $\langle Ga\rangle_\gamma$ and $\{Ga\}$ the parameters are an approximate combined fit to experimental values for *both* phases from 40 K to 302.92 K[18] and from 302.92 K to 2476 K.[19] For $\langle As\rangle_\alpha$ and $\{As\}$ the parameters are an approximate combined fit to experimental values for both phases from 57.2 K to 291 K[20] and from 298.15 K to 1200 K.[21] For $\langle Ga_{0.5}As_{0.5}\rangle_c$, the parameters

C.T. Foxon and B.A. Joyce, "Interaction kinetics of As_2 and Ga on {100} GaAs surfaces," *Surf. Sci.* **64**, 293 (1977).

[17] Bulk $\langle As\rangle_\alpha$ is known to sublime preferentially as As_4; $\langle Ga_{1-x}As_x\rangle_\alpha$ epitaxially oriented to a $\langle Ga_{0.5}As_{0.5}\rangle_c$ substrate may or may not sublime preferentially as As_4.

[18] Y.S. Touloukian and E.H. Buyco, *Thermophysical Properties of Matter Vol. 4, Specific Heat of Metallic Elements and Alloys* (IFI/Plenum, New York, 1970).

[19] M.W. Chase, Jr., C.A. Davies, J.R. Downey, Jr., D.J. Frurip, R.A. McDonald and A.N. Syverud, *JANAF Thermochemical Tables*, 3rd Ed., Part II, Cr-Zr, *J. Phys. Chem. Ref. Data* **14**, Suppl. No. 1 (1985), p. 1204.

[20] Y.S. Touloukian and E.H. Buyco, *Thermophysical Properties of Matter, Vol. 4, Specific Heat of Metallic Elements and Alloys* (IFI/Plenum, New York, 1970).

[21] R. Hultgren, P.D. Desai, D.T. Hawkins, M. Gleiser, K.K. Kelley and D.D. Wagman, *Selected Values of the Thermodynamic Properties of the Elements* (American Society for Metals, Metals Park, Ohio, 1973), pp. 204-209.

Phase	$\Theta_{T,tra}$ (K)	$c_{o,tra}$	$\Theta_{T,rot}$ (K)	$c_{o,rot}$	$\Theta_{T,vib}$ (K)	$c_{o,vib}$
(Ga)	$0.125\,(p/760\,\text{Torr})^{0.4}$	$5k/2$	–	0	–	0
$(\frac{1}{2}As_2)$	$0.079\,(p/760\,\text{Torr})^{0.4}$	$5k/4$	0.144	$k/2$	618	$k/2$

Table 3.3: Heat capacity parameters for the vapor phases (Ga) and $(\frac{1}{2}As_2)$.

are a fit to experimental values from 4 K to 1500 K.[22] In all cases, the usual caveat applies — the heat capacities of nonequilibrium phases (e.g., crystals above their melting temperatures or liquids below their freezing temperatures) are estimates only.

For the vapor phases, the translational, rotational, and vibrational contributions to the heat capacities can be described by Equations 2.31, 2.35 and 2.39, using the parameters listed in Table 3.3. The parameters for As_2 are based on a bond length of 2.104 Åand a ground-electronic-state vibrational stretching frequency of 429.55 cm^{-1}.[23] The electronic contributions can be described by Equations 2.34 and 2.20, using the energies and degeneracies listed in Table 3.4. As always, care must be taken to halve thermodynamic quantities having to do with dimers, in order for their units to be per *atom* rather than per dimer.

For all the phases, we use the enthalpy and entropy offsets listed in Table 3.5. Most of the values are those recommended by Tmar and coworkers.[24] Some, though, have been modified slightly according to the methods used in the Si–Ge system discussed in Chapter 2. The principal modification is to the enthalpy of formation of $(\frac{1}{2}As_2)$. We find that it must be approximately 3% lower than Tmar's value in order for the congruent sublimation temperature of GaAs, described later in this section, to agree with its approximate experimental value of 898 K.

Intermediate Compositions

Just as it was in the Si–Ge system, in the Ga–As system the vapor phase is just a mixture of two ideal gases, and hence can be considered a "perfect" solution. However, in this system the vapor phase is somewhat more complicated, as we assume that it is composed of monomeric Ga and *dimeric*

[22] J.S. Blakemore, "Semiconducting and other major properties of gallium arsenide," *J. Appl. Phys.* **53**, R123 (1982).

[23] S.N. Suchard and J.E. Melzer, Eds., *Spectroscopic Data Vol. 2: Homonuclear Diatomic Molecules* (IFI/Plenum, New York, 1976), p. 43.

[24] M. Tmar, A. Gabriel, C. Chatillon, and I. Ansara, "Critical analysis and optimization of the thermodynamic properties and phase diagrams of the III-V compounds. II. The Ga–As and In–As systems," *J. Cryst. Growth* **69**, 421 (1984).

Molecule	Level ($^{(2S+1)}L_J$)	Degeneracy ($\omega_{i,\mathrm{ele}}$)	Relative energy (eV)
Ga	$^3P_{1/2}$	2	0
	$^3P_{3/2}$	4	0.102
	$^2S_{1/2}$	2	3.07
As_2	$^3\Sigma_g^+$	3	0
	$^3\Sigma_u^+$	3	1.81

Table 3.4: Energies and degeneracies of the lowest lying electronic levels of Ga and As_2.

As. Therefore, its molar Gibbs free energy is that given by Equation 3.20:

$$g^{(\mathrm{Ga}_{1-x}\mathrm{As}_x)} =$$

$$(1-x)\left\{g^{(\mathrm{Ga})}(p_o,T) + \frac{2}{5}c_{o,\mathrm{tra}}^{(\mathrm{Ga})}T\ln\left[\frac{(1-x)p}{p_o}\right]\right\}$$
$$+ (x)\left\{g^{(\frac{1}{2}\mathrm{As}_2)}(p_o,T) + \frac{2}{5}c_{o,\mathrm{tra}}^{(\frac{1}{2}\mathrm{As}_2)}T\ln\left[\frac{(x/2)p}{p_o}\right]\right\}. \qquad (3.41)$$

Exactly at their end point compositions ($x = 0$ for $\langle \mathrm{Ga}_{1-x}\mathrm{As}_x\rangle_\gamma$, $x = 1$ for $\langle \mathrm{Ga}_{1-x}\mathrm{As}_x\rangle_\alpha$, and $x = 0.5$ for $\langle \mathrm{Ga}_{1-x}\mathrm{As}_x\rangle_c$), the molar Gibbs free energies of the three crystalline phases can be readily calculated from the parameters in Tables 3.2, 3.3, 3.4 and 3.5. Away from those end point compositions, however, the molar Gibbs free energies increase extremely rapidly — in other words, compositional "defects" in these three phases are energetically quite costly. Therefore, we will treat these three phases, as illustrated in Figures 3.7 and 3.8, as if their Gibbs free energies rise so steeply that their compositions are essentially "pinned" at those end point compositions.

Finally, from the shape of the experimental liquidus boundary dividing $\{\mathrm{Ga}_{1-x}\mathrm{As}_x\}$ from a two-phase mixture of $\{\mathrm{Ga}_{1-x}\mathrm{As}_x\}$ and $\langle \mathrm{Ga}_{1-x}\mathrm{As}_x\rangle_c$, the $\{\mathrm{Ga}_{1-x}\mathrm{As}_x\}$ liquid has long been known to be fairly well represented by a quasi-regular (or "simple") solution model.[25] More recent analyses indicate that a sub-regular solution model improves the representation somewhat, especially in the extremely As-rich portion of the phase diagram.[26] For our purposes, then, we assume such a sub-regular solution model,

$$g^{\{\mathrm{Ga}_{1-x}\mathrm{As}_x\}} =$$

[25] M.B. Panish, "A thermodynamic evaluation of the simple solution treatment of the Ga-P, In-P and Ga-As systems," *J. Cryst. Growth* **27**, 6 (1974).

[26] A.E. Schultz, *The relationship between (gallium, indium) arsenic and its melt for the bulk and thin film cases*, Ph.D Thesis (U. of Wisconsin-Madison, 1988).

Phase	T_o (K)	$h(T_o)$ (eV/at)	T_o (K)	$s(T_o)$ (eV/(at-K))
$\langle Ga \rangle_\gamma$	298	0	0	0
$\{Ga\}$	302.9	$h^{\langle Ga \rangle_\gamma} + 0.0578$	302.9	$s^{\langle Ga \rangle_\gamma} + 0.0578/302.9$
(Ga)	0	$h^{\langle Ga \rangle_\gamma} + 2.70$	0	0
$\langle Ga_{0.5}As_{0.5} \rangle_c$	298	-0.424	0	0
$\langle As \rangle_\alpha$	298	0	0	0
$\{As\}$	1090	$h^{\langle As \rangle_\alpha} + 0.2533$	1090	$s^{\langle As \rangle_\alpha} + 0.2533/1090$
$(\frac{1}{2}As_2)$	0	$h^{\langle As \rangle_\alpha} + 0.952$	0	0

Table 3.5: s and h offsets for the various phases of Ga and As.

$$\begin{aligned}
&(1-x)g^{\{Ga\}} + (x)g^{\{As\}} - Ts_{\text{mix,ideal}} \\
&+ x(1-x)\left(\Omega_{\text{h,sym}}^{\{Ga_{1-x}As_x\}} - T\Omega_{\text{s,sym}}^{\{Ga_{1-x}As_x\}}\right) \\
&+ x(1-x)(x-\frac{1}{2})\left(\Omega_{\text{h,asy}}^{\{Ga_{1-x}As_x\}} - T\Omega_{\text{s,asy}}^{\{Ga_{1-x}As_x\}}\right).
\end{aligned} \tag{3.42}$$

Following Brebrick,[27] we note that of the four interaction parameters (two symmetric and two antisymmetric), only one may be chosen freely to fit the "shape" of the phase boundaries to experimental data. The other three are then determined by the known enthalpy (Δh_{m}) and entropy ($\Delta s_{\text{m}} = \Delta h_{\text{m}}/T_{\text{m}}$) of fusion of the zincblende compound, and by the known eutectic temperature ($T_{\text{e}} = 1079$ K) and composition ($x_{\text{e}} = 0.976$) at which the As-rich liquid coexists in equilibrium with the zincblende compound and As-rich solid:

$$\begin{aligned}
&\frac{1}{4}\Omega_{\text{h,sym}}^{\{Ga_{1-x}As_x\}} = \\
&\qquad h_{T_{\text{m}}}^{\langle Ga_{0.5}As_{0.5}\rangle_c} + \Delta h_{\text{m}} - 0.5\left(h_{T_{\text{m}}}^{\{Ga\}} + h_{T_{\text{m}}}^{\{As\}}\right) \\
&\frac{1}{4}\left(\Omega_{\text{h,sym}}^{\{Ga_{1-x}As_x\}} - T_{\text{m}}\Omega_{\text{s,sym}}^{\{Ga_{1-x}As_x\}}\right) = \\
&\qquad g_{T_{\text{m}}}^{\langle Ga_{0.5}As_{0.5}\rangle_c} + T_{\text{m}}s_{\text{mix,ideal}}(0.5) - 0.5\left[g_{T_{\text{m}}}^{\{Ga\}} + g_{T_{\text{m}}}^{\{As\}}\right]
\end{aligned} \tag{3.43}$$

$$\begin{aligned}
&\left[\Omega_{\text{h,asy}}^{\{Ga_{1-x}As_x\}} - T_{\text{e}}\Omega_{\text{s,asy}}^{\{Ga_{1-x}As_x\}}\right] x_{\text{e}}(1-x_{\text{e}})(x_{\text{e}} - \frac{1}{2}) = \\
&\qquad 2(1-x_{\text{e}})g_{T_{\text{e}}}^{\langle Ga_{0.5}As_{0.5}\rangle_c} + 2(x_{\text{e}} - \frac{1}{2})g_{T_{\text{e}}}^{\langle As\rangle_\alpha}
\end{aligned} \tag{3.44}$$

[27] R.F. Brebrick, "Quantitative fits to the liquidus line and high temperature thermodynamic data for InSb, GaSb, InAs and GaAs," *Met. Trans.* **8A**, 403 (1977).

$\Omega_{\mathrm{h,sym}}^{\{\mathrm{Ga}_{1-x}\mathrm{As}_x\}}$ (eV/atom)	$\Omega_{\mathrm{s,sym}}^{\{\mathrm{Ga}_{1-x}\mathrm{As}_x\}}$ (meV/(atomK))	$\Omega_{\mathrm{h,asy}}^{\{\mathrm{Ga}_{1-x}\mathrm{As}_x\}}$ (eV/atom)	$\Omega_{\mathrm{s,asy}}^{\{\mathrm{Ga}_{1-x}\mathrm{As}_x\}}$ (meV/(atomK))
0.031	0.27	-0.011	0

Table 3.6: Mixing enthalpies and entropies for liquid $\{\mathrm{Ga}_{1-x}\mathrm{As}_x\}$.

$$- \left[\Omega_{\mathrm{h,sym}}^{\{\mathrm{Ga}_{1-x}\mathrm{As}_x\}} - T_{\mathrm{e}}\Omega_{\mathrm{s,sym}}^{\{\mathrm{Ga}_{1-x}\mathrm{As}_x\}}\right] x_{\mathrm{e}}(1-x_{\mathrm{e}})$$
$$+ T_{\mathrm{e}} s_{\mathrm{mix,ideal}}(x_{\mathrm{e}}) - (x_{\mathrm{e}}) g_{T_{\mathrm{e}}}^{\{\mathrm{As}\}} + (1-x_{\mathrm{e}}) g_{T_{\mathrm{e}}}^{\{\mathrm{Ga}\}}. \qquad (3.45)$$

The values found for the four interaction parameters are listed in Table 3.6. It was not found necessary to invoke a nonzero $\Omega_{\mathrm{s,asy}}^{\{\mathrm{Ga}_{1-x}\mathrm{As}_x\}}$ in order to reproduce the experimentally measured liquidus boundaries dividing the one-phase liquid from the two-phase liquid plus solid regions, as can be seen in the left side of Figure 3.7.

From these molar Gibbs free energies, we can now calculate, using the common tangent construction, which phase or combination of phases minimizes the total Gibbs free energy (and hence represents the equilibrium configuration of the system) for a given *overall* system composition, pressure and temperature. As is customary, we will illustrate the resulting phase diagrams as x-T cuts at fixed p, as x-p cuts at fixed T, and as p-T cuts at fixed x.[28]

3.4.2 Phase Equilibria

In Subsection 3.4.1, we estimated the composition, pressure, and temperature dependences of the molar Gibbs free energies of the various alloy phases of Ga and As. In this subsection, we discuss the x-T phase diagrams that are deduced from those free energy functions using the common tangent construction. For example, Figures 3.7 and 3.8 illustrate x-T phase diagrams at four fixed pressures ranging from 1.5×10^3 Torr to 10^{-9} Torr. To make the derivation of the diagrams more concrete, we also show the molar Gibbs free energies of the five phases of interest at 1000 K and 850 K, and the common tangents at those temperatures.

At the highest pressure, 1.5×10^3 Torr, there should, for each of the five phases, be a distinct region in which, at equilibrium, the system contains only that phase ($(\mathrm{Ga}_{1-x}\mathrm{As}_x)$, $\{\mathrm{Ga}_{1-x}\mathrm{As}_x\}$, $\langle\mathrm{Ga}_{1-x}\mathrm{As}_x\rangle_{\mathrm{c}}$, $\langle\mathrm{Ga}_{1-x}\mathrm{As}_x\rangle_\alpha$ or $\langle\mathrm{Ga}_{1-x}\mathrm{As}_x\rangle_\gamma$). As expected, the $(\mathrm{Ga}_{1-x}\mathrm{As}_x)$ phase can be seen in Figure 3.7 at the highest temperatures, and the $\{\mathrm{Ga}_{1-x}\mathrm{As}_x\}$ phase can be

[28] J. van den Boomgaard and K. Schol, "The p-T-x phase diagrams of the systems In–As, Ga–As and In–P," *Philips Res. Rep.* **12**, 127 (1957).

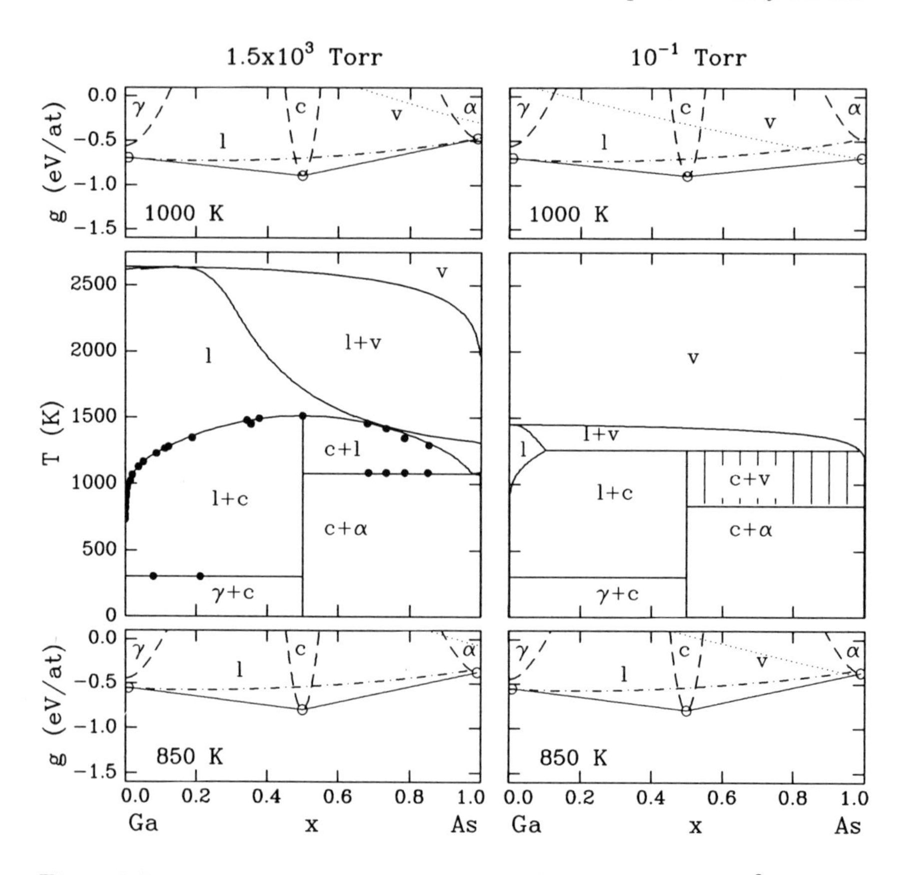

Figure 3.7: x-T phase diagrams for $Ga_{1-x}As_x$ at pressures of 1.5×10^3 Torr (left) and 10^{-1} Torr (right). Above and below each diagram are also shown the molar Gibbs free energies of the various phases at 1000 K and 850 K, and their common tangents. The solid circles in the left diagram represent experimental measurements.[a]

[a] J.C. DeWinter and M.A. Pollack, "Ga-As liquidus at temperatures below 650 C," *J. Appl. Phys.* **58**, 2410 (1985); R.N. Hall, "Solubility of III-V compound semiconductors in column III liquids," *J. Electrochem. Soc.* **110**, 385 (1963); and V.W. Köster and B. Thoma, "Aufbau der Systeme Gallium-Antimon, Gallium-Arsen and Aluminium-Arsen," *Z. Metallkd.* **46**, 291 (1955).

seen at intermediate temperatures. The three crystalline phase regions, however, are infinitesimally narrow, as we have forced their compositions to be pinned at their end point compositions.[29] The $\langle Ga_{1-x}As_x \rangle_\gamma$ single-

[29] On the scale of Figures 3.7 and 3.8, the three crystalline phase regions would have

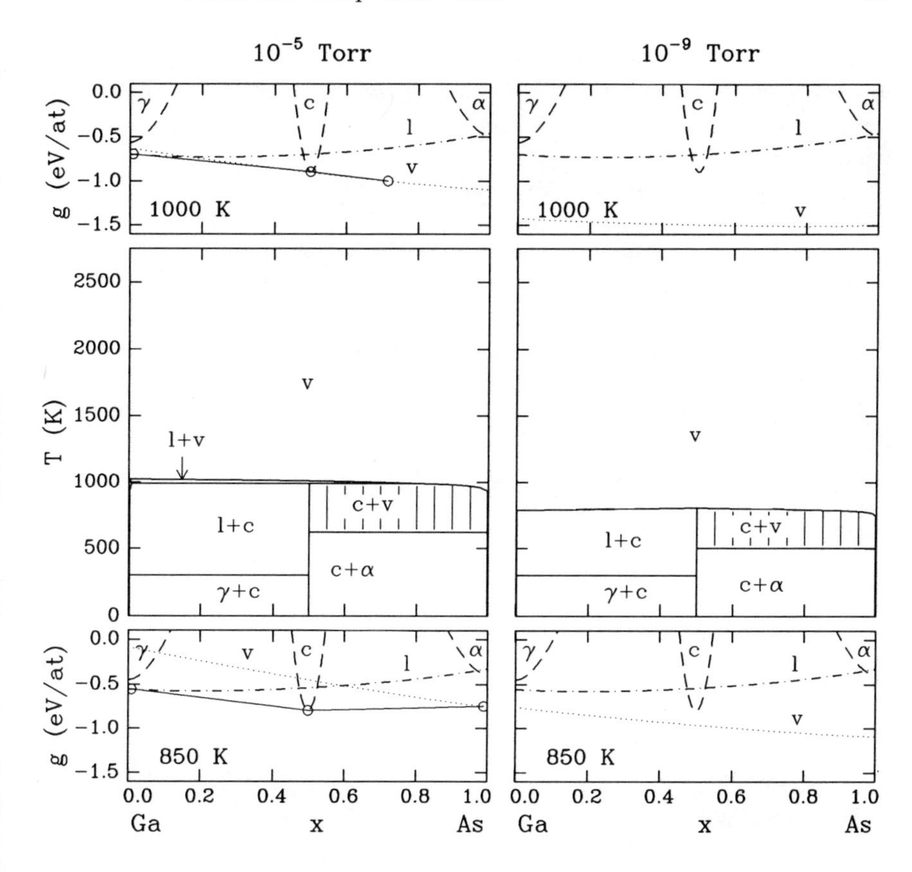

Figure 3.8: x-T phase diagrams for $Ga_{1-x}As_x$ at pressures of 10^{-6} Torr (left) and 10^{-9} Torr (right). Above and below each diagram are also shown the molar Gibbs free energies of the various phases at 1000 K and 850 K, and their common tangents.

phase "region" is essentially a vertical line at $x = 0$ from 0 K to 302.9 K (the melting temperature of $\langle Ga \rangle_\gamma$). The $\langle Ga_{1-x}As_x \rangle_\alpha$ single-phase "region" is essentially a vertical line at $x = 1$ from 0 K to approximately 1090 K (the melting temperature of $\langle As \rangle_\alpha$). The $\langle Ga_{1-x}As_x \rangle_c$ single-phase "region" is essentially a vertical line at $x = 0.5$ from 0 K to 1513.5 K (the melting temperature of $\langle Ga_{0.5}As_{0.5} \rangle_c$).

As we saw in the Si–Ge system, as the pressure decreases, the vapor

appeared infinitesimally narrow even had we not forced their compositions to be pinned at their end point compositions, but had instead used realistic composition dependences for their molar Gibbs free energies.

becomes increasingly stable relative to the liquid and crystalline phases. Then, the vapor-phase regions in the x-T diagrams move downward in temperature with decreasing pressure, impinging first on the liquid-phase regions and then on the crystalline-phase regions.

To see how, consider the top panels of Figures 3.7 and 3.8, which show the molar Gibbs free energies of the various phases at 1000 K. At 1.5×10^3 Torr the molar Gibbs free energy of the vapor is so high that it is visible only in the upper right portion of the figure. At 10^{-1} and 10^{-5} Torr it has moved downward far enough to intersect the molar Gibbs free energy of the liquid; a two-phase compound plus vapor region then opens up at compositions straddling that intersection. Below 10^{-5} Torr, it lies just below the molar Gibbs free energies of both compound *and* liquid, so that only the vapor phase is stable.

A similar behavior can be seen in the bottom panels of Figures 3.7 and 3.8, which show the molar Gibbs free energies of the various phases at 850 K. At this temperature, though, the molar Gibbs free energies of the vapor are all higher than they were at 1000 K, both absolutely, and relative to the molar Gibbs free energies of the crystal and liquid. Therefore, the equivalent intersections between the molar Gibbs free energies of the vapor and the liquid occur at lower pressures.

3.4.3 The MBE "Window"

In Subsection 3.4.2, we discussed x-T phase diagrams in the Ga–As alloy system. In this subsection, we use these as well as x-p and p-T diagrams to understand the preferred environmental conditions for GaAs MBE.

To do so, we note that, by definition, GaAs MBE is condensation of the vapor into $\langle Ga_{0.5}As_{0.5}\rangle_c$. At the same time, however, condensation of the vapor into other condensed phases such as $\{Ga_{1-x}As_x\}$ or $\langle Ga_{1-x}As_x\rangle_\alpha$ must be avoided. Avoiding such condensation can only be guaranteed, however, if the two-phase mixture of $\langle Ga_{0.5}As_{0.5}\rangle_c$ and $(Ga_{1-x}As_x)$, both of which must be present during MBE, actually minimizes the molar Gibbs free energy. Otherwise, another mix of phases will have some tendency to form. In other words, *MBE will be thermodynamically preferred if it occurs at an overall system composition, temperature and pressure for which the equilibrium mix of phases is* $\langle Ga_{0.5}As_{0.5}\rangle_c$ *and* $(Ga_{1-x}As_x)$.

These compound plus vapor two-phase "windows" in which MBE is preferred are shown as shaded regions in Figures 3.7 and 3.8. Consider, e.g., the 10^{-5} Torr diagram shown in the left half of Figure 3.8. The MBE window is bounded on the left by a Ga-rich liquid plus GaAs compound two-phase region. In other words, the overall system must be As rich; otherwise it will tend to "decompose" into a Ga-rich liquid and GaAs compound, as

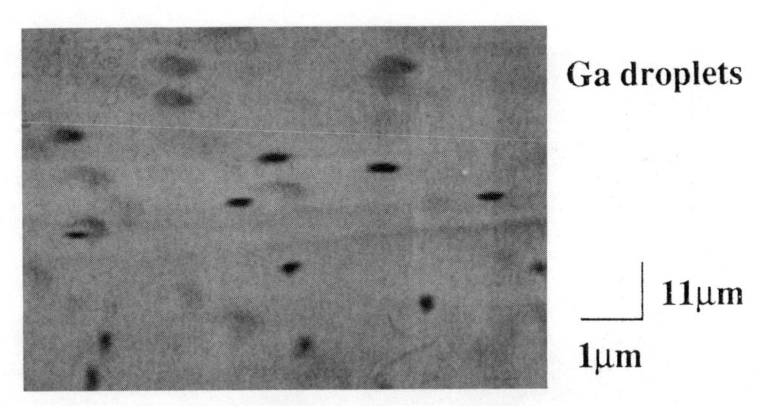

Figure 3.9: *In situ* scanning electron microscope image of Ga droplets (black ovals) formed[a] during GaAs MBE under Ga-rich growth conditions at 610°C. The gray ovals are GaAs mounds formed by previous experimental sequences. The image is foreshortened by about 20 times in the vertical direction.

[a]N. Inouye, "MBE monolayer growth control by in-situ electron microscopy," *J. Cryst. Growth* **111**, 75 (1991).

illustrated in Figure 3.9, into Ga-rich liquid plus As-rich vapor. The reason is that at these temperatures the vapor pressure of Ga over liquid Ga is so low that excess Ga cannot re-evaporate into the vapor. Note, though, that since the compound is very nearly stoichiometric, a Ga-rich overall system can be avoided by maintaining an overpressure of As-rich vapor.

The MBE window is bounded on the bottom by a GaAs compound plus As-rich crystal two-phase region. The reason is that at temperatures so low that the vapor pressure of $\langle As\rangle_\alpha$ is lower than the impinging As pressure, excess As will tend to condense into crystalline As, rather than sublime from the growing surface,[30] as illustrated in Figure 3.10. Finally, the MBE window is bounded on the top by a Ga-rich liquid plus As-rich vapor two-phase region. The reason is that at temperatures so high that the vapor pressure of Ga over the GaAs compound is higher than the vapor pressure of Ga over the Ga-rich liquid, the GaAs compound, even as it grows, will itself tend to decompose.

These MBE windows can also be seen in x-p phase diagrams. Consider, e.g., the 1000 K diagram shown in the right half of Figure 3.11. Again,

[30]Note, though, that this low-temperature boundary to the MBE window is not as well defined as the others, due to the possibility that the As-rich crystal sublimes as As tetramers, which we have disallowed. Because $(\frac{1}{4}As_4)$ is more stable than $(\frac{1}{2}As_2)$, the boundary will move down in temperature, and the MBE windows would be even wider than those shown.

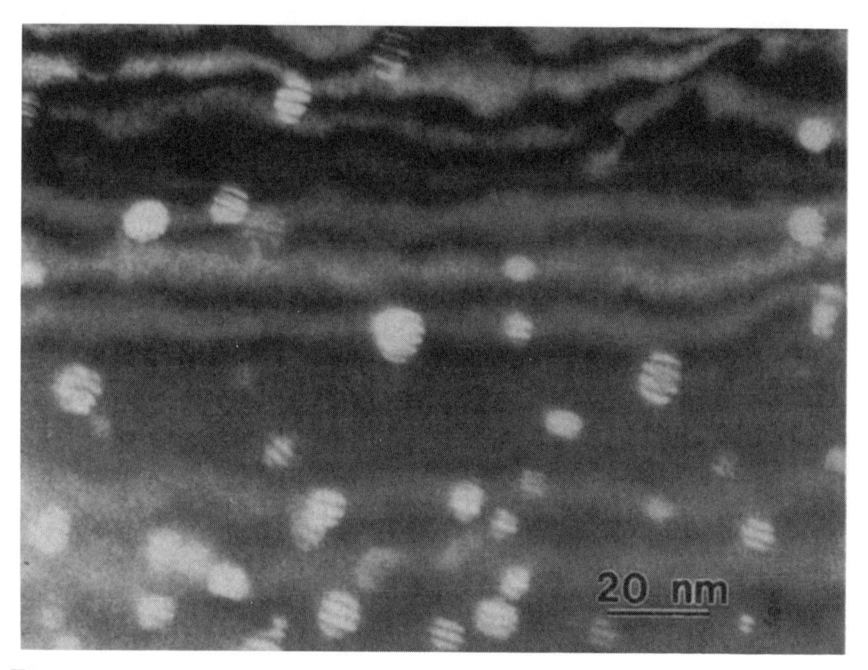

Figure 3.10: Plan-view transmission electron micrograph of hexagonal As precipitates formed[a] after a 20 min. 620°C anneal of a GaAs layer grown by MBE at 190°C.

[a]Z. Liliental-Weber, G. Cooper, R. Mariella, Jr., and C. Kocot, "The role of As in molecular-beam epitaxy GaAs layers grown at low temperature," *J. Vac. Sci. Technol.* **B9**, 2323 (1991).

the MBE window is bounded on the left by a Ga-rich liquid plus GaAs compound two-phase region. The reason is the same as that given above: at these high pressures, excess Ga condenses into Ga-rich liquid, rather than reevaporating into the vapor. The origins of the top and bottom boundaries are also the same as before, but reversed: the window is bounded on the top by a GaAs compound plus As-rich crystal two-phase region, and on the bottom by a Ga-rich liquid plus As-rich vapor two-phase region.

Notice that the MBE windows in this system are actually quite large. As illustrated in Figure 3.8, at a typical overpressure of 10^{-5} Torr, the temperature may range from roughly 350°C to 730°C. As illustrated in Figure 3.11, at a typical growth temperature of $\approx 577°C$, the overpressure may range from roughly 3×10^{-7} Torr to 3×10^{-1} Torr. In terms of MBE growth windows, GaAs is a relatively forgiving compound, a fact that is

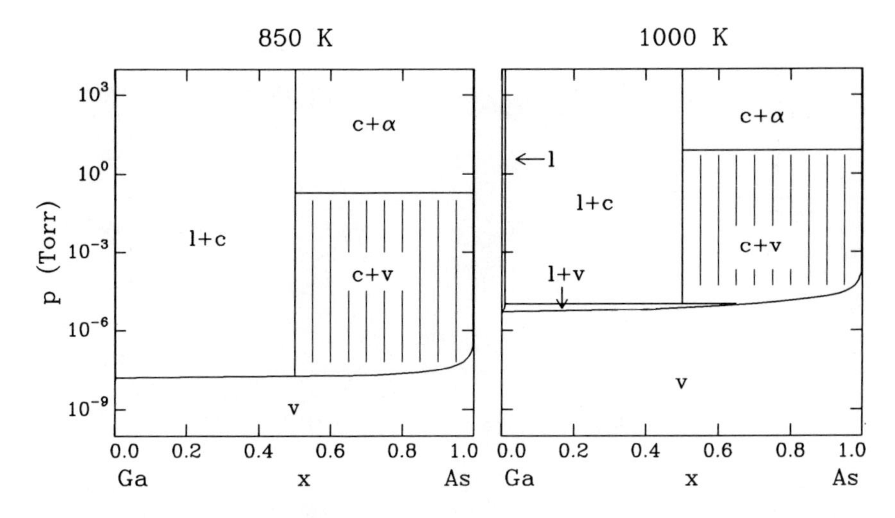

Figure 3.11: x-p phase diagrams for $Ga_{1-x}As_x$ at temperatures of 850 K (left) and 1000 K (right).

in part responsible for the ease and success with which it can be grown. Other III-V compounds alloys are not so forgiving. For example, the MBE growth window for InSb epitaxy is much narrower,[31] due to the lower vapor pressure of solid Sb than of solid As, and the higher vapor pressure of liquid In than of liquid Ga.

Consider, finally, the p-T phase diagram shown in Figure 3.12 for a fixed overall system composition of $x = 0.51$. Note that although the overall system composition is just slightly As rich, the composition of the equilibrium vapor at the various phase boundaries is nearly unity, and *the vapor can be considered nearly pure* As_2. In other words, the phase boundaries are essentially the critical As_2 overpressures at which various phase mixtures coexist. These p-T diagrams can therefore be of great practical use to the crystal grower, because substrate temperature and As_2 vapor overpressure can both be directly and readily controlled.

As in the other diagrams, the MBE window is the GaAs compound plus As-rich vapor shaded region of the diagram. The upper $\alpha + c/c + v$ boundary defines the As_2 overpressure beyond which solid As will tend to form. As illustrated in the upper left part of Figure 3.12, that boundary is defined by the pressure at which the molar Gibbs free energy of the vapor just intersects that of the solid at the As-rich side of the diagram. That

[31] J.Y. Tsao, "Phase equilibria during InSb molecular beam epitaxy," *J. Cryst. Growth* **110**, 595 (1991).

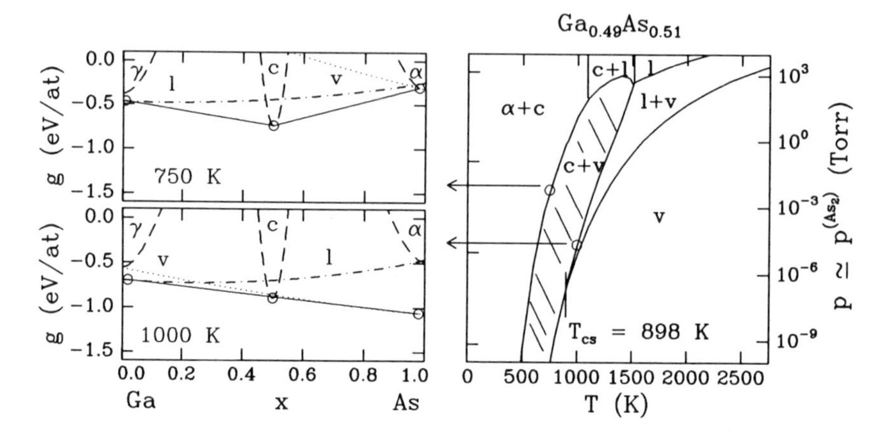

Figure 3.12: p-T phase diagram for $Ga_{1-x}As_x$ at an overall system composition of $x = 0.51$. To the left are the molar Gibbs free energies at the pressures and temperatures indicated by the open circles.

boundary is the vapor pressure of As_2 over $\langle As \rangle_\alpha$.

The lower $c + v/l + v$ boundary defines the As_2 overpressure below which Ga-rich liquid will tend to form. As illustrated in the lower left part of Figure 3.12, that boundary is defined by the pressure at which all three phases coexist, and hence share a common tangent. Below that critical pressure, the molar Gibbs free energy of the vapor decreases below that of the compound, and the equilibrium state of a system at $x = 0.51$ becomes a mixture of Ga-rich liquid and As-rich vapor. Above that critical pressure, the molar Gibbs free energy of the vapor increases above that of the compound, and the equilibrium state of a system at $x = 0.51$ becomes a mixture of GaAs compound and As-rich vapor.

Finally, the lower $c + v/v$ boundary defines the As_2 overpressure above which the GaAs will tend to grow by condensation *from* the vapor, rather than shrink by sublimation *into* the vapor.

3.4.4 Congruent and Incongruent Sublimation

In Subsection 3.4.3, we discussed the window in temperature and As_2 overpressure within which the GaAs compound coexists stably with an As-rich vapor. On the p-T diagram of Figure 3.12, the window is bounded at the bottom either by a liquid plus vapor region above 898 K, or by a pure vapor region below 898 K. This critical temperature is known as the *congruent sublimation temperature*.

In this subsection, we discuss the origin of this critical tempeature. We

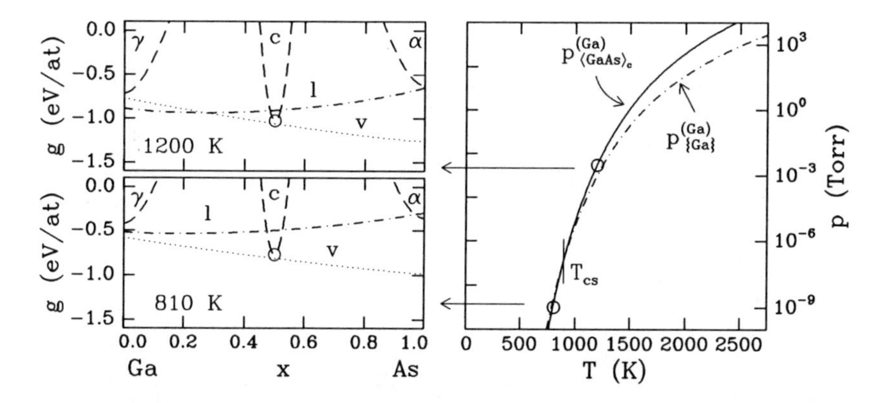

Figure 3.13: Vapor pressures of (Ga) over $\langle Ga_{0.5}As_{0.5}\rangle_c$ and {Ga}. Molar Gibbs free energies of the subliming vapor over the compound at 1200 K (top left) and 810 K (bottom left).

begin by supposing that growth is terminated, the usual overpressure of vapor As is taken away, and the GaAs compound is held at constant temperature in ultra-high-vacuum. We then ask whether, under such nonequilibrium, "Langmuir" evaporation conditions, it is still possible to apply equilibrium phase diagrams.

In fact, it *is* possible. Recall our arguments at the end of our discussion of the Si–Ge system. There, we adopted a *kinetic* point of view, in which the equilibrium between crystal and vapor was considered to be a dynamic competition between simultaneous, independent processes of condensation and sublimation. At equilibrium, condensation and sublimation are balanced; but away from equilibrium, they are imbalanced, and net condensation or sublimation occurs.

Imagine that our system is at 810 K, bathed in an ambient As_2 pressure of 10^{-9} Torr, so that, as indicated in the lower left panel of Figure 3.13, the molar Gibbs free energies of the compound and the vapor at $x = 0.5$ are equal, and the compound and vapor are in equilibrium with each other. Then, the "condensation" and "sublimation" pressures are just balanced, and no net growth occurs. Importantly, though, the two pressures are *independent* of each other. If we suddenly remove the "condensation" pressure, the "sublimation" pressure persists, and the compound will shrink. If we suddenly decrease the temperature to decrease the "sublimation" pressure, the condensation pressure persists, and the compound will grow. For example, Figure 3.14 shows direct measurements of GaAs sublimation after GaAs condensation is terminated.

Here, we imagine removing the condensation pressure. Then, under such

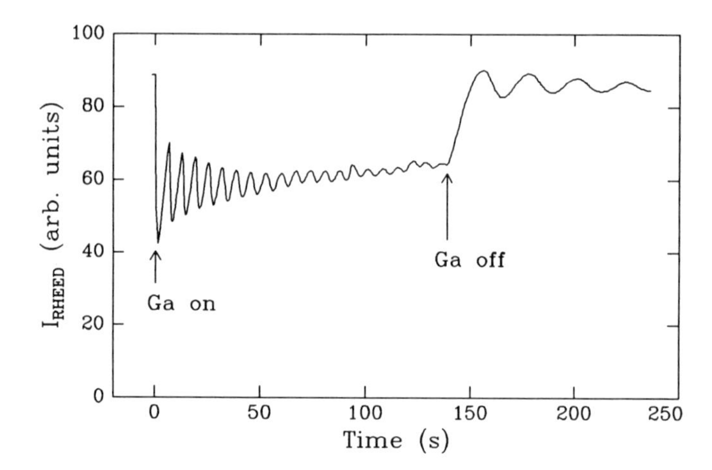

Figure 3.14: Reflection high-energy electron diffraction (RHEED) intensity measurements during a GaAs MBE growth and sublimation sequence[a] at 938 K. In the presence of a Ga flux, the oscillations in time indicate bilayer-by-bilayer condensation; in the absence of a Ga flux, the oscillations indicate bilayer-by-bilayer sublimation (see Chapter 6 for a discussion of RHEED oscillations). An As flux was maintained throughout the growth and sublimation sequence. Note that at this temperature and growth rate sublimation is much slower than growth.

[a] J.M. Van Hove and P.I. Cohen, "Mass-action control of AlGaAs and GaAs growth in molecular beam epitaxy," *Appl. Phys. Lett.* **47**, 726 (1985).

Langmuir, free evaporation conditions, the sublimation pressure persists, and will be the same as the condensation pressure at which the compound and vapor *would* have been in equilibrium. The molar Gibbs free energy of the vapor that sublimes is the same as that of the vapor that would have condensed in equilibrium, and, as indicated in the left panels of Figure 3.13, just touches that of the compound.

At 810 K, it can be seen that at the sublimation pressure (10^{-9} Torr) for which the vapor is in equilibrium with the compound, the molar Gibbs free energy of the vapor lies safely below that of the liquid. Therefore, at this temperature, the compound will sublime *congruently* into the stoichiometric vapor.

At 1200 K, however, it can be seen that at the sublimation pressure (3×10^{-3} Torr) for which the vapor is in equilibrium with the compound, the molar Gibbs free energy of the vapor *intersects* that of the Ga-rich liquid, and on the Ga-rich side lies *above* it. That means that the vapor, once formed by sublimation from the compound, itself is unstable with respect to decomposition into a Ga-rich liquid and an As-rich vapor. At this temperature, the compound will *not* sublime congruently, but rather will decompose "incongruently" into Ga-rich liquid and As-rich vapor.

Notice that at higher temperatures, the molar Gibbs free energy of the liquid decreases relative to that of the crystal, because of its higher entropy. Therefore, at higher temperatures the molar Gibbs free energy of the vapor subliming from the compound will always cut across that of the liquid, and sublimation will always be incongruent. At lower temperatures, however, the molar Gibbs free energy of the liquid increases relative to that of the crystal. Therefore, at lower temperatures the molar Gibbs free energy of the vapor subliming from the compound will never cut across that of the liquid, and sublimation will always be congruent. The critical temperature below which sublimation is congruent is what is known as the *congruent sublimation temperature*, T_{cs}.

Another way to understand the origin of the congruent sublimation temperature is to consider the temperature-dependent sublimation pressures of Ga over $\{Ga\}$ liquid and over the $\langle Ga_{0.5}As_{0.5}\rangle_c$ compound, as shown in the right half of Figure 3.13. The sublimation pressure of Ga over $\{Ga\}$ is determined by the pressure at which the molar Gibbs free energy of the vapor just intersects that of the liquid on the Ga-rich side. From Equation 2.48, that pressure can be written as

$$\ln\left(\frac{p^{(Ga)}_{\{Ga\}}}{p_o}\right) = \frac{g^{\{Ga\}}(T) - g^{(Ga)}(p_o,T)}{kT}. \quad (3.46)$$

The sublimation pressure of Ga over $\langle Ga_{0.5}As_{0.5}\rangle_c$ is determined by the pressure at which the molar Gibbs free energy of the vapor just intersects that of the stoichiometric GaAs compound. Therefore, using Equation 3.20,

$$\begin{aligned} g^{\langle Ga_{0.5}As_{0.5}\rangle_c} &= \frac{kT}{2}\ln\left[\frac{p^{(Ga)}_{\langle Ga_{0.5}As_{0.5}\rangle_c}}{p_o}\right] + \frac{kT}{4}\ln\left[\frac{p^{(Ga)}_{\langle Ga_{0.5}As_{0.5}\rangle_c}}{2p_o}\right] \\ &\quad + \frac{1}{2}\left[g^{(Ga)}(p_o,T) + g^{(\frac{1}{2}As_2)}(p_o,T)\right], \end{aligned} \quad (3.47)$$

where we have made use of the fact that for stoichiometric sublimation, $p^{(Ga)} = 2p/3$ and $p^{(As_2)} = p/3$. Solving for the Ga vapor pressure then gives

$$\ln\left[\frac{p^{(Ga)}_{\langle Ga_{0.5}As_{0.5}\rangle_c}}{2^{1/3}p_o}\right] = \frac{g^{\langle Ga_{0.5}As_{0.5}\rangle_c}(T) - \frac{1}{2}\left[g^{(Ga)}(p_o,T) + g^{(\frac{1}{2}As_2)}(p_o,T)\right]}{3kT/4}. \quad (3.48)$$

At high temperatures Ga has a higher vapor pressure over $\langle Ga_{0.5}As_{0.5}\rangle_c$ than over $\{Ga\}$; therefore, (Ga), once formed by sublimation from the compound, will have a tendency to condense into the liquid. At low temperatures, Ga has a lower vapor pressure over $\langle Ga_{0.5}As_{0.5}\rangle_c$ than over $\{Ga\}$; therefore, (Ga), once formed by sublimation from the compound, will *not* have a tendency to condense into the liquid. Physically, the reason is that disordered $\{Ga\}$ has a higher entropy than ordered $\langle Ga_{0.5}As_{0.5}\rangle_c$. Consequently, $g^{\{Ga\}}$, which enters into Equation 3.46, decreases faster with increasing temperature than does $g^{\langle Ga_{0.5}As_{0.5}\rangle_c}$, which enters into Equation 3.48.

Finally, to determine quantitatively the congruent sublimation temperatures, we can equate Equations 3.46 and 3.48, giving, after some algebra,

$$\begin{aligned} kT_{cs}\ln(2^{1/3}) &= \frac{2}{3}[g^{\{Ga\}}(T) + g^{(\frac{1}{2}As_2)}(p_o,T) - 2g^{\langle Ga_{0.5}As_{0.5}\rangle_c}(T)] \\ &\quad + \frac{1}{3}[g^{\{Ga\}}(T) - g^{(Ga)}(p_o,T)]. \end{aligned} \tag{3.49}$$

This equation determines the congruent sublimation temperature in terms of the molar Gibbs free energy of evaporation of $\{Ga\}$, and the molar Gibbs free energy of formation of $\langle Ga_{0.5}As_{0.5}\rangle_c$ from $\{Ga\}$ and $(\frac{1}{2}As_2)$.

The *practical* significance of the congruent sublimation temperature is illustrated in Figure 3.15. There, we show experimental measurements of the ratio between the As and Ga fluxes leaving a GaAs (001) surface under Langmuir evaporation conditions. At temperatures below the congruent sublimation temperature, Ga and As leave the surface in equal amounts; therefore, the surface does *not* need to be bathed in an As ambient to avoid formation of Ga-rich liquid. At temperatures above the congruent sublimation temperature, however, As preferentially leaves the surface, leaving behind Ga-rich liquid droplets; therefore, the surface *must* be bathed in an As ambient to avoid formation of Ga-rich liquid.

Note that both the pressure of the subliming vapor [essentially the vapor pressure of $(Ga_{1-x}As_x)$ along the Ga-rich part of the $\{Ga_{1-x}As_x\} + \langle Ga_{0.5}As_{0.5}\rangle_c$ liquidus] and its composition can be calculated fairly straightforwardly. At a given temperature, one draws the common tangent between the molar Gibbs free energies of the Ga-rich liquid and the GaAs compound. Then, the pressure of the vapor is adjusted so that the molar Gibbs free energy of the vapor shares this common tangent. At that pressure, all three phases can coexist in equilibrium. The composition, x^v_{out}, of the vapor is then given by the composition at which the common tangent just touches the molar Gibbs free energy of the vapor, and the ratio between the As and Ga fractions in the vapor is $x^v_{out}/(1-x^v_{out})$. The results of such a calculation, as shown in Figure 3.15, agree with measurements within the

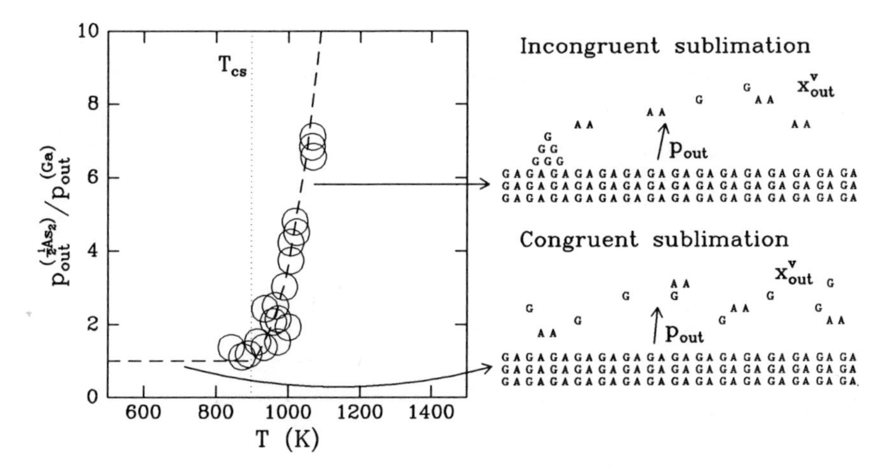

Figure 3.15: Left: Ratio between As and Ga fluxes desorbing from GaAs (001) during Langmuir evaporation. The open circles are from mass spectrometry measurements;[a] the dashed line is the prediction of equilibrium thermodynamics. The sizes of the data points reflect uncertainties in the relative sensitivities of the mass spectrometer to Ga and As_2 fluxes. Right: Schematic illustrations of congruent sublimation of GaAs compound into pure vapor (bottom) and incongruent sublimation of GaAs compound into Ga-rich liquid and As-rich vapor (top).

[a] C.T. Foxon, J.A. Harvey, and B.A. Joyce, "The evaporation of GaAs under equilibrium and nonequilibrium conditions using a modulated beam technique," *J. Phys. Chem. Solids* **34**, 1693 (1973).

uncertainty of the data.

3.4.5 Vapor Pressures

In Subsection 3.4.4, we discussed the vapor pressures of (Ga) over the pure compound $\langle Ga_{0.5}As_{0.5}\rangle_c$ and pure liquid {Ga} phases. At temperatures higher than the congruent sublimation temperature, (Ga) subliming from $\langle Ga_{0.5}As_{0.5}\rangle_c$, if uncompensated by incoming As, will condense into {Ga}. At temperatures lower than the congruent sublimation temperature, (Ga) subliming from $\langle Ga_{0.5}As_{0.5}\rangle_c$, even if uncompensated by incoming As, will not condense into {Ga}.

In this subsection, we discuss the equilibrium vapor pressures of both As_2 and Ga over various *mixtures* of phases, i.e., along the the phase *boundaries* shown in Figure 3.12. The phase boundaries of particular interest are those that bound the MBE window. The vapor pressures along those boundaries are shown in Figure 3.16. We plot in all cases the pressures

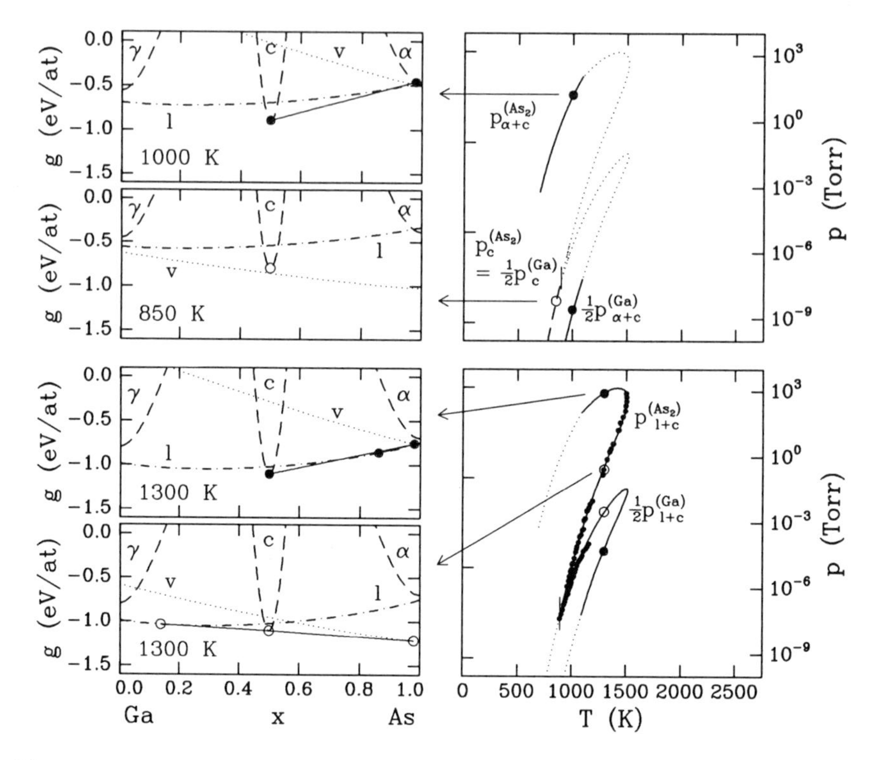

Figure 3.16: Vapor pressures of (As_2) and (Ga) along the phase-coexistence lines bounding the MBE window. The molar Gibbs free energy diagrams on the left correspond, from top to bottom, to conditions for which the equilibrium mix of condensed phases are: $\langle As\rangle_\alpha$ and $\langle Ga_{0.5}As_{0.5}\rangle_c$; $\langle Ga_{0.5}As_{0.5}\rangle_c$; As-rich $\{Ga_{1-x}As_x\}$ and $\langle Ga_{0.5}As_{0.5}\rangle_c$; and Ga-rich $\{Ga_{1-x}As_x\}$ and $\langle Ga_{0.5}As_{0.5}\rangle_c$. The small dots in the lower right panel are experimental measurements of Arthur[a] and Panish.[b]

[a] J.R. Arthur, "Vapor pressures and phase equilibria in the Ga-As system," *J. Phys. Chem. Solids* **22**, 2257 (1967).

[b] M.B. Panish, "A thermodynamic evaluation of the simple solution treatment of the Ga-P, In-P and Ga-As systems," *J. Cryst. Growth* **27**, 6 (1974).

$p^{(As_2)}$ and $\frac{1}{2}p^{(Ga)}$, rather than $p^{(As_2)}$ and $p^{(Ga)}$, so as to better illustrate that $p^{(As_2)} = \frac{1}{2}p^{(Ga)}$ at (and below) the congruent sublimation temperature.

For example, the vapor pressures over the two-phase mixture $\alpha + c$ are indicated by the solid lines in the top right panel, and were determined by the graphical constructions shown in the top left panel. In order for pure $(\frac{1}{2}As_2)$ to be in equilibrium with (nearly) pure $\langle As\rangle_\alpha$, their molar Gibbs free energies must be equal:

$$g^{(\frac{1}{2}\mathrm{As}_2)} = g^{\langle\mathrm{As}\rangle_\alpha}. \tag{3.50}$$

Then, through use of Equation 2.47, the As_2 vapor pressure must be

$$\ln\left[\frac{p_{\alpha+c}^{(\mathrm{As}_2)}}{p_o}\right] = \frac{g^{\langle\mathrm{As}\rangle_\alpha}(T) - g^{(\frac{1}{2}\mathrm{As}_2)}(p_o,T)}{\frac{2}{5}c_{o,\mathrm{tra}}^{(\frac{1}{2}\mathrm{As}_2)}T}. \tag{3.51}$$

To deduce the Ga vapor pressure, consider the common tangent between the molar Gibbs free energies of nearly pure $\langle\mathrm{As}\rangle_\alpha$ and stoichiometric $\langle\mathrm{Ga}_{0.5}\mathrm{As}_{0.5}\rangle_c$. The intersection of that common tangent with $x = 0$ must be the molar Gibbs free energy of pure (Ga):

$$\frac{g^{\langle\mathrm{As}\rangle_\alpha} - g^{\langle\mathrm{Ga}_{0.5}\mathrm{As}_{0.5}\rangle_c}}{1.0 - 0.5} = \frac{g^{\langle\mathrm{Ga}_{0.5}\mathrm{As}_{0.5}\rangle_c} - g^{(\mathrm{Ga})}}{0.5 - 0.0}. \tag{3.52}$$

Then, again through use of Equation 2.47, the Ga vapor pressure must be

$$\ln\left[\frac{p_{\alpha+c}^{(\mathrm{Ga})}}{p_o}\right] = \frac{2g^{\langle\mathrm{Ga}_{0.5}\mathrm{As}_{0.5}\rangle_c} - g^{\langle\mathrm{As}\rangle_\alpha}(T) - g^{(\mathrm{Ga})}(p_o,T)}{\frac{2}{5}c_{o,\mathrm{tra}}^{(\mathrm{Ga})}T}. \tag{3.53}$$

Likewise, the vapor pressures over the two-phase mixture $c + l$ are indicated by the solid lines in the bottom right panel, and were determined by the graphical constructions shown in the lower left two panels. Consider again the common tangent, this time between the molar Gibbs free energies of the $\{\mathrm{Ga}_{1-x}\mathrm{As}_x\}$ liquid and stoichiometric $\langle\mathrm{Ga}_{0.5}\mathrm{As}_{0.5}\rangle_c$. The intersections of that common tangent with $x = 0$ and $x = 1$ must be the molar Gibbs free energies of pure (Ga) and $(\frac{1}{2}\mathrm{As}_2)$:

$$\begin{aligned} \frac{g^{(\frac{1}{2}\mathrm{As}_2)} - g^{\langle\mathrm{Ga}_{0.5}\mathrm{As}_{0.5}\rangle_c}}{1.0 - 0.5} &= \frac{g^{\langle\mathrm{Ga}_{0.5}\mathrm{As}_{0.5}\rangle_c} - g^{\{\mathrm{Ga}_{1-x}\mathrm{As}_x\}}}{0.5 - x^{\{\mathrm{Ga}_{1-x}\mathrm{As}_x\}}} \\ \frac{g^{\{\mathrm{Ga}_{1-x}\mathrm{As}_x\}} - g^{(\mathrm{Ga})}}{x^{\{\mathrm{Ga}_{1-x}\mathrm{As}_x\}} - 0.0} &= \frac{g^{\langle\mathrm{Ga}_{0.5}\mathrm{As}_{0.5}\rangle_c} - g^{\{\mathrm{Ga}_{1-x}\mathrm{As}_x\}}}{0.5 - x^{\{\mathrm{Ga}_{1-x}\mathrm{As}_x\}}}. \end{aligned} \tag{3.54}$$

Then, again through use of Equation 2.47, the Ga and As_2 vapor pressures must be given by

$$\frac{2}{5}c_{o,\mathrm{tra}}^{(\frac{1}{2}\mathrm{As}_2)}T\ln\left[\frac{p_{l+c}^{(\mathrm{As}_2)}}{p_o}\right] =$$

$$g^{\langle\mathrm{Ga}_{0.5}\mathrm{As}_{0.5}\rangle_c} + \frac{0.5\left[g^{\langle\mathrm{Ga}_{0.5}\mathrm{As}_{0.5}\rangle_c} - g^{\{\mathrm{Ga}_{1-x}\mathrm{As}_x\}}\right]}{0.5 - x^{\{\mathrm{Ga}_{1-x}\mathrm{As}_x\}}} - g^{(\frac{1}{2}\mathrm{As}_2)}(p_o,T)$$

$$\frac{2}{5}c_{o,\text{tra}}^{(\text{Ga})}T\ln\left(\frac{p_{l+c}^{(\text{Ga})}}{p_o}\right) =$$

$$g^{\langle\text{Ga}_{0.5}\text{As}_{0.5}\rangle_\text{c}} + \frac{0.5\left(g^{\{\text{Ga}_{1-x}\text{As}_x\}} - g^{\langle\text{Ga}_{0.5}\text{As}_{0.5}\rangle_\text{c}}\right)}{0.5 - x^{\{\text{Ga}_{1-x}\text{As}_x\}}} - g^{(\text{Ga})}(p_o, T). \tag{3.55}$$

Below the congruent sublimation temperature, it is no longer possible for $(\text{Ga}_{1-x}\text{As}_x)$ to be in equilibrium with both $\langle\text{Ga}_{0.5}\text{As}_{0.5}\rangle_\text{c}$ and $\{\text{Ga}_{1-x}\text{As}_x\}$. Instead, $\{\text{Ga}_{1-x}\text{As}_x\}$, if formed, will have a tendency to evaporate, leaving only $\langle\text{Ga}_{0.5}\text{As}_{0.5}\rangle_\text{c}$. If the vapor has the same composition as the stoichiometric compound, then the two vapor pressures are determined by the intersection of the molar Gibbs free energy curves of $\langle\text{Ga}_{0.5}\text{As}_{0.5}\rangle_\text{c}$ and $(\text{Ga}_{1-x}\text{As}_x)$, as illustrated in the graphical constructions shown in the second panel on the left of Figure 3.16. Therefore,

$$\frac{g^{(\text{Ga})} + g^{(\frac{1}{2}\text{As}_2)}}{2} = g^{\langle\text{Ga}_{0.5}\text{As}_{0.5}\rangle_\text{c}}. \tag{3.56}$$

Together with Equation 2.47 and the condition $p^{(\text{Ga})} = 2p^{(\text{As}_2)}$ for congruent sublimation, the Ga and As_2 vapor pressures can then be deduced to be

$$\ln\left[\frac{p_c^{(\text{As}_2)}}{p_o}\right] = \ln\left[\frac{p_c^{(\text{Ga})}}{2p_o}\right] =$$

$$\frac{2g^{\langle\text{Ga}_{0.5}\text{As}_{0.5}\rangle_\text{c}} - g^{(\text{Ga})}(p_o,T) - g^{(\frac{1}{2}\text{As}_2)}(p_o,T) - \frac{2}{5}c_{o,\text{tra}}^{(\text{Ga})}T\ln(2)}{\frac{2}{5}\left[c_{o,\text{tra}}^{(\text{Ga})} + c_{o,\text{tra}}^{(\frac{1}{2}\text{As}_2)}\right]T}, \tag{3.57}$$

and are indicated by the dashed lines in the top right panel.

If the vapor does *not* have the same composition as the stoichiometric compound, then the vapor may still be in equilibrium with the compound, although its overall pressure will be different. Indeed, inside the MBE window shown in Figure 3.12, and away from the phase coexistence boundaries, there will be a range of partial pressures of (Ga) and (As_2) with which the pure compound can coexist. The compound can coexist with every pair of partial pressures determined by the tangents that pass through the minimum of $g^{\langle\text{Ga}_{0.5}\text{As}_{0.5}\rangle_\text{c}}$, provided those tangents lie below all other molar Gibbs free energy curves. As illustrated in Figure 3.17, as these tangents pivot around the minimum of the molar Gibbs free energy of $\langle\text{Ga}_{0.5}\text{As}_{0.5}\rangle_\text{c}$, their intercepts with the $x = 0$ and $x = 1$ axes trace out the relationship

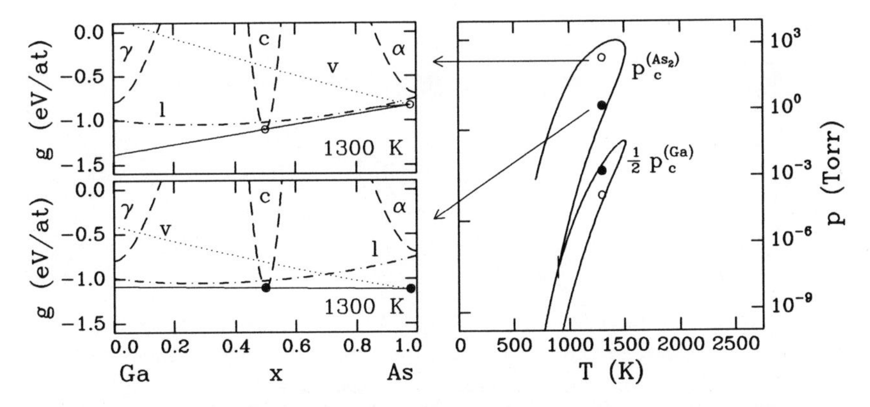

Figure 3.17: Vapor pressures of (As_2) and (Ga) over $\langle Ga_{0.5}As_{0.5}\rangle_c$. As the common tangent between $(Ga_{1-x}As_x)$ and $\langle Ga_{0.5}As_{0.5}\rangle_c$ pivots around the minimum of the molar Gibbs free energy of $\langle Ga_{0.5}As_{0.5}\rangle_c$, the intercepts of the common tangent trace out the relationship between the molar Gibbs free energies (and vapor pressures) of (As_2) and (Ga).

between the molar Gibbs free energies (and vapor pressures) of (As_2) and (Ga).

Algebraically, the constraint on the allowed tangents is

$$g^{(\frac{1}{2}\mathrm{As_2})} + g^{(\mathrm{Ga})} = 2g^{\langle \mathrm{Ga_{0.5}As_{0.5}}\rangle_c}. \tag{3.58}$$

Combined with Equation 2.47, this constraint forms the basis for what is known as the "law of mass action:"

$$\ln\left\{\left[\frac{p_c^{(\mathrm{As_2})}}{p_o}\right]^{c_{o,tra}^{(\frac{1}{2}\mathrm{As_2})}}\left[\frac{p_c^{(\mathrm{Ga})}}{p_o}\right]^{c_{o,tra}^{(\mathrm{Ga})}}\right\} = \frac{2g^{\langle \mathrm{Ga_{0.5}As_{0.5}}\rangle_c} - \left[g^{(\frac{1}{2}\mathrm{As_2})}(p_o,T) + g^{(\mathrm{Ga})}(p_o,T)\right]}{T}. \tag{3.59}$$

Finally, the As_2 vapor pressures along the various phase boundaries can also be interpreted as the critical As_2 overpressures at which various phase mixtures become stable. Therefore, those vapor pressures can be used to plot an As_2-overpressure/temperature phase stability diagram, as shown in Figure 3.18. That diagram is analogous to the *p*-*T* phase diagram of Figure 3.12, except that it is in terms of the As_2 overpressure rather than the total pressure. As discussed above, the diagrams are very nearly identical except near and below the congruent sublimation pressure.

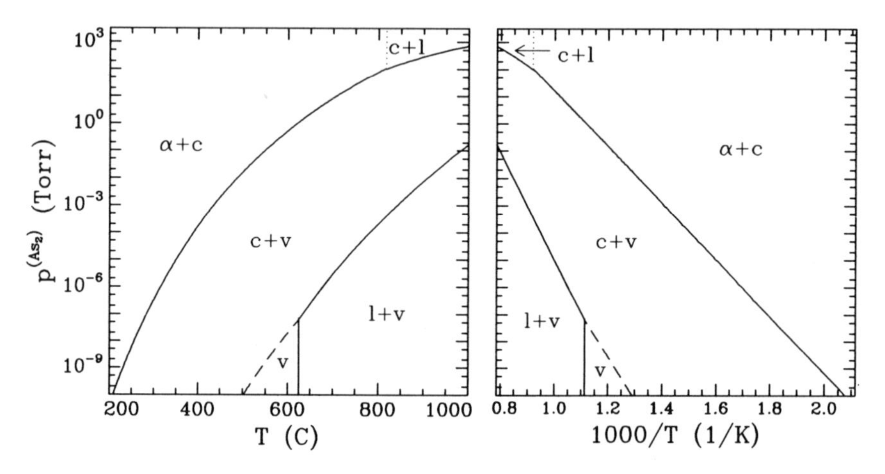

Figure 3.18: p–T (left) and p–$1/T$ (right) plots of critical As_2 overpressures at which various phase mixtures are stable.

Suggested Reading

1. J.W. Christian, *The Theory of Transformations in Metals and Alloys Part I: Equilibrium and General Kinetic Theory*, 2nd Ed. (Pergamon Press, Oxford, 1975).

2. P. Haasen, *Physical Metallurgy* (Cambridge Univ. Press, Cambridge, 1978).

3. F.A. Kroger, *The Chemistry of Imperfect Solids*, 2nd ed. (North-Holland, Amsterdam, 1973).

4. A. Prince, *Alloy Phase Equilibria* (Elsevier, Amsterdam, 1966).

Exercises

1. Verify the derivation of Equations 3.3 and 3.4 from Equations 3.2 and 3.5.

2. Derive the molar Gibbs free energy of a perfect solution consisting of a mixture of monomeric and tetrameric ideal gases. By factoring out the pressure dependences of the molar Gibbs free energies, derive an explicit form for the molar Gibbs free energy analogous to Equation 3.16.

3. A positive entropy of mixing (over and above the ideal entropy of mixing) might arise if the two components of a binary condensed-phase alloy did not mix fully randomly. How might a negative entropy of mixing arise?

4. Suppose the equilibrium vapor pressures of Si and Ge had been measured over the high-temperature $\{Si_{1-x}Ge_x\}$ liquid. Show how such a measurement could be used to deduce the interaction parameter for $\{Si_{1-x}Ge_x\}$. Suppose this measurement were made through a small hole in a 1" diameter, 3" long Knudsen cell. How small would the hole need to be for the measurement itself not to perturb the pressure in the cell? How long could the measurement go on before the composition of the alloy in the cell itself began to change significantly?

5. Verify Equations 3.26 and 3.29 for the chemical potentials of the components of ideal and regular solutions.

6. Verify Equation 3.38 for the critical temperature at which the miscibility gap of a strictly regular solution vanishes.

7. Strictly speaking, condensation and sublimation pressures are only equivalent up to a correction factor that depends on the sticking coefficient, i.e., the probability that an impinging molecule from the vapor will actually " stick" to the solid. What are those correction factors for elemental and alloy vapors?

8. How might the congruent sublimation temperature (or measurements of the congruent sublimation temperature by Langmuir free evaporation studies) depend on the orientation of the surface?

9. Along which of the phase boundaries shown in Figures 3.12 and 3.18 will the total pressure deviate most from the As_2 pressure, and by how much?

10. The usual form for the law of mass action is

$$\left[p^{(\mathrm{Ga})}\right]\left[p^{(\mathrm{As}_2)}\right]^{\frac{1}{2}} = K_p(T), \tag{3.60}$$

where $K_p(T)$ is a temperature-dependent equilibrium constant. Using Equation 3.59, derive an expression for that equilibrium constant.

Part II

Thin Film Structure and Microstructure

In Part I, we described the thermodynamic properties of bulk condensed and vapor phases. These properties determine whether epitaxy will occur at all, and are of primary importance in choosing growth conditions.

In this part, we describe modifications to the thermodynamic properties of bulk phases when the phases are constrained to grow as epitaxial thin films. These modifications are of secondary importance in choosing growth conditions, but are nevertheless crucial in determining the detailed structure and microstructure of the epitaxial phases as they condense. Indeed, even if a coarse view reveals only that the desired phase is condensing, a finer view may reveal a wide range of properties.

We begin, in Chapter 4, by discussing the tendency of epitaxial alloy phases to order and cluster. Then, in Chapter 5, we discuss the tendency of lattice-mismatched epitaxial phases to at first grow coherently with their underlying substrate, but then later to grow semi-coherently, through the introduction of misfit dislocations.

Chapter 4

Ordering and Clustering

In this chapter, we discuss the tendency of alloy phases, constrained to grow as epitaxial thin films, to order and cluster. We would like to know whether, during MBE of alloy phases, the individual components will tend on a microscopic scale to attract or repel each other, so that there is short-range order. We would also like to know whether the individual components will tend on a macroscopic scale to cluster into ordered or disordered phases of particular stoichiometries.

For concreteness, our discussion will center on "pseudobinary" III/V alloys — alloys composed of binary mixtures of two distinct III/V compounds. These alloys are exceedingly useful to device engineers because their lattice constants and electronic properties can be tuned continuously by adjusting the relative fractions of the two III/V compounds. These alloys are also characterized by positive enthalpies of mixing, and hence have a tendency to "unmix."[1] Those enthalpies of mixing originate mainly from microscopic strain caused by the different bond lengths of the two III/V compounds. Therefore, we begin the chapter by describing, in Section 4.1, how to estimate the strain in microscopic clusters using what are known as "valence force field" (VFF) models. If these microscopic clusters are embedded in an epitaxial thin film on a substrate with a different lattice constant, then they will also be "externally" strained. In Section 4.2, we discuss how to estimate that external strain.

In Section 4.3, we introduce a powerful technique, the cluster variation method, for building a macroscopic description of alloy thermodynamics from statistical combinations of such microscopic clusters. In Section 4.4, we apply this method in an approximate way to $In_{1-x}Ga_xAs$, a pseudobi-

[1] E.K. Müller and J.L. Richards, "Miscibility of III-V semiconductors studied by flash evaporation," *J. Appl. Phys.* **35**, 1233 (1964).

nary alloy of current technological interest. We will find that the thermodynamic properties of $In_{1-x}Ga_xAs$ depend greatly on whether the alloy is coherent or incoherent with the substrate, i.e., on whether the interface between the epitaxial film and the substrate is crystallographically perfect or not.[2] If the alloy is incoherent with the substrate, then it is free to adopt the in-plane lattice constant that minimizes its free energy. If the alloy is coherent with the substrate, then it must adopt the in-plane lattice constant of the substrate; the resulting elastic strain energy can increase its overall free energy significantly.

In fact, such coherency constraints greatly suppress the tendency for alloys to separate into their pure-component "endpoint" phases, and at the same time greatly enhance their tendency to form ordered compounds at certain stoichiometric compositions. These tendencies can be understood quantitatively from the full cluster variation method calculation, but they can also be understood semiquantitatively through simpler semi-empirical models. We end the chapter, therefore, with a simple analytical treatment in Section 4.5 of coherency-constrained clustering and ordering.

4.1 Microscopic Strain

Let us start, in this section, by discussing microscopic strain in pseudobinary III/V alloys. We begin, in Subsection 4.1.1, by introducing a simple bond stretching and bond bending force field model for calculating the equilibrium atomic positions of a small alloy cluster. Then, in Subsection 4.1.2, we use those atomic positions to estimate the strain energy, which is the dominant contribution to the enthalpy of mixing.

4.1.1 Virtual Crystals and Covalent Radii

Let us begin, in this subsection, by calculating the microscopic bond distortions that occur when two III/V compounds are mixed. For concreteness, let us consider GaAs and InAs. Bulk alloys in this system are known to obey Vegard's law quite accurately: their *overall* lattice constants are the averages of the bulk GaAs and InAs lattice constants, weighted by mole fraction. If we imagine the alloy to be a "virtual crystal," in that each atom sits on geometrically perfect zincblende lattice sites,[3] then its lattice

[2]D.M. Wood and A. Zunger, "Epitaxial effects on coherent phase diagrams of alloys," *Phys. Rev.* **B40**, 4062 (1989).

[3]L. Nordheim, "Electron theory of metals," *Ann. Phys. (Leipzig)* **9**, 607 and 641 (1931).

constant can be expressed as

$$a_{VC} = (1-x)a_{GaAs,o} + xa_{InAs,o}, \tag{4.1}$$

where x is the InAs fraction in the alloy, and $a_{GaAs,o}$ and $a_{InAs,o}$ are the equilibrium lattice constants of (unstrained) bulk GaAs and InAs, respectively. Indeed, measurements[4] show that the *second*-nearest-neighbor distances between group III atoms (or between group V atoms) in the lattice are very nearly those — $a_{VC}/\sqrt{2}$ — expected for such virtual crystals.

In contrast, however, *first*-nearest-neighbor distances between group III and group V atoms deviate significantly from those — $\sqrt{3}a_{VC}/4$ — expected for such virtual crystals. Instead, Ga–As bonds are shorter, and In–As bonds are longer, than the virtual crystal bonds. That this is so is not unexpected, since the Ga–As bond in bulk GaAs is shorter than the In–As bond in bulk InAs, so in some sense the As "prefers" to be nearer to Ga than to In atoms. Indeed, one might imagine that, instead of occupying virtual crystal lattice sites, the atoms would occupy sites such that the bulk Ga–As and In–As bond lengths, and the associated "covalent radii" of the Ga, As, and In atoms, were preserved.[5]

To see which extreme of behavior is closer to the truth, consider the 5-atom $In_{0.5}Ga_{0.5}As$ tetrahedron at the right of Figure 4.1. This tetrahedron is one of the five distinct tetrahdra shown at the top of Figure 4.2 from which, as discussed in Section 4.4, an $In_{1-x}Ga_xAs$ alloy of arbitrary composition may be constructed. On the one hand, if the central As atom occupies the geometric center of the tetrahedron, then the tetrahedral bond angles associated with the sp^3 hybridized bonds can be preserved, but at the expense of InAs bonds that are too short and GaAs bonds that are too long. On the other hand, if the central As atom moves down slightly, then the InAs and GaAs bonds can approach their bulk equilibrium values, but at the expense of In–As–In bond angles that are too acute and Ga–As–Ga bond angles that are too obtuse.

The "elastic" energies associated with these kinds of distortions are often quantified using what are known as valence force field (VFF) models,[6] in which the energies of individual bonds and bond angles are considered to be independent of each other. In the most popular representation for diamond-

---

[4] J.C. Mikkelsen, Jr., and J.B. Boyce, "Extended x-ray-absorption fine-structure study of GaInAs random solid solutions," *Phys. Rev.* **B28**, 7130 (1983).

[5] L. Pauling and M.L. Huggins, "Covalent radii of atoms and interatomic distances in crystals containing electron-pair bonds," *Z. Kristallogr. Kristallgeom. Kristallphys. Kristallchem.* **87**, 205 (1934).

[6] M.J.P. Musgrave and J.A. Pople, "A general valence force field for diamond," *Proc. Roy. Soc. London* **A268**, 474 (1962).

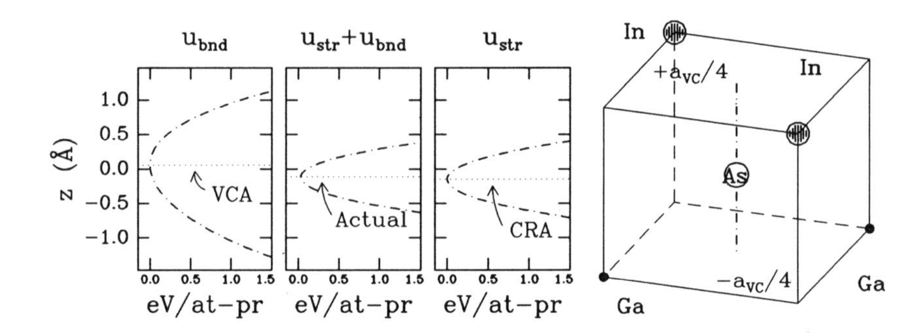

Figure 4.1: Valence force field potential energies versus As position within a 5-atom $In_{0.5}Ga_{0.5}As$ tetrahedron. Left and right panels show contributions due to bond bending and bond stretching forces; center panel shows the sum of the two contributions. The geometric center of the tetrahedron is at $z = 0$; the actual As position is shifted downward toward the Ga atoms. The predictions of the virtual crystal (VCA) and covalent radius (CRA) approximations discussed in the text are also shown.

structure semiconductors, the "Keating potential,"[7] the stretching energy associated with bond i is proportional to the squared deviations of the squared actual bond length from the squared equilibrium length,

$$u_{\mathrm{str},i} = \frac{3}{8}\alpha_i \frac{\left(d_i^2 - d_{i,\mathrm{o}}^2\right)^2}{d_{i,\mathrm{o}}^2}, \tag{4.2}$$

and the bending energy associated with adjacent bonds i and j is proportional to the squared deviations of the dot products of actual adjacent bonds from the dot products of the equilibrium bonds,

$$u_{\mathrm{bnd},ij} = \frac{3}{8}\frac{\beta_i + \beta_j}{2}\frac{\left(\mathbf{d}_i \cdot \mathbf{d}_j - \mathbf{d}_{i,\mathrm{o}} \cdot \mathbf{d}_{j,\mathrm{o}}\right)^2}{d_{i,\mathrm{o}} d_{j,\mathrm{o}}}. \tag{4.3}$$

The two microscopic stretching and bending force constants, α and β, are assumed sufficient to characterize completely the microscopic elastic behavior of both the pure and mixed III-V compounds. Moreover, they can be used to predict various macroscopic elastic phenomena, and hence can be deduced from bulk elastic constants. The most commonly used values are listed in Table 4.1 for a number of diamond-structure materials.

[7] P.N. Keating, "Effect of invariance requirements on the elastic strain energy of crystals with application to the diamond structure," *Phys. Rev.* **145**, 637 (1966).

Material	α (N/m)	β (N/m)	β/α	C_{11} (10^{10} N/m^2)	C_{12} (10^{10} N/m^2)	C_{44} (10^{10} N/m^2)
C	129.33	84.76	0.655	107.6	12.50	57.68
Si	48.50	13.81	0.285	16.57	6.39	7.96
Ge	38.67	11.35	0.294	12.89	4.83	6.71
AlSb	35.35	6.77	0.192	8.94	4.43	4.16
GaP	47.32	10.44	0.221	14.12	6.25	7.05
GaAs	41.19	8.95	0.217	11.81	5.32	5.92
GaSb	33.16	7.22	0.218	8.84	4.03	4.32
InP	43.04	6.24	0.145	10.22	5.76	4.60
InAs	35.18	5.50	0.156	8.33	4.53	3.96
InSb	29.61	4.77	0.161	6.67	3.65	3.02
ZnS	44.92	4.78	0.107	10.40	6.50	4.62
ZnSe	35.24	4.23	0.120	8.10	4.88	4.41
ZnTe	31.35	4.45	0.142	7.13	4.07	3.12
CdTe	29.02	2.43	0.084	5.35	3.68	1.99
CuCl	12.60	1.00	0.079	2.72	1.87	1.57

Table 4.1: Microscopic bond stretching (α) and bond bending (β) force constants deduced from macroscopic elastic constants (C_{11}, C_{12}, and C_{44}) of various cubic semiconducting materials.[a]

[a] R.M. Martin, "Elastic properties of ZnS structure semiconductors," *Phys. Rev.* **B1**, 4005 (1970).

To calculate the stretching energy of tetrahedra such as that shown at the right of Figure 4.1, we sum Equation 4.2 over the four bonds to the central As atom, divide by two because each bond is shared by two atoms, then multiply by two because there is a pair of atoms per tetrahedron:

$$u_{\text{str}} = \sum_{i=1}^{4} u_{\text{str},i}. \tag{4.4}$$

To calculate the bending energy of the tetrahedron, we sum Equation 4.3 over each distinct pair of adjacent bonds, and multiply by two because we have only accounted for the bonds centered on the group V atoms, but not those centered on the group III atoms:

$$u_{\text{bnd}} = 2\sum_{i=1}^{4}\sum_{j<i} u_{\text{bnd},ij}. \tag{4.5}$$

For the particular tetrahedron shown at right in Figure 4.1, the total elastic energy, per atom-pair, is then

$$\begin{aligned} u_{(\blacktriangle)} &= u_{\rm str} + u_{\rm bnd} \\ &\approx \frac{3}{4}\left[\alpha_{\rm GaAs}\frac{\left(d_{\rm GaAs}^2 - d_{\rm GaAs,o}^2\right)^2}{d_{\rm GaAs,o}^2} + \alpha_{\rm InAs}\frac{\left(d_{\rm InAs}^2 - d_{\rm InAs,o}^2\right)^2}{d_{\rm InAs,o}^2}\right] \\ &+ \frac{3}{4}\left[\beta_{\rm GaAs}\frac{\left(d_{\rm GaAs}^2\cos^2\theta_{\rm GaAsGa} - d_{\rm GaAs,o}^2\cos^2\theta_{\rm T}\right)^2}{d_{\rm GaAs,o}^2}\right. \\ &\left. + \beta_{\rm InAs}\frac{\left(d_{\rm InAs}^2\cos^2\theta_{\rm InAsIn} - d_{\rm InAs,o}^2\cos^2\theta_{\rm T}\right)^2}{d_{\rm InAs,o}^2}\right]. \end{aligned} \tag{4.6}$$

In this equation, the actual and equilibrium GaAs and InAs bond lengths are denoted $d_{\rm GaAs}$, $d_{\rm GaAs,o}$, $d_{\rm InAs}$ and $d_{\rm InAs,o}$; the actual Ga–As–Ga and In–As–In bond angles are denoted $\theta_{\rm GaAsGa}$ and $\theta_{\rm InAsIn}$; and the ideal tetrahedral bond angle is $\theta_{\rm T} = 2\tan^{-1}(1/\sqrt{2}) \approx 109.47°$. Note that we have used the symmetry of the tetrahedron to set $\theta_{\rm GaAsGa} = \theta_{\rm AsGaAs}$, $\theta_{\rm InAsIn} = \theta_{\rm AsInAs}$ and $\theta_{\rm GaAsIn} = \theta_{\rm InAsGa} \approx \theta_{\rm T}$.

In terms of the vertical displacement, z, of the As atom from the geometric center of the tetrahedron, the actual GaAs and InAs bond lengths can be written as

$$\begin{aligned} d_{\rm GaAs}^2 &= \left(\frac{a_{\rm VC}}{2\sqrt{2}}\right)^2 + \left(\frac{a_{\rm VC}}{4} + z\right)^2 \\ d_{\rm InAs}^2 &= \left(\frac{a_{\rm VC}}{2\sqrt{2}}\right)^2 + \left(\frac{a_{\rm VC}}{4} - z\right)^2, \end{aligned} \tag{4.7}$$

and the actual Ga–As–Ga and In–As–In bond angles can be written as

$$\begin{aligned} \cos(\theta_{\rm GaAsGa}/2) &\approx \frac{(a_{\rm VC}/4) + z}{(\sqrt{3}a_{\rm VC}/4) + z/\sqrt{3}} \\ \cos(\theta_{\rm InAsIn}/2) &\approx \frac{(a_{\rm VC}/4) - z}{(\sqrt{3}a_{\rm VC}/4) - z/\sqrt{3}}, \end{aligned} \tag{4.8}$$

where $a_{\rm VC}/2$ is the length of an edge of the cube circumscribing the tetrahedron.

Then, substituting back into Equation 4.6, we can calculate, as shown in the left three panels of Figure 4.1, the distortion energies as a function of z. The left and right panels show only the bending and stretching energies, respectively; the center panel shows their total. Those panels illustrate how the actual position of the As atom at the center of the tetrahedron is determined by a competition between bending and stretching forces. Given only bending forces, the virtual crystal approximation (VCA) holds, and bond angles are nearly ideally tetrahedral.[8] Given only stretching forces, the covalent radius approximation (CRA) holds, and bond lengths are undistorted from the bulk pure component compounds. Given both forces, neither holds exactly, but, as can be seen, the CRA is the better approximation. In this pseudobinary III/V system, bending forces are about 5 times weaker than stretching forces, and bond lengths are very nearly preserved upon mixing. They deviate slightly, however, due to the "steric" constraints provided by bending forces.

To obtain an analytic form for the position of the As atom, we can expand Equations 4.6, 4.7, and 4.8 to second order in z, giving

$$\begin{aligned} u_{(\blacktriangle)} \approx\; & \frac{3}{2}\alpha_{\mathrm{GaAs}}\left[\frac{\sqrt{3}}{4}a_{\mathrm{VC}} - d_{\mathrm{GaAs,o}} + \frac{z}{\sqrt{3}}\right]^2 \\ & + \frac{3}{2}\alpha_{\mathrm{InAs}}\left[\frac{\sqrt{3}}{4}a_{\mathrm{VC}} - d_{\mathrm{InAs,o}} - \frac{z}{\sqrt{3}}\right]^2 \\ & + \frac{3}{8}\beta_{\mathrm{GaAs}}\left[\frac{-2}{3}\left(\frac{\sqrt{3}}{4}a_{\mathrm{VC}} - d_{\mathrm{GaAs,o}}\right) + \frac{2}{\sqrt{3}}z\right]^2 \\ & + \frac{3}{8}\beta_{\mathrm{InAs}}\left[\frac{-2}{3}\left(\frac{\sqrt{3}}{4}a_{\mathrm{VC}} - d_{\mathrm{InAs,o}}\right) - \frac{2}{\sqrt{3}}z\right]^2 . \end{aligned} \tag{4.9}$$

Then, solving for $\partial u_{(\blacktriangle)}/\partial z = 0$, the equilibrium position can be deduced to be

$$z_{\mathrm{equ}} = \left(\frac{-\sqrt{3}}{2}\right)\frac{\alpha_{\mathrm{GaAs}} + \alpha_{\mathrm{InAs}} - \beta_{\mathrm{GaAs}}/3 + \beta_{\mathrm{InAs}}/3}{\alpha_{\mathrm{GaAs}} + \alpha_{\mathrm{InAs}} - \beta_{\mathrm{GaAs}} + \beta_{\mathrm{InAs}}}(d_{\mathrm{InAs,o}} - d_{\mathrm{GaAs,o}}). \tag{4.10}$$

In the limit $\beta \to 0$, $z_{\mathrm{equ}} \to -(\sqrt{3}/2)(d_{\mathrm{InAs,o}} - d_{\mathrm{GaAs,o}})$, and the VCA holds; in the limit $\alpha \to 0$, $z_{\mathrm{equ}} \to -(d_{\mathrm{InAs,o}} - d_{\mathrm{GaAs,o}})/(2\sqrt{3})$, and the CRA holds.

To see how the bond lengths in these alloys depend on composition, similar calculations can be performed for 5-atom GaAs, $\mathrm{In_{0.25}Ga_{0.75}As}$,

[8] The angles are not exactly tetrahedral because the Keating representation of the "valence forces" does not cleanly separate stretching from bending motions, since Equation 4.3 consists of deviations of dot products (rather than of angles) between adjacent bonds. Other representations do, but at the expense of not appearing to predict distortion energies as accurately [W.A. Harrison, *Electronic Structure and the Properties of Solids* (W.H. Freeman, San Francisco, 1980), pp. 193-197].

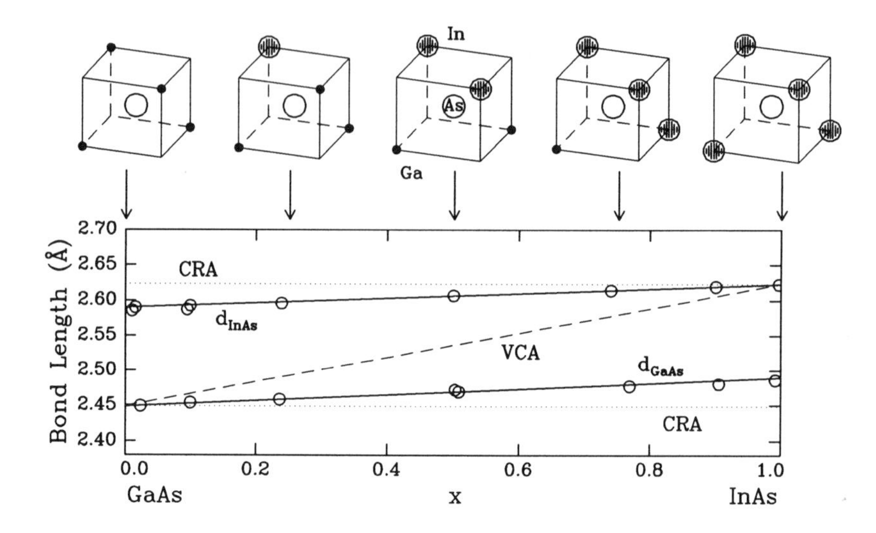

Figure 4.2: Measured and calculated Ga–As and In–As bond lengths in $In_{1-x}Ga_xAs$ alloys. Experimental data are from (open circles) X-ray-absorption fine structure (EXAFS)[a] measurements; calculations are based on valence-force-field potentials using either bond bending forces in the virtual crystal approximation (dashed line), bond stretching forces in the covalent radius approximation (dotted lines), or both (solid lines).[b]

[a] J.C. Mikkelsen, Jr., and J.B. Boyce, "Extended x-ray-absorption fine-structure study of GaInAs random solid solutions," *Phys. Rev.* **B28**, 7130 (1983).

[b] T. Fukui, "Calculation of bond length in InGaAs ternary semiconductors," *Jpn. J. Appl. Phys.* **23**, L208 (1984).

$In_{0.75}Ga_{0.25}As$ and InAs tetrahedra, which are the most probable tetrahedra in the corresponding GaAs, $In_{0.25}Ga_{0.75}As$, $In_{0.75}Ga_{0.25}As$ and InAs alloys. The results are shown in Figure 4.2. The dotted and dashed lines are the stretching-force-only (CRA) and bending-force-only (VCA) bond lengths; the solid lines are the stretching-force plus bending-force bond lengths. The predictions agree extremely well with the measurements shown as open circles. That agreement indicates that, consistent with more complete calculations,[9] elastic energies dominate chemical energies in this alloy system. Indeed, this dominance appears to hold for most isovalent, though not for heterovalent, mixtures of semiconductors.[10]

[9] T. Ito, "A pseudopotential approach to mixing enthalpies of III-V ternary semiconductor alloys," *Jpn. J. Appl. Phys.* **26**, 256 (1987).

[10] W.A. Harrison and E.A. Kraut, "Energies of substitution and solution in semiconductors," *Phys. Rev.* **B37**, 8244 (1988).

4.1.2 Mixing Enthalpies

In Subsection 4.1.1, we calculated the microscopic distortions that minimize the sum of the bond stretching and bond bending energies of an As-centered cluster containing both Ga and In. In this subsection, we make use of those distortions to calculate the strain energy associated with the cluster, and then to estimate the mixing enthalpy associated with the alloy as a whole.

To obtain a simplified formula for the energy of the $In_{0.5}Ga_{0.5}As$ tetrahedron, we insert the equilibrium position of the As atom given by Equation 4.10 into Equation 4.9. Then, approximating the individual bond stretching and bending force constants by their averages, $\overline{\alpha} \equiv (\alpha_{\mathrm{GaAs}} + \alpha_{\mathrm{InAs}})/2$ and $\overline{\beta} \equiv (\beta_{\mathrm{GaAs}} + \beta_{\mathrm{InAs}})/2$, we obtain, after some algebra,

$$u_{(\blacktriangle)} \approx \frac{\overline{\alpha}\overline{\beta}}{2(\overline{\alpha} + \overline{\beta})}(\Delta a_o)^2, \tag{4.11}$$

where

$$\Delta a_{\mathrm{o}} \equiv a_{\mathrm{InAs,o}} - a_{\mathrm{GaAs,o}} = \frac{4}{\sqrt{3}}(d_{\mathrm{InAs,o}} - d_{\mathrm{GaAs,o}}). \tag{4.12}$$

The distortion energy of the tetrahedron calculated in this way is listed in Table 4.7 on page 132. The energy is proportional to the square of the difference, Δa_o, between the lattice parameters of the component compounds, precisely what one expects from a model based on linear elasticity. The effective spring constant, $1/[(1/\overline{\alpha}) + (1/\overline{\beta})]$, is the "parallel" sum of the individual stretching and bending force constants. Since, as mentioned above, $\overline{\beta}$ is approximately 5 times weaker than $\overline{\alpha}$, the effective spring constant is dominated by $\overline{\beta}$. In other words, as with all coupled spring systems, most of the energy is stored in the weaker and more deformed spring.

If we now imagine building a lattice solely out of $In_{0.5}Ga_{0.5}As$ tetrahedra, then Equation 4.11 can also be used to estimate the enthalpy of mixing of the $In_{0.5}Ga_{0.5}As$ alloy. On the one hand, it will be an overestimate: our simple calculation did not account for relaxation of the corner group III atoms of the tetrahedron away from their virtual crystal positions, which would decrease the tetrahedron energy. On the other hand, it will be an underestimate: as discussed later in Section 4.4, a real $In_{0.5}Ga_{0.5}As$ alloy at finite temperature would also contain some fraction of more highly deformed tetrahedra of other compositions, which would increase the energy of the alloy as a whole.

To see how well this estimate works, let us approximate the alloy as a strictly regular solution, and identify its interaction enthalpy at $x = 1/2$ with the VFF elastic energy of the $In_{0.5}Ga_{0.5}As$ tetrahedron: $\Omega_{\mathrm{VFF}} \approx$

Material	Crystal Structure	a_o or a/c (Å or Å/Å)	$\alpha_T = \partial \ln a_o/\partial T$ ($10^{-6}K^{-1}$)
C	Diamond	3.56683	$0.87 + 0.0092(T - 273)$
Si	Diamond	5.43095	$3.08 + 0.0019(T - 273)$
Ge	Diamond	5.64613	$6.05 + 0.0036(T - 273)$
α-Sn	Diamond	6.48920	
SiC	Wurtzite	3.086/15.117	
BN	Zincblende	3.6150	
BP	Zincblende	4.5380	
AlP	Zincblende	5.4510	
AlAs	Zincblende	5.6605	$3.40 + 0.0064(T - 273)$
AlSb	Zincblende	6.1355	
GaN	Zincblende	3.189/5.185	
GaP	Zincblende	5.4512	5.81
GaAs	Zincblende	5.6533	$5.35 + 0.0080(T - 273)$
GaSb	Zincblende	6.0959	6.7
InP	Zincblende	5.8686	
InAs	Zincblende	6.0584	$4.33 + 0.0038(T - 273)$
InSb	Zincblende	6.4794	
ZnO	Rock Salt	4.580	
ZnS	Zincblende	5.420	$6.70 + 0.0128(T - 313)$
ZnS	Wurtzite	3.82/6.26	
CdS	Zincblende	5.8320	
CdS	Wurtzite	4.16/6.756	
CdTe	Zincblende	6.482	
CdSe	Zincblende	6.050	
PbS (Galena)	Rock Salt	5.9362	$18.81 + 0.0074(T - 273)$
PbTe (Altaite)	Rock Salt	6.4620	19.80

Table 4.2: Crystal structures, room-temperature lattice parameters and thermal expansion coefficients of various semiconductors.[a]

[a] Adapted from S.M. Sze, *Physics of Semiconductor Devices*, 2nd Ed. (John Wiley & Sons, New York, 1981), and R.S. Krishnan, R. Srinivasan and S. Devanarayanan, *Thermal Expansion of Crystals* (Pergamon Press, Oxford, 1979).

$4u_{(\blacktriangle)}$. Then

$$\Omega_{\mathrm{VFF}} \approx \frac{2\overline{\alpha}\overline{\beta}}{\overline{\alpha} + \overline{\beta}} (\Delta a_o)^2. \tag{4.13}$$

This equation can be used to estimate the elastic part of the regular solution interaction parameter for any pseudobinary mixture whose microscopic elastic constants and lattice parameters are known. Its predictions are shown in Figure 4.3 for a number of alloys, using the lattice parameters listed in Table 4.2 and the bond stretching and bending force constants listed in Table 4.1. Within the (fairly large) uncertainty in the values deduced from experimental measurements, the equation predicts the regular solution parameters surprisingly accurately. It represents the physical basis[11] for what is known as the Delta-Lattice-Parameter (DLP) model, originally based on the empirical observation that heats of mixing are approximately proportional to the squared mismatches between the lattice parameters of the constituent components.[12]

4.2 Macroscopic Strain

In Section 4.1, we noted that, from a microscopic point of view, pseudobinary III-V alloys can be viewed as a collection of elementary tetrahedra such as those shown in Figure 4.2. Except for the pure-component tetrahedra, none are perfectly tetrahedral: their bond lengths and angles deviate from the CRA lengths and VCA angles, respectively. These *internal* distortions give rise to the elastic strain energies listed in Table 4.7 on page 132 even in tetrahedra embedded in bulk alloys of the same overall composition as the tetrahedra themselves.

Superimposed on these internal distortions, however, are distortions due to *externally* imposed constraints on the dimensions of the tetrahedra. These constraints arise because the tetrahedra, each with an "ideal" dimension or shape, are all embedded in a macroscopic lattice whose unit cells have their own (and possibly different) average dimension or shape. In this section, we discuss these externally imposed distortions. Conceptually, they can be decomposed into two components: one that is mainly volumetric and one that is mainly distortional.

The volumetric component comes about either when alloys are grown in bulk form, or when epitaxial films are grown coherently on a lattice-*matched* substrate. Consider such an alloy, whose overall composition is $x_{\mathrm{epi}} = 0.5$, and whose mean (or virtual crystal) lattice parameter is given, using Equation 4.1, by

$$a_{\mathrm{epi,o}} = 0.5a_{\mathrm{GaAs,o}} + 0.5a_{\mathrm{InAs,o}}. \qquad (4.14)$$

[11] P.A. Fedders and M.W. Muller, "Mixing enthalpy and composition fluctuations in ternary III-V semiconductor alloys," *J. Phys. Chem. Solids* **45**, 685 (1984); J.L. Martins and A. Zunger, "Bond lengths around isovalent impurities and in semiconductor solid solutions," *Phys. Rev.* **B30**, 6217 (1984)

[12] G.B. Stringfellow, "Calculation of regular solution interaction parameters in semiconductor solid solutions," *J. Phys. Chem. Solids* **34**, 1749 (1973).

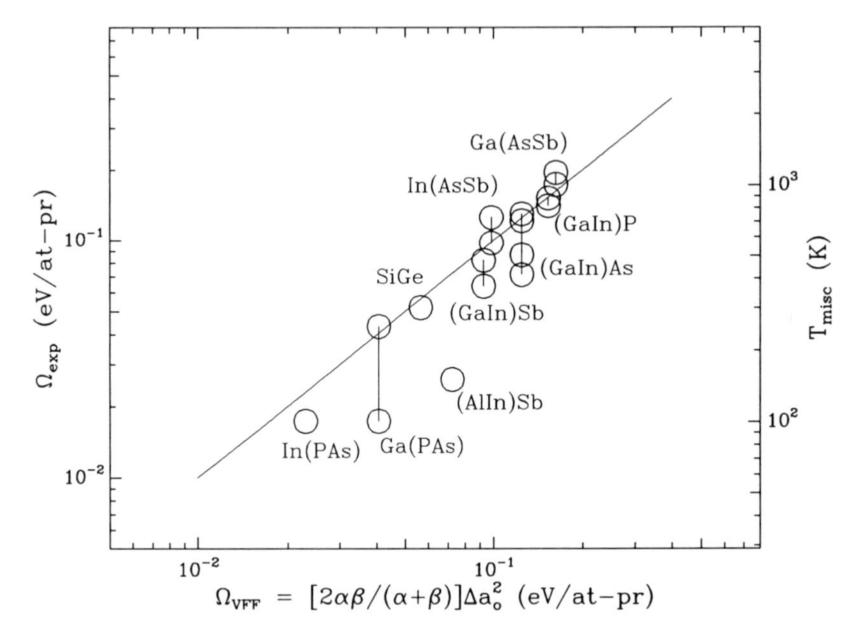

Figure 4.3: Regular solution parameters for various pseudobinary alloys. Values plotted along the bottom axis were calculated using Equation 4.13; values plotted along the left axis are experimental measurements[a]; values plotted along the right axis are the critical temperatures, deduced from Equation 3.38, above which the constituent components are fully miscible.

[a] Adapted from G.B. Stringfellow, "Calculation of ternary and quaternary III-V phase diagrams," *J. Cryst. Growth* **27**, 21 (1974).

The only tetrahedron whose "ideal" dimension is also given by Equation 4.14 is the $In_{0.5}Ga_{0.5}As$ tetrahedron, which will not be externally strained, and whose excess energy will be due solely to internal distortions. All other tetrahedra will have "ideal" dimensions different from that given by Equation 4.14. If embedded in the $x_{epi} = 0.5$ alloy, they will be constrained to occupy volumes different from their ideal volume, and will have additional energies due to the externally imposed volumetric distortions. For example, an $In_{0.75}Ga_{0.25}As$ tetrahedron has an ideal dimension (neglecting relaxations of corner atoms) of $0.25a_{GaAs,o} + 0.75a_{InAs,o}$, and must be compressed before it can fit into a $In_{0.5}Ga_{0.5}As$ lattice.

The distortional component comes about when epitaxial films are grown coherently on a lattice-*mismatched* substrate. Suppose, for example, that the substrate is a single (unstrained) crystal of bulk $In_{1-x_{sub}}Ga_{x_{sub}}As$ itself, whose In composition is x_{sub} and whose mean (or virtual crystal) lattice parameter is a weighted average of the two endpoint lattice parameters,

$$a_{\mathrm{sub}} = (1 - x_{\mathrm{sub}})a_{\mathrm{GaAs,o}} + x_{\mathrm{sub}}a_{\mathrm{InAs,o}} \tag{4.15}$$

As illustrated in the right half of Figure 4.4, if the epitaxial film is coherent with the substrate, then its lattice parameter parallel to the interface must be the same as that of the substrate, independent of the composition of the epitaxial film itself:

$$a_{\mathrm{epi},\|} = a_{\mathrm{sub}} = (1 - x_{\mathrm{sub}})a_{\mathrm{GaAs,o}} + x_{\mathrm{sub}}a_{\mathrm{InAs,o}}. \tag{4.16}$$

In other words, there will be a parallel strain in the film of

$$\epsilon_{\mathrm{epi},\|} \equiv 2\frac{a_{\mathrm{epi},\|} - a_{\mathrm{epi,o}}}{a_{\mathrm{epi},\|} + a_{\mathrm{epi,o}}}, \tag{4.17}$$

where

$$a_{\mathrm{epi,o}} = (1 - x_{\mathrm{epi}})a_{\mathrm{GaAs,o}} + x_{\mathrm{epi}}a_{\mathrm{InAs,o}}, \tag{4.18}$$

is the equilibrium (unstrained) lattice parameter of the epitaxial film.

As illustrated in the left half of Figure 4.4, however, its lattice parameter in a direction perpendicular to the interface will not be the same as the equilibrium lattice parameters of either the substrate or the epitaxial film. If the film is locked to a substrate with a smaller lattice parameter, then the in-plane compressional "squeezing" will force its perpendicular lattice parameter to increase in order to preserve (approximately) its unit cell volume. If the film is locked to a substrate with a larger lattice parameter, then the in-plane tensile "stretching" will force its perpendicular lattice parameter to decrease, again in order to preserve (approximately) its unit cell volume.

To understand both the volumetric and distortional components of the externally imposed strains quantitatively, we write what is known as the generalized Hooke's law for cubic crystals,[13]

$$\begin{pmatrix} \sigma_x \\ \sigma_y \\ \sigma_z \\ \tau_{xy} \\ \tau_{yz} \\ \tau_{zx} \end{pmatrix} = \begin{pmatrix} C_{11} & C_{12} & C_{12} & 0 & 0 & 0 \\ C_{12} & C_{11} & C_{12} & 0 & 0 & 0 \\ C_{12} & C_{12} & C_{11} & 0 & 0 & 0 \\ 0 & 0 & 0 & C_{44} & 0 & 0 \\ 0 & 0 & 0 & 0 & C_{44} & 0 \\ 0 & 0 & 0 & 0 & 0 & C_{44} \end{pmatrix} \begin{pmatrix} \epsilon_x \\ \epsilon_y \\ \epsilon_z \\ \gamma_{xy} \\ \gamma_{yz} \\ \gamma_{zx} \end{pmatrix}, \tag{4.19}$$

[13]See, e.g., A.J. Durelli, E.A. Phillips, and C.H. Tsao, *Introduction to the Theoretical and Experimental Analysis of Stress and Strain* (McGraw-Hill, New York, 1958), Chap. 4.

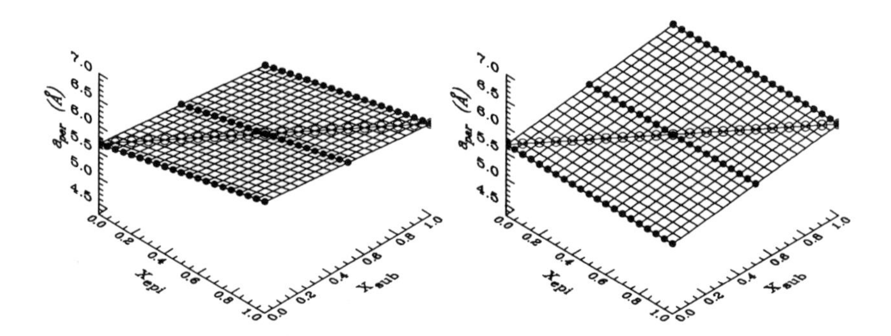

Figure 4.4: Perpendicular (left) and parallel (right) lattice parameters of $In_{1-x}Ga_xAs$ at composition $x_{\rm epi}$ grown coherently on substrates having the lattice parameters of bulk $In_{1-x}Ga_xAs$ at composition $x_{\rm sub}$. The filled circles represent $In_{1-x}Ga_xAs$ grown on substrates with compositions $x_{\rm sub} = 0, 1/2, 1$. The open circles represent $In_{1-x}Ga_xAs$ grown on "lattice-matched" substrates with compositions $x_{\rm sub} = x_{\rm epi}$, or, alternatively, to incoherent growth.

where the ϵ_i's and σ_i's are the normal strains and stresses, respectively, and the γ_{ij}'s and τ_{ij}'s are the shear strains and stresses, respectively.

If the epitaxial film and its substrate are oriented along one of the $\langle 100 \rangle$ cubic symmetry directions, then this equation reduces to

$$\begin{pmatrix} \sigma_{\rm epi,\|} \\ \sigma_{\rm epi,\perp} \end{pmatrix} = \begin{pmatrix} C_{11}+C_{12} & C_{12} \\ 2C_{12} & C_{11} \end{pmatrix} \begin{pmatrix} \epsilon_{\rm epi,\|} \\ \epsilon_{\rm epi,\perp} \end{pmatrix}. \tag{4.20}$$

If, in addition, the epitaxial film has a free surface, such that perpendicular stresses vanish, then

$$\sigma_{\rm epi,\perp} = 2C_{12}\epsilon_{\rm epi,\|} + C_{11}\epsilon_{\rm epi,\perp} = 0, \tag{4.21}$$

and the perpendicular strain and lattice parameter of the film are

$$\begin{aligned} \epsilon_{\rm epi,\perp} &= \frac{-2C_{12}}{C_{11}}\epsilon_{\rm epi,\|} \\ a_\perp(x_{\rm epi}, x_{\rm sub}) &= a_{\rm epi,o}\frac{1+\epsilon_{\rm epi,\perp}/2}{1-\epsilon_{\rm epi,\perp}/2}. \end{aligned} \tag{4.22}$$

On average, then, the unit cell of the epitaxial film has parallel and perpendicular dimensions given by Equations 4.16 and 4.22, respectively.

Now consider the microscopic tetrahedra that are embedded within this epitaxial film. On average, they must be constrained to the same dimensions as the unit cell.[14] However, each individual tetrahedron has its own "ideal" size, given approximately by Vegard's law:

$$a_{i,\mathrm{o}} = (1 - x_i)a_{\mathrm{GaAs,o}} + x_i a_{\mathrm{InAs,o}}, \tag{4.23}$$

where x_i is the composition of the ith elementary tetrahedron. If they are all constrained to the average dimension of the unit cell of the epitaxial film, then they will be strained, according to

$$\begin{aligned} \epsilon_{i,\|} &= 2\frac{a_{\mathrm{epi},\|} - a_{i,\mathrm{o}}}{a_{\mathrm{epi},\|} + a_{i,\mathrm{o}}} \\ \epsilon_{i,\perp} &= 2\frac{a_{\mathrm{epi},\perp} - a_{i,\mathrm{o}}}{a_{\mathrm{epi},\perp} + a_{i,\mathrm{o}}}. \end{aligned} \tag{4.24}$$

The resulting strain energies of the various tetrahedra (per atom pair) due to these external constraints can then be approximated, through use of Equation 4.20, by[15]

$$\begin{aligned} u_{i,\mathrm{ext}} &= \frac{1}{2}\left[2\sigma_{i,\|}\epsilon_{i,\|} + \sigma_{i,\perp}\epsilon_{i,\perp}\right] \\ &= (C_{i,11} + C_{i,12})\epsilon_{i,\|}^2 + 2C_{i,12}\epsilon_{i,\perp}\epsilon_{i,\|} + \frac{1}{2}C_{i,11}\epsilon_{i,\perp}^2, \end{aligned} \tag{4.25}$$

where the elastic constants of the individual tetrahedra can be taken to be Vegard's law averages of the elastic constants of the pure component binary alloys:

$$\begin{aligned} C_{i,11} &= (1 - x_i)C_{\mathrm{GaAs,o},11} + x_i C_{\mathrm{InAs,o},11} \\ C_{i,12} &= (1 - x_i)C_{\mathrm{GaAs,o},12} + x_i C_{\mathrm{InAs,o},12} \end{aligned} \tag{4.26}$$

4.3 The Cluster Variation Method

In Sections 4.1 and 4.2, we explored the origin of elastic distortion energies in small microscopic tetrahedra such as those shown in Figure 4.2. In

[14] Note that the tetrahedra with more In atoms will be somewhat larger than the average, and those with fewer will be somewhat smaller. Nevertheless, we make the simplifying approximation, as we did in Section 4.1.1, that the virtual crystal approximation holds for *second*-nearest-neighbor distances, and that all tetrahedra have the same dimensions.

[15] We neglect, in this simple treatment, the strain-induced-splitting of the degeneracies of tetrahedra differing only by permutations of group III atoms, and treat all tetrahedra having the same number of In and Ga atoms to be the same.

this section, we ask: how can we use such microscopic information to deduce macroscopic quantities of interest, such as enthalpies and entropies of mixing, or tendencies toward short- and long-range ordering?

One classic approach to this problem is the cluster variation method (CVM),[16] in which solids are built by statistically combining a finite number of independent, elementary clusters. In principal, the method may be made arbitrarily accurate by choosing arbitrarily large clusters. In practice, actual implementations of the method represent trade-offs between accuracy and speed.[17] The larger the clusters, the less important the intercluster interaction energies are relative to intracluster energies, and the more accurate the assumption of cluster independence becomes. However, the larger the clusters, the more types of elementary clusters (of different composition) there will be, and the more time-consuming the combinatorics become.

In this section, we give a brief introduction to the cluster variation method. The method can be viewed as an increasingly accurate sequence of approximations, and so it is convenient to illustrate it by applying it to successively more complex structures: first alloys on 1D linear (in Subsection 4.3.1), then 2D triangular (in Subsection 4.3.2) and finally 3D zincblende (in Subsection 4.3.3) lattices.

The introduction given in this section is somewhat lengthy, both because the cluster variation method gives insight into so many aspects of alloy thermodynamics and because a comparable introductory treatment does not appear to exist elsewhere. However, it will not be necessary to understand the cluster variation method in detail in order to follow its application in Sections 4.4 and 4.5 to $In_{1-x}Ga_xAs$, a prototypical pseudobinary III/V alloy. The casual reader is advised to begin with Section 4.5.

4.3.1 1D Linear Lattice

We start, in this subsection, by illustrating the cluster variation method using a simple one-dimensional linear lattice. We consider, in turn, two possible ways of constructing this lattice. In the first way, the lattice is constructed from uncorrelated "points" of atoms, as shown in Figure 4.5. In the second way, the lattice is constructed from correlated "pairs" of atoms, as shown in Figure 4.6.

[16] R. Kikuchi, "A theory of cooperative phenomena," *Phys. Rev.* **81**, 988 (1951).

[17] D.M. Burley, "Closed form approximations for lattice systems," in C. Domb and M.S. Green, Eds., *Phase transitions and critical phenomena* (Academic Press, London, 1972), Vol. 2, Chap. 9.

Figure 4.5: Construction of a 1D linear chain of points by the addition of a new node (open circle) to an existing lattice (filled circles).

Points

Consider first the lowest order "point" approximation, in which the largest clusters are the individual atoms themselves. We imagine building an ensemble of n linear chains, each composed of nodes which are either type A or type B atoms. If the overall fractions of A and B atoms in the ensemble of chains are x_0 and $x_1 = 1 - x_0$, then each node of the ensemble of chains will have $x_0 n$ A atoms and $x_1 n$ B atoms.

Now suppose we wish to add another node to this ensemble of n chains. Since the nodes are all independent, we are free to add A atoms to $x_0 n$ nodes of the ensemble in any order, and then to add B atoms to the rest of the $x_1 n$ nodes of the ensemble, again in any order. The number of distinguishable ways the atoms may be added is $W = n!/[(x_0 n)!(x_1 n)!]$. If we introduce the CVM notation shown in Table 4.3,

$$() = n! \tag{4.27}$$

$$(\bullet) = \prod (x_i n)!, \tag{4.28}$$

then we have the compact expression

$$W = ()/(\bullet). \tag{4.29}$$

The entropy per node and per chain in the ensemble can then be calculated, using Stirling's formula, to be

$$s = \frac{k}{n} \ln W = -k \sum x_i \ln x_i. \tag{4.30}$$

As expected, this equation reproduces the entropy of a random mixture of noninteracting components.

Since, by assumption, the nodes do not interact, the energy per node and per chain in the ensemble is just a weighted sum of the energies of the individual A and B atoms:

$$u = \sum x_i u_i. \tag{4.31}$$

Largest cluster	Combinatorial Factor	Uncorrelated cluster Identity
Space	$() = n!$	—
Point	$(\bullet) = \prod(x_i n)!$	—
Pair	$(-) = \prod(y_i n)!^{\beta_i}$	$(\not-) = (\bullet)^2/()$
Triangle	$(\triangle) = \prod(z_i n)!^{\gamma_i}$	$(\not\triangle) = (\bullet)^3/()^2$
Tetrahedron	$(\blacktriangle) = \prod(w_i n)!^{\delta_i}$	$(\not\blacktriangle) = (\bullet)^4/()^3$

Table 4.3: Heirarchy of CVM approximations showing combinatorial factors and uncorrelated cluster identities.

Note, though, that although we have assumed that the nodes do not interact *directly*, we may still allow them to interact *indirectly* by allowing the energies u_i to depend on the mean composition. For example, if u_0 is proportional to the average concentration of B, $u_0 = \Omega x_1/2$, and u_1 is proportional to the average concentration of A, $u_1 = \Omega x_0/2$, then the molar energy becomes

$$u = \Omega x_0 x_1, \tag{4.32}$$

which reproduces the strictly regular solution model for alloys.

Finally, the free energy of the system, $f = u - Ts$, can be seen to be a function of two parameters, x_0 and x_1. Only one can be chosen freely, however, since, as listed in Table 4.4, they must together obey the constitutive "space" relationship

$$x_0 + x_1 = 1. \tag{4.33}$$

Therefore, given the overall composition, $x \equiv x_1$, the free energy is given directly by Equations 4.30 and 4.31.

Pairs

Consider now the next CVM approximation, in which the largest clusters are pairs of atoms. Again imagine building an ensemble of n linear chains, whose nodes have $x_0 n$ A atoms and $x_1 n$ B atoms. This time, however, we include only those chains for which the overall fractions of AA, AB, BA, and BB atom pairs (or bonds) assume particular values, say, y_0, y_1, y_1, and y_2.

Note that we have assumed that y_1 is, by symmetry, the number of both the AB and the BA atom pairs. Then, as listed in Table 4.4, the degeneracies of the configurations are $\beta_1 = 2$ and $\beta_0 = \beta_2 = 1$. Also note that the atom pair fractions y_i are not independent of the atom fractions

Configuration	Fraction	Degeneracy	Constitutive Relation
$()$	1	—	$1 = x_0 + x_1$
$(\bullet)_A$	x_0	—	$x_0 = y_0 + y_1$
$(\bullet)_B$	x_1	—	$x_1 = y_1 + y_2$
$(-)_{A_2}$	y_0	$\beta_0 = 1$	$y_0 = z_0 + z_1$
$(-)_{AB}$	y_1	$\beta_1 = 2$	$y_1 = z_1 + z_2$
$(-)_{B_2}$	y_2	$\beta_2 = 1$	$y_2 = z_2 + z_3$
$(\triangle)_{A_3}$	z_0	$\gamma_0 = 1$	$z_0 = w_0 + w_1$
$(\triangle)_{A_2B}$	z_1	$\gamma_1 = 3$	$z_1 = w_1 + w_2$
$(\triangle)_{AB_2}$	z_2	$\gamma_2 = 3$	$z_2 = w_2 + w_3$
$(\triangle)_{B_3}$	z_3	$\gamma_3 = 1$	$z_3 = w_3 + w_4$
$(\blacktriangle)_{A_4}$	w_0	$\delta_0 = 1$	
$(\blacktriangle)_{A_3B}$	w_1	$\delta_1 = 4$	
$(\blacktriangle)_{A_2B_2}$	w_2	$\delta_2 = 6$	
$(\blacktriangle)_{AB_3}$	w_3	$\delta_3 = 4$	
$(\blacktriangle)_{B_4}$	w_4	$\delta_4 = 1$	

Table 4.4: Configurations, fractions, degeneracies and constitutive relations for empty, point, pair, triangular and tetrahedral clusters.

x_i, but must obey the constitutive "point" relations

$$\begin{aligned} x_0 &= y_0 + y_1 \\ x_1 &= y_1 + y_2. \end{aligned} \tag{4.34}$$

These relations arise because all AA and AB pairs are associated on the left with an A atom, and all BA and BB pairs are associated on the left with a B atom.

Now suppose we wish to add another node to this ensemble of n chains. In this case, the nodes are *not* independent, so we are not free to add A atoms to x_0n nodes of the ensemble in any order, nor to add B atoms to the rest of the x_1n nodes of the ensemble in any order. Instead, we must add them in such a way that the fractions of new atom-pairs are also y_0, y_1, y_1, and y_2.

A convenient way of doing this is illustrated in Figure 4.6. To the x_0n chains in the ensemble having A atoms as their last node we add y_0n A atoms and y_1n B atoms. The number of distinguishable ways these additions can be done is $(x_0n)!/[(y_0n)!(y_1n)!]$. Then, to the remaining x_1n chains in the ensemble having B atoms as their last node we add y_1n A atoms and y_2n B atoms. The number of distinguishable ways these

Figure 4.6: Construction of a 1D linear chain of pairs by the addition of a new node (open circle) to an existing lattice (filled circles).

additions can be done is $(x_1n)!/[(y_1n)!(y_2n)!]$. The total number of ways is the product, or $W = [(x_0n)!(x_1n)!]/[(y_0n)!(y_1n)!^2(y_2n)!]$.

If we introduce the CVM notation

$$(-) = \prod [(y_i n)!]^{\beta_i} , \tag{4.35}$$

then we can again write more compactly

$$W = \frac{(\bullet)}{(-)}. \tag{4.36}$$

Equations 4.29 and 4.36 are now seen to take the same form, which by induction can be written

$$W = \frac{\text{The part already filled}}{\text{The whole to be completed}}. \tag{4.37}$$

This rule generalizes and simplifies the calculation of combinatoric factors for even the most complicated lattice and cluster topologies.

The entropy per node and per chain in the ensemble can now be deduced, again using Stirling's formula, to be

$$s = \frac{k}{n} \ln W = k \left(\sum x_i \ln x_i - \sum \beta_i y_i \ln y_i \right). \tag{4.38}$$

Note that if the atom pairs were randomly distributed, then $y_0 = x_0^2$, $y_1 = x_0 x_1$, and $y_2 = x_1^2$. Then, we would have $\sum \beta_i y_i \ln y_i = 2 \sum x_i \ln x_i$, and Equation 4.38 would reduce to Equation 4.30, the entropy of mixing in the point approximation. In a more compact notation, we can write

$$(\not-) \equiv (\bullet)^2/(), \tag{4.39}$$

where $(\not-)$ denotes a pair of "uncorrelated" points. Then,

$$W = \frac{(\bullet)}{(\not-)} = \frac{(\bullet)}{(\bullet)^2/()} = \frac{()}{(\bullet)}, \tag{4.40}$$

which again is the point approximation result.

Since, by assumption, individual atoms *do* interact in the pair approximation, the energy per node and per chain in the ensemble is written as a weighted sum of the energies of the various kinds of *pairs* of A and B atoms:

$$u = \sum \beta_i y_i u_i. \tag{4.41}$$

The free energy of the system, $f = u - Ts$, is then seen to be a function of five fractions, x_0, x_1, y_0, y_1, and y_2. As before, of the two "point" fractions, at most one can be chosen freely, due to the constitutive "space" relationship of Equation 4.33. In addition, of the three "pair" fractions, only one as well can be chosen freely, due to the constitutive "point" relationships of Equation 4.34.

Now, if the overall composition, $x \equiv x_1$, were free to vary, then the equilibrium value of the free energy would be determined by minimizing f with respect to both x and one of the pair probabilities, say, y_1. This might be the case, e.g., if the lattice were composed not of atoms whose overall numbers we know, but of spins which are free to flip, as in an Ising model. Then, x would play the role of the overall magnetization.

For problems in alloy thermodynamics, however, $x \equiv x_1$ is usually fixed, and is not free to vary. Then, the equilibrium value of the free energy is determined by minimizing f with respect to one of the pair probabilities, usually taken to be the unlike pair probability, y_1. In other words, we wish to minimize

$$\begin{aligned} f &= y_0 u_0 + 2y_1 u_1 + y_2 u_2 \\ &\quad + kT\left[y_0 \ln y_0 + 2y_1 \ln y_1 + y_2 \ln y_2 - (1-x)\ln(1-x) - x\ln(x)\right] \end{aligned} \tag{4.42}$$

with respect to y_1, where $y_0 = 1 - x - y_1$ and $y_2 = x - y_1$. Taking the derivative and setting it equal to zero then gives

$$\frac{\partial f}{\partial y_1} = 2u_1 - u_0 - u_2 + kT \ln\left(\frac{y_1^2}{y_0 y_2}\right) = 0. \tag{4.43}$$

This expression can be recast, again using the constitutive point relations of Equation 4.34, into the form

$$\frac{y_1^2}{y_0 y_2} = e^{-(2u_1 - u_0 - u_2)/kT}. \tag{4.44}$$

If each atom pair is considered, in a loose sense, to be a molecule, then the equilibrium ratio between the number of AB or BA molecules and the product of the numbers of AA and BB molecules is seen to be given by a Boltzmann factor. This is exactly the "mass-action" law expected for the chemical reaction

$$AA + BB \rightleftharpoons 2AB, \tag{4.45}$$

which can be derived by equating a forward rate, proportional to the product of the concentrations of the AA and BB species, to a backward rate, proportional to the concentration of the AB or BA species. In this way, the pair approximation is equivalent[18] to what is known as the "quasi-chemical" treatment[19] of alloy thermodynamics, for which Equation 4.44 is the central assumption.

Equation 4.44 has two limiting behaviors. On the one hand, if $2u_1 \ll u_0 + u_2$, then AB pairs are highly favored over AA and BB pairs, the A and B atoms tend to arrange themselves next to each other, and the pair probability y_1 approaches $(1/2) - \sqrt{(1/4) - x(1-x)}$. On the other hand, if $2u_1 \gg u_0 + u_2$, then AA and BB pairs are highly favored over AB pairs, the A and B atoms tend to segregate away from each other, and the pair probability y_1 approaches 0. In between, if $2u_1 = u_0 + u_2$, then AB pairs are neither favored nor unfavored over AA and BB pairs, the A and B atoms tend to arrange themselves randomly, and the pair probability y_1 approaches $x(1-x)$.

Often, it is useful to characterize these behaviors by a short-range "order parameter,"

$$\sigma^{\mathrm{SRO}} \equiv \frac{y_1 - y_1^{\mathrm{ran}}}{y_1^{\mathrm{ord}} - y_1^{\mathrm{ran}}} = \frac{y_1 - x(1-x)}{(1/2) - \sqrt{(1/4) - x(1-x)} - x(1-x)}, \tag{4.46}$$

which is zero if the atoms are arranged randomly, one if the atoms are ordered, and minus one if the atoms are "anti-ordered." For the special case of $x = 1/2$, Equations 4.38 and 4.41 can be recast, after some algebra, into the forms

$$\begin{aligned} s &= -k\left[\frac{1-\eta}{2}\ln\left(\frac{1-\eta}{2}\right) + \frac{1+\eta}{2}\ln\left(\frac{1+\eta}{2}\right)\right] \\ u &= \frac{1}{4}(u_0 + 2u_1 + u_2) + \frac{\sigma^{\mathrm{SRO}}}{4}(2u_1 - u_0 - u_2). \end{aligned} \tag{4.47}$$

If the resulting free energy is minimized with respect to η, then one finds

$$\sigma^{\mathrm{SRO}} = \frac{1 - e^{(2u_1 - u_0 - u_2)/2kT}}{1 + e^{(2u_1 - u_0 - u_2)/2kT}}. \tag{4.48}$$

[18] R. Kikuchi, "Theory of ternary III-V semiconductor phase diagrams," *Physica* **103B**, 41 (1981).

[19] E.A. Guggenheim, "The statistical mechanics of regular solutions," *Proc. Roy. Soc. (London)* **A148**, 304 (1935).

For negative $2u_1 - u_0 - u_2$, $\sigma^{\text{SRO}} > 0$, and A and B atoms *order* on a microscopic scale; for positive $2u_1 - u_0 - u_2$, $\sigma^{\text{SRO}} < 0$, and A and B atoms *anti-order* on a microscopic scale.

4.3.2 2D Triangular Lattice

In Subsection 4.3.1, we illustrated the cluster variation method using a simple 1D linear lattice. In this subsection, we illustrate the cluster variation method using the more complicated 2D triangular lattice shown in Figure 4.7. This lattice may be constructed either from points, pairs or triangles.

In the lowest order point approximation, the entropies and energies are the same as those for the 1D linear lattice, and Equations 4.30 and 4.31 for the entropies and energies can be carried over without modification. In the pair and triangle approximations, however, the topology of the lattice must be taken into account, because it imposes correlations *between* the various pairs and triangles of atoms. We consider, in turn, these two possible ways of constructing this lattice.

Pairs

Consider first the pair approximation. As before, we assume that individual atoms interact pairwise, so that the energy per node and per chain in the ensemble can, as in Equation 4.41, still be written as a weighted sum of the energies of the various kinds of pairs of A and B atoms. Also, as before, we imagine building a large ensemble of n lattices, whose nodes have $x_0 n$ A atoms and $x_1 n$ B atoms, and for which the overall fractions of AA, AB, BA, and BB atom pairs are y_0, y_1, y_1, and y_2.

Suppose we wish to add another node to this ensemble of lattices, in such a way that the node contains $x_0 n$ A atoms and $x_1 n$ B atoms, and each new ensemble of bonds, a–b, a–c and a–d, contains y_0 AA pairs, y_1 AB pairs, y_1 BA pairs, and y_2 BB pairs. This we can do in three steps.

First, add node a with respect to node b without regard to correlations with nodes c and d. The number of ways this can be done is the same as that for the 1D linear lattice, namely,

$$W' = \frac{(\bullet)}{(-)}. \tag{4.49}$$

Second, correct (approximately) for the correlation between a and c by multiplying by the factor

$$W'' = \frac{(\bullet)/(-)}{()/(\bullet)}. \tag{4.50}$$

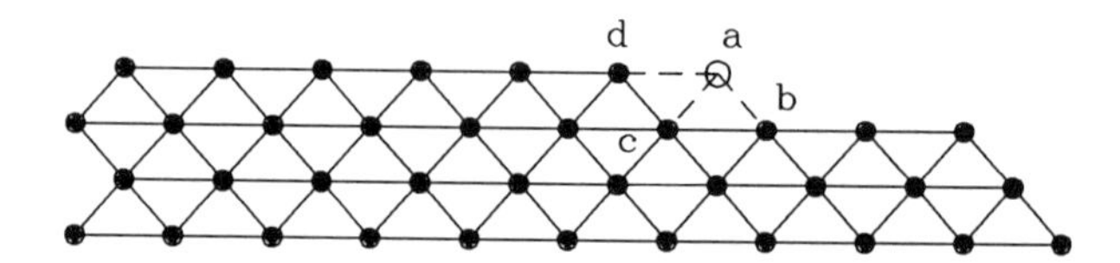

Figure 4.7: Construction of a 2D triangular lattice by the addition of a new node (open circle) to an existing lattice (filled circles).

This factor is the ratio between the number of ways atoms *should* have been placed on node a with respect to node c, $(\bullet)/(-)$, and the number of ways atoms *actually were* placed on node a with respect to node c, $()/(\bullet)$. Third, correct (approximately) for the correlation between a and d by multiplying by the same factor

$$W''' = \frac{(\bullet)/(-)}{()/(\bullet)}. \tag{4.51}$$

Another way to look at these two correction factors is to use Equation 4.39 to rewrite them as

$$W'' = W''' = \frac{(\bullet)/(-)}{()/(\bullet)} = \frac{(\neq)}{(-)}, \tag{4.52}$$

so that, in the spirit of Equation 4.37, they carry the physical meaning that correlated pairs are being built from uncorrelated pairs. Indeed, the first combinatorial factor can itself be rewritten as

$$W' = \frac{(\bullet)}{(-)} = \left[\frac{()}{(\bullet)}\right]\left[\frac{(\neq)}{(-)}\right], \tag{4.53}$$

which carries the physical meaning that an uncorrelated point is first added, and then a correlated pair is built from an uncorrelated pair.

The overall number of ways of adding atoms to node a then becomes

$$W = W'W''W''' = \frac{()}{(\bullet)}\left[\frac{(\neq)}{(-)}\right]^3 = \frac{(\bullet)^5}{(-)^3()^2}. \tag{4.54}$$

In other words, again in the spirit of Equation 4.37, we first add an uncorrelated point, then correlate the resulting three uncorrelated pairs.

The entropy per node and per chain in the ensemble can now be calculated, using Stirling's formula, to be

$$s = \frac{k}{n}\ln W = k\left(5\sum x_i \ln x_i - 3\sum \beta_i y_i \ln y_i\right). \tag{4.55}$$

Aside from the different numerical factors for the point and pair sums in Equations 4.38 and 4.55, all the arguments in Section 4.3.1 hold.

Triangles

Consider now the triangle approximation. In this case, we assume that the energies of atoms can be expressed as sums over triangular triplets of atoms, so that the energy per node and per chain in the ensemble can be written as a weighted sum of the energies of the various kinds of triangles:

$$u = \sum \gamma_i z_i u_i, \tag{4.56}$$

Here, the overall fractions of A_3, A_2B, AB_2 and B_3 triplets are $\gamma_0 z_0$, $\gamma_1 z_1$, $\gamma_2 z_2$ and $\gamma_3 z_3$, with the degeneracies, γ_i, listed in Table 4.4 on page 111.

Now suppose we wish to add another node to this ensemble of n lattices, in such a way that (1) the node contains $x_0 n$ A atoms and $x_1 n$ B atoms, (2) each new ensemble of pairs, a–b, a–c and a–d, contains y_0 AA pairs, y_1 AB pairs, y_1 BA pairs and y_2 BB pairs, and (3) each new ensemble of triangles, a–b–c and a–c–d, contains $\gamma_0 z_0$ A_3 triangles, $\gamma_1 z_1$ A_2B triangles, $\gamma_2 z_2$ AB_2 triangles, and $\gamma_3 z_3$ B_3 triangles. Again, we proceed in steps.

First, we add node a with respect to the pair b–c without regard to correlations with node d. In the spirit of Equation 4.37, the number of ways this can be done is

$$W' = \frac{(-)}{(\triangle)}, \tag{4.57}$$

where

$$(\triangle) \equiv \prod [(z_i n)!]^{\gamma_i} . \tag{4.58}$$

Second, correct for the correlation within the triangle a–c–d by multiplying by the factor

$$W'' = \frac{(-)/(\triangle)}{(\bullet)/(-)}. \tag{4.59}$$

The numerator of this factor is the ratio between the number of ways atoms should have been placed on node a with respect to the pair c–d. The denominator is the number of ways atoms actually were placed on node a with respect to the pair c–d, namely, the number of ways the correlated pair a–c forming one side of the correlated triangle a–b–c could be formed from the point c.

Again, it is useful to rewrite W' and W'' as

$$W' = \frac{(-)}{(\triangle)} = \frac{()}{(\bullet)} \left[\frac{(-)}{(\not-)}\right] \left[\frac{(\not\triangle)}{(\triangle)}\right]$$

$$W'' = \frac{(-)/(\triangle)}{(\bullet)/(-)} = \left[\frac{(-)}{(\not{-})}\right]^2 \left[\frac{(\not\triangle)}{(\triangle)}\right], \tag{4.60}$$

where

$$(\not\triangle) \equiv \frac{(\bullet)^3}{()^2} \tag{4.61}$$

generalizes Equation 4.39 to uncorrelated triangles.

The physical meaning of W' can now be seen to be the addition of an uncorrelated point, the *decorrelation* of the previously correlated pair b–c, and the correlation of the now uncorrelated triangle a–b–c. The physical meaning of W'' is seen to be the decorrelation of the pairs a–c and c–d, which had been previously correlated, followed by the correlation of the now uncorrelated triangle a–c–d.

Finally, then, the overall number of ways of adding atoms to node a becomes

$$W = W'W'' = \frac{()}{(\bullet)}\left[\frac{(-)}{(\not{-})}\frac{(\not\triangle)}{(\triangle)}\right]\left[\frac{(-)^2}{(\not{-})^2}\frac{(\not\triangle)}{(\triangle)}\right] = \frac{(-)^3}{(\triangle)^2(\bullet)}. \tag{4.62}$$

In other words, we first form an uncorrelated point, then for each of the two triangles that the point belongs to, we uncorrelate all previously correlated pairs in the triangles and then correlate the triangles.

The entropy per node and per chain in the ensemble can now be calculated, using Stirling's formula, to be

$$s = \frac{k}{n}\ln W = k\left(3\sum \beta_i y_i \ln y_i - 2\sum \gamma_i z_i \ln z_i - \sum x_i \ln x_i\right). \tag{4.63}$$

The free energy of the system, $f = u - Ts$, is a function of nine parameters, x_0, x_1, y_0, y_1, y_2, z_0, z_1, z_2, and z_3. As before, however, of the two point parameters, only one can be chosen freely, due to the constitutive space relationship. In addition, none of the three pair parameters can be chosen freely, because they must obey the constitutive pair relationships

$$\begin{aligned} y_0 &= z_0 + z_1 \\ y_1 &= z_1 + z_2 \\ y_2 &= z_2 + z_3. \end{aligned} \tag{4.64}$$

These equations express the fact that each A_3 or A_2B triangle is formed by coupling an atom to an A_2 pair, that each A_2B or AB_2 triangle is formed by coupling an atom to an AB pair, and that each AB_2 or B_3 triangle is formed by coupling an atom to a B_2 pair.

Finally, of the four triangle parameters, only two can be chosen freely, because of the constitutive point relations listed in Table 4.4 on page 111.

Therefore, for a fixed overall composition, $x \equiv x_1$, the equilibrium value of the free energy is determined by minimizing f with respect to two of the triangle probabilities, which can be taken to be the mixed triangle probabilities, z_1 and z_2.

4.3.3 3D Zincblende Lattice

In Subsections 4.3.1 and 4.3.2, we illustrated the cluster variation method using first a 1D linear lattice and then a 2D triangular lattice. In this subsection, we illustrate the cluster variation method using a 3D zincblende lattice, whose projection onto an (001) plane is shown in Figure 4.8. For a III-V semiconductor, such a lattice would be built from a superposition of two face-centered-cubic sublattices, one containing group III species and the other containing group V species. Since we are ultimately interested in treating pseudobinary III-III-V alloys, we are interested in the entropy of mixing of group III species on the group III sublattice. Note, though, that these group III species do not form nearest-neighbor bonds with each other; instead, they form next-nearest-neighbor bonds mediated by the group V atoms on the group V sublattice. Therefore, two, three, or four group III atoms can be considered to form a pair, triangle, or tetrahedron if and only if they are all bonded to the same group V atom.

Triangles

Consider first the triangle approximation, in which we assume that the energies of atoms can be expressed as sums over triangular triplets of atoms, as given by Equation 4.56. Suppose we wish to add another node to an ensemble of zincblende lattices, in such a way that all the point, pair and triangle probabilities are preserved. To do so, we use the following simplified rules,[20] generalized from Section 4.3.2:

1. Add an uncorrelated point via the combinatorial factor $()/(\bullet)$.

2. Enumerate all the largest clusters created by adding that point, regardless of overlap.

3. For each such cluster: (a) uncorrelate all (previously correlated) subclusters via the combinatorial factors $(\triangle)/(\not\triangle)$, $(-)/(\not-)$, etc., starting from large to small; and (b) correlate the cluster itself via the combinatorial factors $(\not\blacktriangle)/(\blacktriangle)$, $(\not\triangle)/(\triangle)$, $(\not-)/(-)$, etc.

[20] The rules are not exact, but must be made recursive when clusters overlap in subclusters larger than pairs.

We start, then, by adding an uncorrelated point a, via the combinatorial factor

$$W' = ()/(\bullet). \tag{4.65}$$

Then, we note that by adding point a, we have formed three new triangles, a–b–c, a–b–d, and a–e–f, and one new pair, a–g. We do not include the triangles a–d–g and a–f–g, because these clusters of group III atoms are not all bonded to a common group V atom. Within triangle a–e–f, we must uncorrelate the pair e–f and then correlate the triangle via the combinatorial factor

$$W'' = \left[\frac{(-)}{(\neq)}\right]\left[\frac{(\not\triangle)}{(\triangle)}\right]. \tag{4.66}$$

Similarly, within triangle a–b–c, we must uncorrelate the pair b–c and then correlate the triangle via the combinatorial factor

$$W''' = \left[\frac{(-)}{(\neq)}\right]\left[\frac{(\not\triangle)}{(\triangle)}\right]. \tag{4.67}$$

Within triangle a–b–d, we must now uncorrelate *two* pairs, a–b (which we just correlated in correlating the triangle a–b–c) and b–d, before correlating the triangle:

$$W'''' = \left[\frac{(-)}{(\neq)}\right]^2 \frac{(\not\triangle)}{(\triangle)}. \tag{4.68}$$

Finally, we must correlate the pair a–g via the combinatorial factor

$$W''''' = (\neq)/(-). \tag{4.69}$$

Altogether, the number of ways of adding an atom at a is

$$W = W'W''W'''W''''W''''' = \frac{(-)^3(\bullet)^2}{(\triangle)^3()^2}, \tag{4.70}$$

and the entropy is

$$s = k\left(3\sum \beta_i y_i \ln y_i + 2\sum x_i \ln x_i - 3\sum \gamma_i z_i \ln z_i\right). \tag{4.71}$$

As before, the free energy of the system, $f = u - Ts$, is a function of nine parameters, x_0, x_1, y_0, y_1, y_2, z_0, z_1, z_2 and z_3. However, for a fixed overall composition, $x \equiv x_1$, the constitutive relations eliminate all but two. The equilibrium value of the free energy is then determined by minimizing f with respect to two of the triangle probabilities, e.g., the mixed triangle probabilities z_1 and z_2.

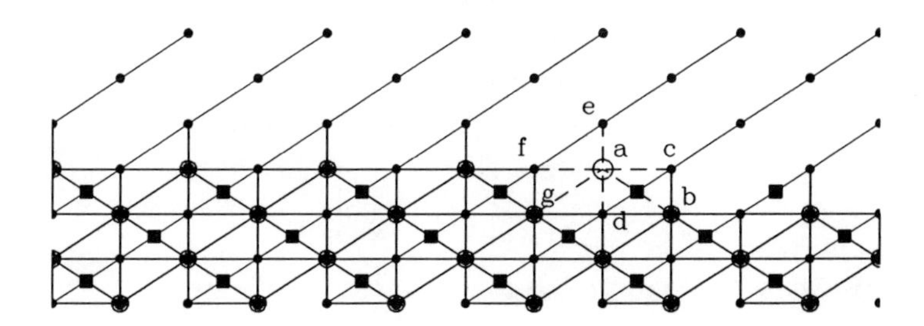

Figure 4.8: Top view of the construction of a 3D zincblende lattice by the addition of a new node (open circle) to an existing lattice (filled circles). For a III-III-V alloy, the filled circles and squares would correspond to group III and group V atoms, respectively. The atoms in each (001) sheet are represented by symbols of the same size; the smaller the symbol the deeper the sheet. The solid lines represent next-nearest-neighbor bonds between group III atoms mediated by group V atoms.

Tetrahedra

In the tetrahedron approximation we assume that the energies of atoms can be expressed as sums over tetrahedral quadruplets of atoms,

$$u = \sum \delta_i w_i u_i, \tag{4.72}$$

Here, the overall fractions of A_4, A_3B, A_2B_2, A_1B_3, and B_4 quadruplets are $\delta_0 w_0$, $\delta_1 w_1$, $\delta_2 w_2$, $\delta_3 w_3$, and $\delta_4 w_4$, respectively, with the degeneracies, δ_i, listed in Table 4.4 on page 111.

Suppose we wish to add another node to this ensemble of lattices, in such a way that all the point, pair, triangle, *and* tetrahedron probabilities are preserved. To do so, we again use the rules outlined in Section 4.3.3.

We start by adding an uncorrelated point a, via the combinatorial factor

$$W' = ()/(\bullet). \tag{4.73}$$

Then, we note that by adding point a, we have formed one new tetrahedron, a–b–c–d, one new triangle, a–e–f, and one new pair, a–g. Again, we include neither the tetrahedron a–d–g–f nor the triangles a–d–g and a–f–g, because these groups of atoms are not all bonded to a common group V atom.

Within tetrahedron a–b–c–d, we must uncorrelate the triangle b–c–d and then correlate the tetrahedron via the combinatorial factor

$$W'' = \left[\frac{(\triangle)}{(\not\triangle)}\right]\left[\frac{(\not\blacktriangle)}{(\blacktriangle)}\right], \tag{4.74}$$

where

$$(\blacktriangle) \equiv \frac{(\bullet)^4}{()^3} \tag{4.75}$$

generalizes Equation 4.39 to uncorrelated tetrahedra. Within triangle a–e–f, we must uncorrelate the pair e–f before correlating the triangle:

$$W''' = \left[\frac{(-)}{(\not-)}\right]\left[\frac{(\not\triangle)}{(\triangle)}\right]. \tag{4.76}$$

Finally, we must correlate the pair a–g via the combinatorial factor

$$W'''' = (\not-)/(-). \tag{4.77}$$

Altogether, the number of ways of adding an atom at a is

$$W = W'W''W'''W'''' = \frac{(\bullet)^3}{(\blacktriangle)()^2}, \tag{4.78}$$

and the entropy is

$$s = k\left(3\sum x_i \ln x_i - \sum \delta_i w_i \ln w_i\right). \tag{4.79}$$

Note that, for the peculiar topology of the zincblende lattice, the free energy of the system,

$$f = \sum \delta_i w_i u_i + kT\left(\sum \delta_i w_i \ln w_i - 3\sum x_i \ln x_i\right), \tag{4.80}$$

contains no pair and triangle probabilities. If it had, though, they could have beeen eliminated either through the constitutive pair relations listed in Table 4.4 on page 111 or the constitutive "triangle" relations

$$\begin{aligned} z_0 &= w_0 + w_1 \\ z_1 &= w_1 + w_2 \\ z_2 &= w_2 + w_3 \\ z_3 &= w_2 + w_3. \end{aligned} \tag{4.81}$$

The equilibrium value of the free energy is determined by minimizing f with respect to the five tetrahedron probabilities w_i and the two point probabilities x_i, subject to the three constraints embodied in the two constitutive pair relations and the constitutive space relation. In general, this minimization can be performed through standard techniques based on Lagrange multipliers, one of which can be identified with the chemical potential for species B. This leads to a set of seven nonlinear equations that can be solved through a compact procedure called the "natural iteration method."[21] In essence, that method begins by guessing values for the point probabilities, using those guesses to calculate the tetrahedron probabilities, from which the point probabilities can be recalculated, etc.

For the zincblende lattice, however, it is simpler to eliminate directly two of the five tetrahedra probabilities using the constitutive pair and space relations. Taking these to be the "pure" cluster probabilities, we then have

$$\begin{aligned} w_0 &= 1 - x - (3w_1 + 3w_2 + w_3) \\ w_4 &= x - (w_1 + 3w_2 + 3w_3) \end{aligned} \tag{4.82}$$

Therefore, for a fixed overall composition, $x \equiv x_1$, the equilibrium value of the free energy is determined by minimizing f with respect to the remaining three "mixed" tetrahedron probabilities w_1, w_2 and w_3.

Taking derivatives of Equation 4.80 with respect to w_1, w_2, and w_3 gives, after some algebra,

$$\begin{aligned} \frac{w_1^4}{w_0^3 w_4} &= e^{-(4u_1 - 3u_0 - u_4)/kT} \\ \frac{w_2^4}{w_0^2 w_4^2} &= e^{-(4u_2 - 2u_0 - 2u_4)/kT} \\ \frac{w_3^4}{w_0 w_4^3} &= e^{-(4u_3 - u_0 - 3u_4)/kT} \end{aligned} \tag{4.83}$$

Note that Equations 4.83 are in exactly the "mass-action" form expected for chemical reactions between "molecular" tetrahedra:

$$\begin{aligned} 4A_3B &\rightleftharpoons 3A_4 + B_4 \\ 4A_2B_2 &\rightleftharpoons 2A_4 + 2B_4 \\ 4AB_3 &\rightleftharpoons A_4 + 3B_4, \end{aligned} \tag{4.84}$$

and are therefore equivalent, as were Equations 4.44, to a "quasi-chemical" treatment, though of tetrahedra rather than of pairs. In general, chemical reactions between pairs, triplets, and quadruplets form the basis for what are known as the first, second and third quasi-chemical approximations.[22] The tetrahedron approximation of the CVM, applied to a zincblende lattice, is therefore equivalent to the third quasi-chemical approximation. It should be emphasized, though, that CVM calculations are not always equivalent to

[21] R. Kikuchi, "Superposition approximation and natural iteration calculation in cluster-variation method," *J. Chem. Phys.* **60**, 1071 (1974).

[22] E.A. Guggenheim, "Statistical mechanics of regular mixtures," *Proc. Roy. Soc. (London)* **A206**, 335 (1951).

quasi-chemical approximations; in this case the equivalence is a consequence of the peculiar topology of the zincblende lattice, whose combinatorial factor of Equation 4.78 contains no intermediate subclusters such as triangles or pairs. Otherwise, an equivalence can only be established by the additional assumption that those intermediate subclusters are uncorrelated.

Equations 4.83, together with Equations 4.82, form a set of coupled nonlinear equations which can be solved for the tetrahedra probabilities, w_i, in terms of the tetrahedra energies, u_i. To do so, it is convenient to make use of their equivalence to a quasi-chemical treatment by reformulating them as chemical rate equations that can be solved by numerical simulation.

If we rewrite Equations 4.84 in terms of reactions between tetrahedra differing by only the exchange of one atom, then we have

$$\begin{aligned} 4A_3B &\underset{k_1^-}{\overset{k_1^+}{\rightleftharpoons}} 2A_4 + 2A_2B_2 \\ 4A_2B_2 &\underset{k_2^-}{\overset{k_2^+}{\rightleftharpoons}} 2A_3B + 2AB_3 \\ 4AB_3 &\underset{k_3^-}{\overset{k_3^+}{\rightleftharpoons}} 2A_2B_2 + 2B_4. \end{aligned} \tag{4.85}$$

The forward and backward reaction rates can be conveniently chosen to be

$$\begin{aligned} k_1^+ &= w_0 w_2 e^{-(2u_1^* - u_0 - u_2)/kT} \\ k_1^- &= w_1^2 e^{-(2u_1^* - 2u_1)/kT} \\ k_2^+ &= w_1 w_3 e^{-(2u_2^* - u_1 - u_3)/kT} \\ k_2^- &= w_2^2 e^{-(2u_2^* - 2u_2)/kT} \\ k_3^+ &= w_2 w_4 e^{-(2u_3^* - u_2 - u_4)/kT} \\ k_3^- &= w_3^2 e^{-(2u_3^* - 2u_3)/kT}, \end{aligned} \tag{4.86}$$

where u_1^*, u_2^*, and u_3^* are activation energies that can be chosen to match the time-step of the numerical simulation. In practice, the choices

$$\begin{aligned} u_1^* &= \max\{(u_0 + u_2)/2, u_1\} \\ u_2^* &= \max\{(u_1 + u_3)/2, u_2\} \\ u_3^* &= \max\{(u_2 + u_4)/2, u_3\} \end{aligned} \tag{4.87}$$

give convergence to steady-state in a reasonable number of time-steps. Note also that these choices of rate constants guarantee that in the steady-state, defined by setting $k_1^+ = k_1^-$, $k_2^+ = k_2^-$ and $k_3^+ = k_3^-$, Equations 4.83 will be satisfied.

In terms of these rates, the time evolution of the tetrahedra probabilities can be written as

$$\begin{array}{lcl} \gamma_0 \dot{w}_0 & = & -(k_1^+ - k_1^-) \\ \gamma_1 \dot{w}_1 & = & 2(k_1^+ - k_1^-) - (k_2^+ - k_2^-) \\ \gamma_2 \dot{w}_2 & = & -(k_1^+ - k_1^-) + 2(k_2^+ - k_2^-) - (k_3^+ - k_3^-) \\ \gamma_3 \dot{w}_3 & = & - (k_2^+ - k_2^-) + 2(k_3^+ - k_3^-) \\ \gamma_4 \dot{w}_4 & = & - (k_3^+ - k_3^-) \end{array} \quad (4.88)$$

Note that these rate equations are *conservative*, so that an initial probability distribution will remain correctly normalized, and an initial overall composition, $x = w_1 + 3w_2 + 3w_3 + w_4$, will remain constant. In practice, two convenient initial probability distributions are the completely random Bernoullian distribution,

$$\delta_i w_{i,\mathrm{ran}} = \binom{4}{i} x^i (1-x)^{4-i}, \quad (4.89)$$

and the completely nonrandom linear distribution,

$$\delta_i w_{i,\mathrm{ord}} = \max\{0, 1 - 4\,|x - x_i|\}, \quad (4.90)$$

where x_i is the composition of the ith cluster.

4.4 A Pseudobinary III-V Alloy: "InGaAs"

In Section 4.3, we described how, given the energies of various elementary tetrahedra, their occupation statistics could be calculated using the cluster variation method, and the free energy of an alloy as a whole could be determined. In this section, we apply this procedure in an approximate way to the pseudobinary alloy $In_{1-x}Ga_xAs$. The treatment is only semiquantitative, but will include all the most interesting and important features that have been observed in alloys of this type, such as short- and long-range ordering.[23] Tables 4.5 and 4.6, e.g., list the ordered alloys that have been observed thus far in III/V compound semiconductors.

[23] H. Nakayama and H. Fujita, "Direct observation of an ordered phase in a disordered $In_{1-x}Ga_xAs$ alloy," *Inst. Phys. Conf. Ser.* **79**, 289 (1985); H.R. Jen, M.J. Cherng and G.B. Stringfellow, "Ordered structures in GaAsSb alloys grown by organometallic vapor phase epitaxy," *Appl. Phys. Lett.* **48**, 1603 (1986); T.S. Kuan, W.I. Wang and E.L. Wilkie, "Long-range order in $In_{1-x}Ga_xAs$," *Appl. Phys. Lett.* **51**, 51 (1987); and M.A. Shahid and S. Mahajan, "Long-range atomic order in $Ga_xIn_{1-x}As_yP_{1-y}$ epitaxial layers [(x, y) = (0.47,1), (0.37,0.82), (0.34,0.71) and (0.27,0.64)]," *Phys. Rev.* **B38**, 1344 (1988).

Alloy	Growth Technique	Substrate	Structure	Reference
GaPAs	MOVPE	(001)	$L1_1$	H.R. Jen, D.S. Cao and G.B. Stringfellow, *Appl. Phys. Lett.* **54**, 1890 (1989).
InPAs	MOVPE	(001)	$L1_1$	D.H. Jaw, G.S. Chen and G.B. Stringfellow, *Appl. Phys. Lett.* **59**, 114 (1991).
GaPSb	MOVPE	(001)	$L1_1$ (weak)	J.R. Pessetto and G.B. Stringfellow, *J. Cryst. Growth* **62**, 1 (1983).
GaAsSb	MOVPE	(001) (110) (221) (311)	$L1_0$ $E1_1$	H.R. Jen, M.J. Cherng and G.B. Stringfellow, *J. Cryst. Growth* **48**, 1603 (1986).
GaAsSb	MBE	(001)	$L1_1$	I.J. Murgatroyd, A.G. Norman and G.R. Booker, *J. Appl. Phys.* **67**, 2310 (1990); and Y.E. Ihm, N. Otsuka, J.F. Klem and H. Morkoç, *Appl. Phys. Lett.* **51**, 2013 (1987).
InPSb	MOVPE	(001)	$L1_1$	J.R. Pessetto and G.B. Stringfellow, *J. Cryst. Growth* **62**, 1 (1983).
InAsSb	MOVPE	(001)	$L1_1$	H.R. Jen, K.Y. Ma and G.B. Stringfellow, *Appl. Phys. Lett.* **54**, 1154 (1989).

Table 4.5: Ordered III/V-V alloys observed to date in layers formed by any epitaxial growth technique.[a] The growth techniques referred to are molecular beam epitaxy (MBE), metal-organic vapor phase epitaxy (MOVPE), liquid-phase epitaxy (LPE), and vapor-levitation epitaxy (VLE). The structures referred to are illustrated in Figures 4.9 and 4.10.

[a] Adapted from G.B. Stringfellow and G.S. Chen, "Atomic ordering in III/V semiconductor alloys," *J. Vac. Sci. Technol.* **B9**, 2182 (1991).

Alloy	Growth Tech-nique	Sub-strate	Struc-ture	Reference
GaInP	MOVPE	(001)	$L1_1$	J.P. Goral, M.M. Al-Jassim, J.M. Olsen and A. Kibbler, *Mat. Res. Soc. Symp. Proc.* **102**, 583 (1988); T. Suzuki, A. Gomyo, and S. Iijima, *J. Cryst. Growth* **93**, 396 (1988); and O. Ueda, M. Takikawa, J. Komeno, and I. Umebu, *Jpn. J. Appl. Phys.* **26**, L1824 (1987).
AlGaInP	MOVPE	(001)	$L1_1$	G.S. Chen, T.Y. Wang, and G.B. Stringfellow *Appl. Phys. Lett.* **56**, 1463 (1990).
AlGaAs	MOVPE	(001) (110)	$L1_0$	T.S. Kuan, T.F. Kuech, W.I. Wang, and E.L. Wilkie, *Phys. Phys. Lett.* **54**, 201 (1985).
AlInAs	MOVPE	(001)	$L1_0$	A.G. Norman, R.E. Mallard, I.J. Murgatroyd, G.R. Booker, A.H. Moore, and M.D. Scott, *Inst. Phys. Conf. Ser.* **87**, 77 (1987).
InGaAs	LPE	(001)	$L1_0$ $E1_1$ DO_{22}?	H. Nakayama and H. Fujita, *Inst. Phys. Conf. Ser.* **79**, 289 (1985).
InGaAs	MBE	(110)	$L1_0$	T.S. Kuan, W.I. Wang, and E.L. Wilkie, *Appl. Phys. Lett.* **51**, 51 (1987).
InGaAs(P)	VLE	(001)	$L1_1$	M.A. Shahid and S. Mahajan, *Phys. Rev. Lett.* **B38**, 1344 (1988).
InGaAs(Sb)	MOVPE	(001)	$L1_0$ $E1_1$	H.R. Jen, M.J. Cherng, and G.B. Stringfellow, *Inst. Phys. Conf. Ser.* **83**, 159 (1987).

Table 4.6: Ordered III-III/V alloys observed to date in layers formed by any epitaxial growth technique.[a] The growth techniques referred to are molecular beam epitaxy (MBE), metal-organic vapor phase epitaxy (MOVPE), liquid-phase epitaxy (LPE) and vapor-levitation epitaxy (VLE). The structures referred to are illustrated in Figures 4.9 and 4.10.

[a] Adapted from G.B. Stringfellow and G.S. Chen, "Atomic ordering in III/V semiconductor alloys," *J. Vac. Sci. Technol.* **B9**, 2182 (1991).

We begin, in Subsection 4.4.1, by estimating the composition-dependent energies of the various elementary tetrahedra. Then, in Subsection 4.4.2, we apply the cluster variation method to estimate the composition and temperature dependent probabilities of the various elementary tetrahedra. Then, in Subsection 4.4.3, we estimate from these tetrahedra energies and probabilities the composition and temperature dependent molar Gibbs free energy of the alloy as a whole. Finally, in Subsection 4.4.4, we discuss the tendency of these alloys to order, i.e., for the tetrahedra probabilities to be peaked at film compositions that match those of the tetrahedra themselves.

4.4.1 Tetrahedra Energies

Let us start, in this subsection, by describing the energetics of the elementary tetrahedra of which such an alloy is composed. Those energies can be thought of as arising from the two kinds of distortions discussed in Sections 4.1 and 4.2. The first kinds are distortions *internal* to the tetrahedra due to the different equilibrium Ga–As and In–As bond lengths. The second kinds are distortions of the tetrahedra as a whole due to *external* constraints imposed by coherency of the epitaxial film with a substrate. Strictly speaking, these two kinds of distortions are not independent, because various externally imposed distortions may be more or less compatible with particular internal distortions.[24] In this simplified treatment, however, we neglect interactions between the two.

Coherency and External Distortions

First, consider the energies of tetrahedra due to *external* distortions. These distortions arise, as discussed in Section 4.2, because of macroscopic strains imposed by coherency with a substrate. The additional energy due to these distortions is given by Equation 4.25.

Ordering and Internal Distortions

Second, consider the energies of tetrahedra due to *internal* distortions. Those energies were estimated in Section 4.1, in a calculation which assumed that the corner group III atoms were pinned at their virtual crystal positions. In fact, those corner group III atoms will have a tendency to relax away from their virtual crystal positions, thereby decreasing the cluster energy.

[24] A.A. Mbaye, D.M. Wood and A. Zunger, "Stability of bulk and pseudomorphic epitaxial semiconductors and their alloys," *Phys. Rev.* **B37**, 3008 (1988).

Two extremes of behavior can be imagined. On the one hand, if the various tetrahedra were distributed randomly, as in a disordered alloy, then the relaxations of the various corner group III atoms will themselves tend to be random. Then, since each group III atom belongs to four tetrahedra, relaxations that decrease the energy of one tetrahedron will just as likely as not increase the energy of the other three. For this reason, the incoherent superposition of relaxations of group III atoms characteristic of a disordered alloy is not expected to greatly reduce the internal distortional energy from those estimated in Section 4.1 and listed in the first row of Table 4.7.

On the other hand, if the various tetrahedra were distributed in an ordered arrangement, then the relaxations of the various corner group III atoms will themselves tend to be ordered. Relaxations that decrease the energy of one tetrahedron may be exactly the relaxations required to reduce the internal distortion of the adjacent tetrahedra, and so on. For this reason, the coherent superposition of relaxations of group III atoms characteristic of an ordered alloy *is* expected to reduce the internal distortional energy from those estimated in Section 4.1. For example, for the GaAsSb alloy, calculations indicate that the chalcopyrite and famatinite structures illustrated in Figure 4.10 may be the least distorted,[25] although the layered tetragonal and layered trigonal ordered compounds are experimentally more commonly observed (see Tables 4.5 and 4.6). Note also that surface thermodynamics and kinetics effects not taken into account here may influence which of the ordered structure actually appears.[26]

For the disordered alloy, then, we would like to use the cluster energies calculated in Section 4.1 and listed in the first row of Table 4.7; for the ordered alloys, we would like to use the reduced values listed in the second row of Table 4.7; and for partially ordered alloys, we would like to use values somewhere in between. To incorporate these ideas in a semiquantitative way, we assume that the energies of the various tetrahedra depend on the occupation probability of the tetrahedra themselves:

$$u_{i,\text{int}} = u_{i,\text{int,dis}} + (u_{i,\text{int,ord}} - u_{i,\text{int,dis}})(\delta_i w_i)^{\lambda}. \tag{4.91}$$

In other words, as the probability of particular clusters increases, their tendency to interact coherently and lower their energy also increases. At low enough temperatures, this kind of cooperative interaction ultimately leads to long-range ordering into stoichiometric structures. Note, though, that only a few of the "wrong" kind of tetrahedra might be expected to destroy

---

[25] A.A. Mbaye, D.M. Wood and A. Zunger, "Stability of bulk and pseudomorphic epitaxial semiconductors and their alloys," *Phys. Rev.* **B37**, 3008 (1988).

[26] See, e.g., S. Froyen, and A. Zunger, "Surface-induced ordering in GaInP," *Phys. Rev. Lett.* **66**, 2132 (1991).

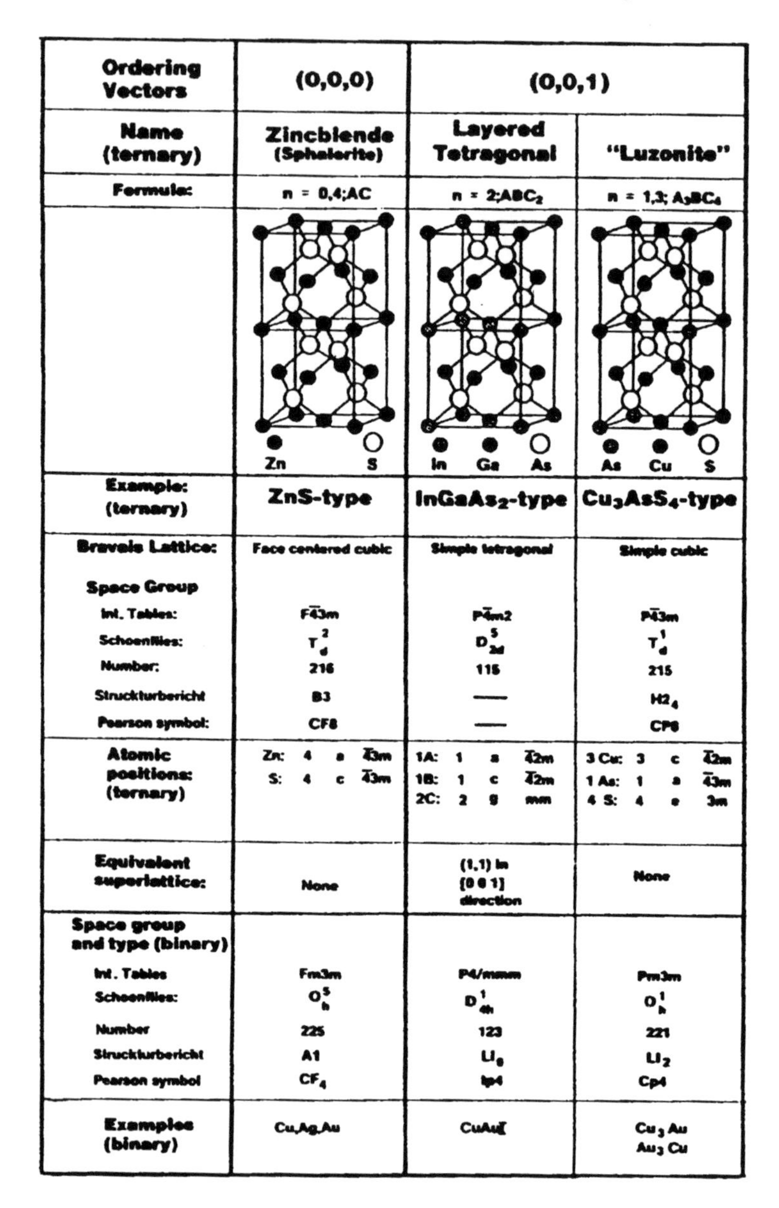

Ordering Vectors	(0,0,0)	(0,0,1)	
Name (ternary)	Zincblende (Sphalerite)	Layered Tetragonal	"Luzonite"
Formula:	n = 0,4;AC	n = 2;ABC_2	n = 1,3; A_3BC_4
	Zn S	In Ga As	As Cu S
Example: (ternary)	ZnS-type	$InGaAs_2$-type	Cu_3AsS_4-type
Bravais Lattice:	Face centered cubic	Simple tetragonal	Simple cubic
Space Group			
Int. Tables:	$F\bar{4}3m$	$P\bar{4}m2$	$P\bar{4}3m$
Schoenflies:	T_d^2	D_{2d}^5	T_d^1
Number:	216	115	215
Struckturbericht	B3	—	$H2_4$
Pearson symbol:	CF8	—	CP8
Atomic positions: (ternary)	Zn: 4 a $\bar{4}3m$ S: 4 c $\bar{4}3m$	1A: 1 a $\bar{4}2m$ 1B: 1 c $\bar{4}2m$ 2C: 2 g mm	3 Cu: 3 c $\bar{4}2m$ 1 As: 1 a $\bar{4}3m$ 4 S: 4 e 3m
Equivalent superlattice:	None	(1,1) in [0 0 1] direction	None
Space group and type (binary)			
Int. Tables	Fm3m	P4/mmm	Pm3m
Schoenflies:	O_h^5	D_{4h}^1	O_h^1
Number	225	123	221
Struckturbericht	A1	$L1_0$	$L1_2$
Pearson symbol	CF_4	tP4	Cp4
Examples (binary)	Cu,Ag,Au	CuAuI	Cu_3Au Au_3Cu

Figure 4.9: Examples of ordered fcc (or pseudobinary zincblende) structures and their space groups.[a]

[a]Reprinted from L.G. Ferreira, S-H Wei and A. Zunger, "First-principles calculation of alloy phase diagrams: the renormalized-interaction approach," *Phys. Rev.* **B40**, 3197 (1989).

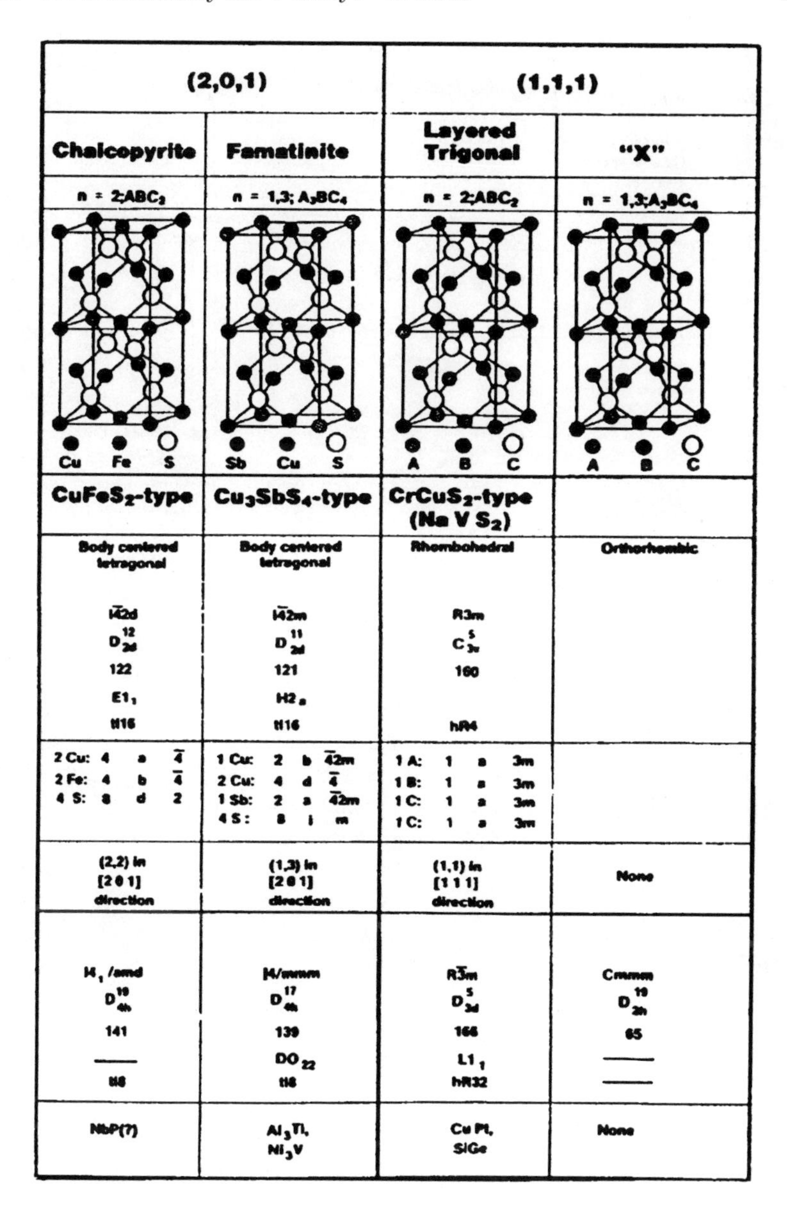

(2,0,1)		(1,1,1)	
Chalcopyrite	Famatinite	Layered Trigonal	"X"
n = 2;ABC$_2$	n = 1,3; A$_3$BC$_4$	n = 2;ABC$_2$	n = 1,3;A$_3$BC$_4$
Cu Fe S	Sb Cu S	A B C	A B C
CuFeS$_2$-type	Cu$_3$SbS$_4$-type	CrCuS$_2$-type (Na V S$_2$)	
Body centered tetragonal I$\bar{4}$2d D^{12}_{2d} 122 E1$_1$ tI16	Body centered tetragonal I$\bar{4}$2m D^{11}_{2d} 121 H2$_6$ tI16	Rhombohedral R3m C^{5}_{3v} 160 hR4	Orthorhombic
2 Cu: 4 a $\bar{4}$ 2 Fe: 4 b $\bar{4}$ 4 S: 8 d 2	1 Cu: 2 b $\bar{4}$2m 2 Cu: 4 d $\bar{4}$ 1 Sb: 2 a $\bar{4}$2m 4 S: 8 i m	1 A: 1 a 3m 1 B: 1 a 3m 1 C: 1 a 3m 1 C: 1 a 3m	
(2,2) in [2 0 1] direction	(1,3) in [2 0 1] direction	(1,1) in [1 1 1] direction	None
I4$_1$/amd D^{19}_{4h} 141 — tI8	I4/mmm D^{17}_{4h} 139 DO$_{22}$ tI8	R$\bar{3}$m D^{5}_{3d} 166 L1$_1$ hR32	Cmmm D^{19}_{2h} 65 — —
NbP(?)	Al$_3$Ti, Ni$_3$V	CuPt, SiGe	None

Figure 4.10: Examples of ordered fcc (or pseudobinary zincblende) structures and their space groups.[a]

[a] Reprinted from L.G. Ferreira, S-H Wei, and A. Zunger, "First-principles calculation of alloy phase diagrams: the renormalized-interaction approach," *Phys. Rev.* **B40**, 3197 (1989).

	$u_{0,\text{int}}$	$u_{1,\text{int}}$	$u_{2,\text{int}}$	$u_{3,\text{int}}$	$u_{4,\text{int}}$
Disordered	0	0.023	0.031	0.023	0
Ordered	0	0.017	0.016	0.017	0

Table 4.7: Estimated internal distortion energies (in eV per atom pair) of the elementary InGaAs tetrahedra shown in Figure 4.2. The energies listed in the first row were estimated for a disordered arrangment of tetrahedra, whose corner atoms, on average, are bound to virtual crystal sites.[a] The energies listed in the second row are those (very roughly) estimated for an ordered arrangement of tetrahedra, whose corner atoms can relax "in-phase" with the corner atoms of adjacent tetrahedra.[b]

[a]M. Ichimura and A. Sasaki, "Short-range order in III-V ternary alloy semiconductors," *J. Appl. Phys.* **60**, 3850 (1986); A. Sher, M. van Schilfgaarde, A.-B. Chen and W. Chen, "Quasi-chemical approximation in binary alloys," *Phys. Rev.* **B36**, 4279 (1987).

[b]Estimated *very* roughly by scaling the results of calculations in the GaAsSb system by L.G. Ferreira, S-H Wei, and A. Zunger, "First-principles calculation of alloy phase diagrams: the renormalized-interaction approach," *Phys. Rev.* **B40**, 3197 (1989).

the coherency of the tetrahedron relaxations. Therefore, we expect the ordering energy to be a highly nonlinear function of the cluster probability itself. In this treatment, we take λ, the nonlinearity parameter, to be eight.

In a sense, we have augmented the tetrahedron approximation of the CVM, which allows different tetrahedra to have different energies, with a point, or mean-field approximation of the CVM to allow each tetrahedron's energy to depend also on the average tetrahedra populations. We must emphasize, though, that this simple, mean-field treatment of long-range order is only a semiquantitative one. To treat long-range order quantitatively within the CVM, it is necessary to distinguish between the (up to) four group III sublattices in the ordered structures, and to account explicitly for the occupation statistics on each sublattice of the (up to) 16 kinds of tetrahedra.[27]

Total Energies

The internal and external strain energies can now be summed to give

$$u_i = u_{i,\text{int}} + u_{i,\text{ext}}. \tag{4.92}$$

[27]W.L. Bragg and E.J. Williams, "The effect of thermal agitation on atomic arrangement in alloys" *Proc. Roy. Soc. (London)* **A145**, 699 (1934); H.A. Bethe, "Statistical theory of superlattices," *Proc. Roy. Soc. (London)* **A150**, 552 (1935); C.M. van Baal, "Order-disorder transformations in a generalized Ising alloy," *Physica* **64**, 571 (1973); and D. de Fontaine, "Configurational thermodynamics of solid solutions," *Solid State Physics* **34**, 73 (1979).

These energies cannot be evaluated directly, since, through Equation 4.91, they depend on the tetrahedron probabilities, which in turn depend (self-consistently) on the energies themselves. However, we can get an approximate idea of how the tetrahedra energies depend on the compositions, x_{epi} and x_{sub}, of the epitaxial film and the substrate by calculating the energies of a completely disordered alloy, so that the tetrahedra probabilities contribute negligibly to Equation 4.91, and $u_{i,\text{int}} \rightarrow u_{i,\text{int,dis}}$. Those energies are plotted in the right column of Figure 4.11 as functions of x_{epi} and x_{sub}.

Consider first the type 0 tetrahedron at the bottom of that column. All of its group III atoms are Ga, and so its externally imposed strain energy is zero when it is embedded in a film of pure GaAs grown undistorted and lattice-matched to a substrate of pure GaAs. Moreover, its internal distortional strain energy is also zero, since the central As atom is symmetrically situated within a tetrahedron of equivalent Ga atoms. Therefore, its total strain energy is zero at $x_{\text{epi}} = x_{\text{sub}} = 0$.

If now we increase x_{epi}, then the average lattice parameter of the epitaxial film increases, and the unit cell of the epitaxial film grows. At the same time, the size of the tetrahedra embedded in the film are tied to those of the unit cell. Therefore, the type 0 tetrahedra themselves must grow, even if they would "prefer" not to, and their strain energies must increase.

Note that even as x_{epi} increases, we can choose either to increase the substrate lattice parameter at the same rate ($x_{\text{sub}} = x_{\text{epi}}$) or to keep it fixed ($x_{\text{sub}} = 0$). If we increase it at the same rate (open circles in Figure 4.11), then the unit cell of the epitaxial film remains an undistorted, albeit larger, cube. The energy of the type 0 tetrahedra increases due to that volume mismatch. If, however, we keep it fixed (filled near circles in Figure 4.11), then the unit cell of the epitaxial film is not only larger, but distorted as well. The energy of the type 0 tetrahedra is therefore also quite high when $x_{\text{epi}} = 1$ and $x_{\text{sub}} = 0$.

Suppose, now, that we fix x_{epi} at zero, but increase x_{sub}. Then, the unit cell of the epitaxial film remains approximately the same size, but it distorts, as its parallel lattice parameter increases and its perpendicular lattice parameter decreases. Therefore, its energy increases, reaching a maximum at $x_{\text{sub}} = 1$. If x_{epi} is now increased, then the volume of the unit cell increases, but the distortion in the unit cell decreases. Initially, the strain energy in the type 0 tetrahedra decreases as the unit cell distortion decreases, but eventually it increases as the volume mismatch between the type 0 tetrahedra and the film unit cell increases.

Similar arguments can be used to understand the dependences of the energies of the other types of tetrahedra on x_{epi} and x_{sub}. In general, the energy minima for the various tetrahedra occur when both x_{epi} and x_{sub} are equal to the composition of the cluster itself. The reasons are that when

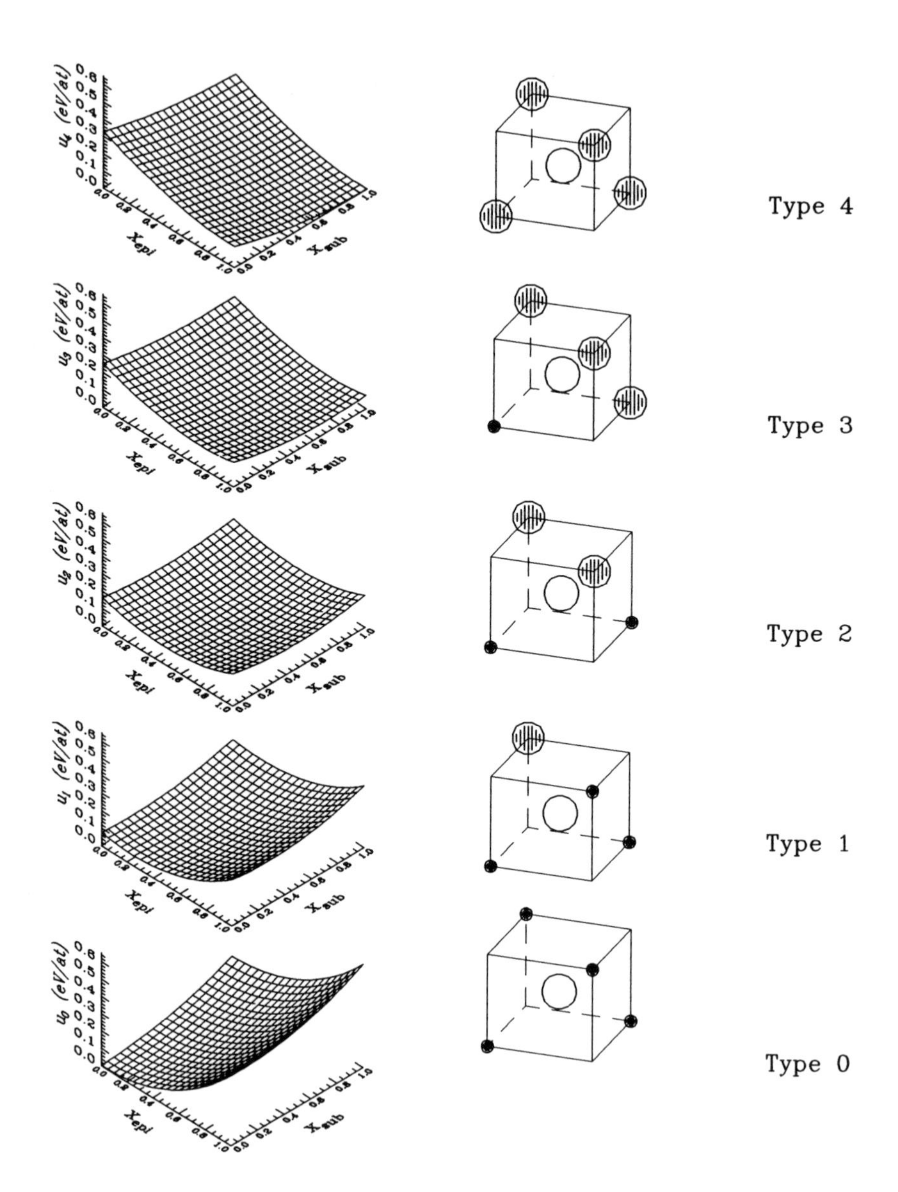

Figure 4.11: Energies of various tetrahedra embedded in disordered $In_{1-x}Ga_xAs$ of composition x_{epi} grown coherently on substrates having the lattice parameters of bulk $In_{1-x}Ga_xAs$ at composition x_{sub}.

$x_i = x_{\rm epi}$, the volume of cluster i is best matched to the volume of the unit cell of the epitaxial film, and when $x_{\rm epi} = x_{\rm sub}$ the unit cell of the film is least distorted. Deviations from $x_{\rm epi} = x_{\rm sub} = x_i$ along the $x_{\rm epi} = x_{\rm sub}$ diagonal lead to volume mismatches and relatively large increases in energy. Deviations from $x_{\rm epi} = x_{\rm sub} = x_i$ through changes in $x_{\rm sub}$ lead to distortions and somewhat smaller increases in energy. Deviations from $x_{\rm epi} = x_{\rm sub} = x_i$ through changes in $x_{\rm epi}$ lead to some of both, and intermediate increases in energy.

Note that Equation 4.92 includes only the elastic potential energy contribution to the energies of each cluster. In principle, the temperature dependences of the molar energies and entropies could also be determined by heat capacity functions for each cluster, via Equations 2.9 and 2.8. However, in this simple treatment, we make the approximation that the various tetrahedra all have the same heat capacities. Then, the temperature dependences to their molar energies and entropies are all the same. Since tetrahedra probabilities depend only on the *relative* energies, we can neglect those temperature dependences.

4.4.2 Tetrahedra Probabilities

In Subsection 4.4.1, we estimated the energetics of the various elementary tetrahedra from which the InGaAs alloy may be constructed. In this subsection, we use these elementary tetrahedra energetics to calculate the tetrahedra probabilities using the rate equation method outlined in Section 4.3.3. These are shown in Figure 4.12 as functions of $x_{\rm epi}$ and $x_{\rm sub}$ at fixed temperatures of 100, 600, and 1100 K. Two opposing tendencies determine the probability distributions.

The first tendency is energy minimization. For a given composition of the epitaxial film, the two tetrahedra whose compositions just straddle $x_{\rm epi}$ will be the least volume mismatched, and will usually have the lowest energies. The film energy will then be minimized if it is composed of a weighted combination of only those two tetrahedra. For example, if $x_{\rm epi} = 3/8$, then the type 1 ($x_1 = 1/4$) and type 2 ($x_2 = 1/2$) tetrahedra will have the lowest energies, and the lowest energy film will be that composed of half type 1 and half type 2 tetrahedra. Therefore, at 100 K (left column of Figure 4.12), where energy minimization is most important, only two kinds of tetrahedra are ever significantly populated, and the probability distribution approaches the linear ramp given by Equation 4.90.

The second, opposing, tendency is entropy maximization. As can be seen in Figure 4.12, as temperature increases and entropy becomes an increasingly important component of the molar Gibbs free energy, the probabilities "diffuse" away from the tetrahedra whose compositions straddle that of the epitaxial film. The probabilities cannot diffuse too far away, however, since the overall composition of the film is still constrained to be $\sum p_i x_i = x_{\text{epi}}$. Ultimately, at 1100 K, the probability distribution approaches the Bernoullian distribution given by Equation 4.89.

4.4.3 Free Energies

In Subsections 4.4.1 and 4.4.2, we estimated the energetics and probabilities of the various elementary tetrahedra from which the InGaAs alloy may be constructed. In this subsection, we use these energetics and probabilities to calculate the molar Gibbs free energy of the film as a whole using Equation 4.80. These free energies are shown in Figure 4.13 as functions of x_{epi} and x_{sub}, again for three fixed temperatures: 100, 600, and 1100 K. These temperatures are representative of three distinct regimes of behavior.

At the highest temperature, 1100 K, the molar Gibbs free energy is everywhere and in every direction concave *up*. Therefore, films cannot lower their molar Gibbs free energies by decomposing spatially into local regions, some having higher x_{epi} and others having lower x_{epi}. Epitaxial films at this temperature are *stable* against such macroscopic compositional clustering.

At the intermediate temperature, 600 K, the molar Gibbs free energy is concave up with respect to horizontal fluctuations in x_{epi} (at fixed x_{sub}), but concave down with respect to diagonal fluctuations in x_{epi} (mimicked by identical flucations in x_{sub}). Therefore, films cannot lower their molar Gibbs free energies by composition fluctuations that preserve x_{sub}, but can by fluctuations that do not preserve x_{sub}. In other words, fluctuations in which the local regions remain coherent with the substrate are suppressed, while fluctuations in which the local regions are incoherent (and hence free to adopt their equilibrium lattice parameter) are not. Epitaxial films at this temperature are stable against *coherent* macrosopic clustering, but unstable against *incoherent* macrosopic clustering.

At the lowest temperature, 100 K, the molar Gibbs free energy is, for some combinations of x_{epi} and x_{sub}, concave down with respect to *both* horizontal and diagonal fluctuations in x_{epi}. Therefore, these films can lower their molar Gibbs free energies both by composition fluctuations that preserve x_{sub}, as well as by fluctuations that do not preserve x_{sub}. These films at this temperature are *not* stable against either coherent or incoherent macroscopic compositional clustering.

Note that the downward concavity of the molar Gibbs free energy at 100 K is most exaggerated at those special compositions (1/4, 1/2, 3/4) for which we have assumed ordering may take place. The sharpness of those cusps is a consequence of the cooperative nature of the ordering process.

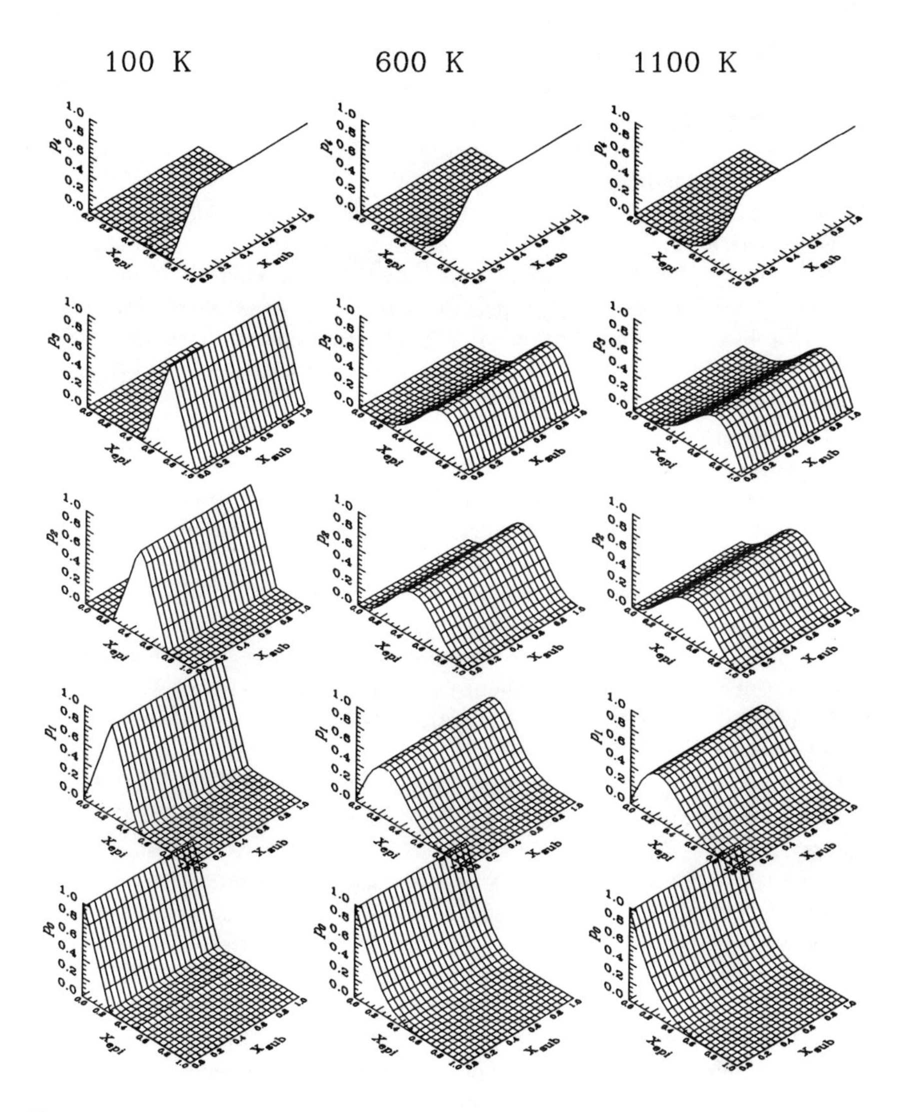

Figure 4.12: 100, 600, or 1100 K probabilities of various tetrahedra embedded in $In_{1-x}Ga_xAs$ of composition x_{epi} grown coherently on substrates having the lattice parameters of bulk $In_{1-x}Ga_xAs$ at composition x_{sub}.

In the $In_{0.25}Ga_{0.75}As$ alloy, for example, the more $In_{0.25}Ga_{0.75}As$ tetrahedra there are, the closer w_1 approaches unity, the lower the energy $u_{1,\text{int}}$ becomes, and the more favored the $In_{0.25}Ga_{0.75}As$ tetrahedra become. At higher temperatures or at compositions slightly off $In_{0.25}Ga_{0.75}As$, there are never enough $In_{0.25}Ga_{0.75}As$ to "get the process going," and the tetrahedra energies are dominated by their disordered values.

Also note that with respect to horizontal fluctuations, the resulting cusps are even, for intermediate substrate compositions, *global* minima in the molar Gibbs free energies. Therefore, films that are constrained to be coherent with a substrate are unstable against clustering into ordered alloys. With respect to diagonal fluctuations, however, the cusps are only *local* minima. Therefore, films not constrained to be coherent with a substrate are unstable against clustering into ordered compounds, but those ordered compounds are themselves unstable against further clustering into (nearly) pure GaAs and (nearly) pure InAs.

To understand these three temperature regimes more concretely, consider an epitaxial film at $x_{\text{epi}} = 0.6$ grown on a substrate also at $x_{\text{sub}} = 0.6$. Because the film is lattice-matched to the substrate, it is free from macroscopic elastic strain. It is, however, also composed preferentially of type 2 and type 3 tetrahedra. Those tetrahedra are internally distorted, and hence, on a microscopic scale, contain a significant amount of internal distortional elastic energy.

Suppose we force the film to decompose into macroscopic clusters, of which in some $x_{\text{epi}} = 0$ and in others $x_{\text{epi}} = 1$. These clusters are composed preferentially of type 0 and type 4 tetrahedra, respectively. Neither type of tetrahedron is internally distorted, and hence both are free of internal distortional elastic energy. However, they may or may not be externally distorted, and hence may or may not be free of external distortional elastic energy.

On the one hand, if the $x_{\text{epi}} = 0$ and $x_{\text{epi}} = 1$ clusters were each free to change their average lattice parameters (i.e., free to change x_{sub}), then the type 0 and type 4 tetrahedra would be free from external distortional elastic energy. Hence, the decomposition of regions having mainly type 2 and 3 tetrahedra into macroscopic clusters having mainly type 0 and type 4 tetrahedra decreases the overall strain energy and will tend to occur.

Note, though, that the number of ways different tetrahedra can be combined to form a macroscopically uniform alloy at $x_{\text{epi}} = 0.6$ is larger than the number of ways they can be combined to form alloys at $x_{\text{epi}} = 0$ and $x_{\text{epi}} = 1$. Since at high enough temperatures, entropic contributions to the molar Gibbs free energies ultimately dominate, there will then be a critical temperature above which mixing will be favored over decomposition.

On the other hand, if the $x_{\text{epi}} = 0$ and $x_{\text{epi}} = 1$ clusters were *not* free to

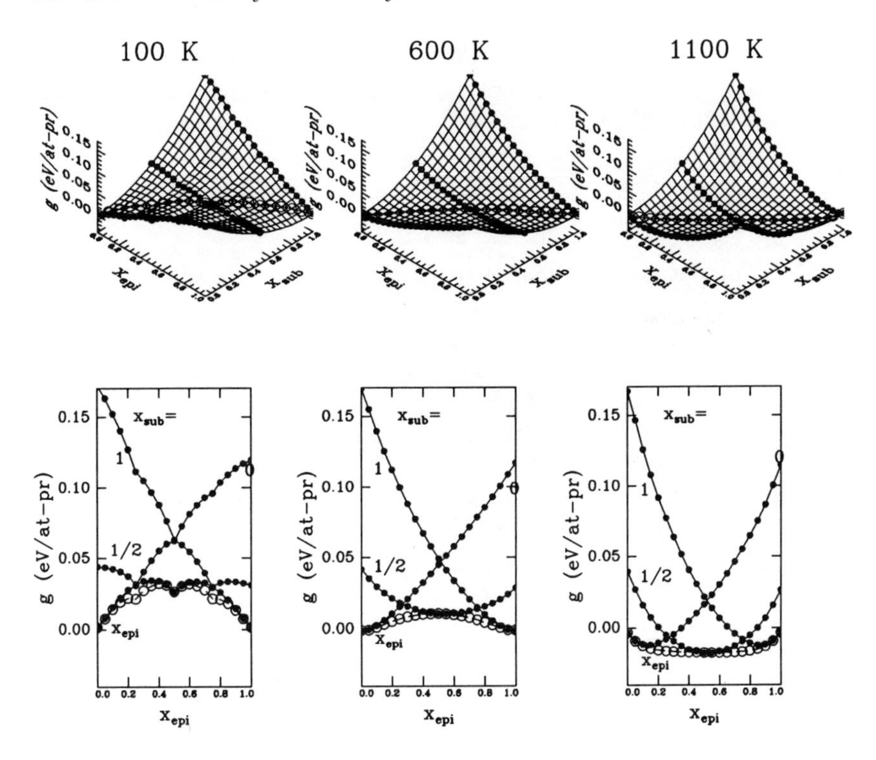

Figure 4.13: 100, 600, or 1100 K molar Gibbs free energies of $In_{1-x}Ga_xAs$ at composition x_{epi} grown coherently on substrates having the lattice parameters of bulk $In_{1-x}Ga_xAs$ at composition x_{sub}. The filled circles represent $In_{1-x}Ga_xAs$ grown on substrates with compositions $x_{\text{sub}} = 0, 1/2, 1$. The open circles represent $In_{1-x}Ga_xAs$ grown on "lattice-matched" substrates with compositions $x_{\text{sub}} = x_{\text{epi}}$, or, alternatively, to incoherent growth.

change their average lattice parameters (i.e., not free to change x_{sub}), then the type 0 and type 4 tetrahedra would not be free from external distortional elastic energy. If that energy is higher than the internal distortional elastic energy of the original type 2 and 3 tetrahedra, then the decomposition is suppressed. Instead, the film will decompose into macroscopic ordered clusters, in some of which $x_{\text{epi}} = 0.5$ and in others $x_{\text{epi}} = 0.75$. These clusters are composed preferentially of type 2 and 3 tetrahedra, respectively, which fit together in such a way as to minimize their internal distortional elastic energy.

Note, though, that just as before, the number of ways different tetrahedra can be combined to form a macroscopically uniform alloy at $x_{\text{epi}} = 0.6$ is larger than the number of ways they can be combined to form ordered alloys at $x_{epi} = 0.5$ and $x_{epi} = 0.75$. Since at high enough temperatures, entropic contributions to the molar Gibbs free energies ultimately dominate, there will then be a critical temperature above which mixing will be favored over ordering.

4.4.4 Short-Range Ordering

In Subsection 4.4.2, we estimated the probabilities of the various elementary tetrahedra from which an InGaAs alloy may be constructed. From Figure 4.12, it can be seen that as the temperature is lowered, the tetrahedra probabilities become less random and more peaked at the film compositions that match those of the tetrahedra themselves. This is a consequence of the fact that the tetrahedra energies are not the same, but are minimum for film compositions that match those of the tetrahedra themselves.

In this subsection, we discuss in more detail this deviation from randomness. Now, first suppose the tetrahedra *were* distributed randomly, according to Equation 4.89. Then, from the constitutive pair and triangle relations listed in Table 4.4 on page 111, the "unlike" pair probability would be

$$\begin{aligned} y_1 &= w_1 + 2w_2 + w_3 \\ &= x(1-x)^3 + 2x^2(1-x)^2 + x^3(1-x) \\ &= x(1-x), \end{aligned} \tag{4.93}$$

as expected. Since the clusters are *not* distributed randomly, we expect deviations from this purely random mixed pair probability.[28]

To quantify these deviations from randomness, we define a short-range order parameter associated with pairs of unlike (next-nearest-neighbor) group III atoms, analogous to that of Equation 4.46,

$$\sigma^{SRO} \equiv \frac{w_1 + 2w_2 + w_3 - x(1-x)}{w_1^{ord} + 2w_2^{ord} + w_3^{ord} - x(1-x)}, \tag{4.94}$$

where the w_i^{ord} are the completely ordered cluster probabilities given by Equation 4.90. σ^{SRO} is unity if every In atom is surrounded by as many Ga atoms as possible, zero if every In atom is surrounded by a random number of Ga atoms, and negative if every In atom is surrounded by as many In atoms as possible.

[28] M.T. Czyżyk, M. Podgórny, A. Balzarotti, P. Letardi, N. Motta, A. Kisiel and M. Kimnal-Starnawska, "Thermodynamic properties of ternary semiconducting alloys," *Z. Phys.* **B62**, 153 (1986).

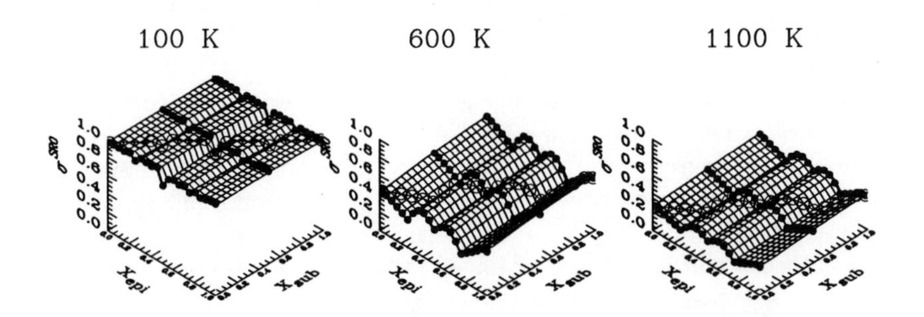

Figure 4.14: 100, 600, or 1100 K short-range order parameters in $In_{1-x}Ga_xAs$ at composition x_{epi} grown coherently on substrates having the lattice parameters of bulk $In_{1-x}Ga_xAs$ at composition x_{sub}. The filled circles represent $In_{1-x}Ga_xAs$ grown on substrates with compositions $x_{\mathrm{sub}} = 0, 1/2, 1$. The open circles represent $In_{1-x}Ga_xAs$ grown on "lattice-matched" substrates with compositions $x_{\mathrm{sub}} = x_{\mathrm{epi}}$, or, alternatively, to incoherent growth.

This short-range order parameter is plotted in Figure 4.14 for the three temperatures 100, 600, and 1100 K. Note that even at the highest temperature, there is a preference toward bonding between unlike group III atoms, although the preference becomes more pronounced at the lower temperatures. Note also that the short-range ordering becomes somewhat less pronounced (cusped downward) at compositions corresponding to the various elementary tetrahedra. This is so even though the unlike pair probability itself decreases smoothly on both sides of $x_{\mathrm{epi}} = 1/2$.

The reason is that exactly at those stoichiometric compositions, it is more difficult to suppress the occupation of composition-straddling tetrahedra. For example, at $x_{\mathrm{epi}} = 1/2$, the film will be dominated by type 2 tetrahedra, but some type 1 and 3 tetrahedra, differing in composition from x_{epi} by only 1/4, will also be present. At $x_{\mathrm{epi}} = 5/8$, the film will be dominated by a mix of type 2 and 3 tetrahedra. The type 1 and 4 tetrahedra, however, differ in composition from x_{epi} by 3/8. Since the elastic distortion energies of the tetrahedra vary with the *square* of the composition mismatch, the type 1 and 4 tetrahedra will be suppressed more effectively for $x_{\mathrm{epi}} = 5/8$ than the type 1 and 3 tetrahedra were for $x_{\mathrm{epi}} = 1/2$.

Finally, we are in a position to understand the microscopic origins of clustering and ordering. In epitaxial films at all temperatures, different tetrahedra have different energies. Usually, tetrahedra that are most nearly volume-matched to the average unit cell volume have the lowest energies, and hence are most favored. Therefore, films of intermediate composition will be preferentially composed of tetrahedra of intermediate composition, and will be short-range ordered in the sense of having an excess of pairs between unlike next-nearest neighbors.

At high temperatures, a homogeneous film of intermediate composition will always be favored over macroscopic clusters at endpoint compositions, because of the increased entropy associated with an increased number of ways of combining tetrahedra of different compositions. Moreover, because of the relatively wide distribution of tetrahedra of different compositions, the tetrahedra will tend to be arranged randomly with respect to one another.

At low temperatures, homogeneous films are no longer favored. As the occupation probabilities become more and more concentrated among those tetrahedra whose compositions straddle the composition of the film, it becomes possible for the tetrahedra to order in such a way that their internal distortions are minimized. Then, films will have a tendency to form macroscopic, ordered clusters composed nearly exclusively of tetrahedra of a certain kind, arranged in a certain way. In *coherent* films, these ordered clusters are the stable state of the system, because tetrahedra in disordered clusters at intermediate compositions have too much internal distortional energy, and tetrahedra in clusters at endpoint compositions have too much external distortional energy. In *incoherent* films, however, the tetrahedra in clusters at the endpoint compositions have no external distortional energy, and hence will ultimately form at the expense of both a homogeneous film or a film composed of ordered clusters.

4.5 Semi-empirical Models

In Sections 4.1–4.4, we have been concerned with developing a microscopic description of the thermodyamics of coherent and incoherent pseudobinary III-V alloys. There is of course no substitute for the physical insight that such a microscopic description gives. However, many of the overall *results* of such a description, such as the molar Gibbs free energy, can be understood using simpler, macroscopic, semi-empirical models. Such models have the advantage, as discussed in Chapter 3, of being described by analytic equations that can be more easily used to calculate phase diagrams and other thermodynamic quantities of interest. In this section, we develop such a semi-empirical model.

We will begin, in Subsection 4.5.1, by describing semi-empirical, physically motivated expressions for the molar Gibbs free energies of disordered and ordered pseudobinary alloys. Then, in Subsection 4.5.2, we use these molar Gibbs free energies to calculate equilibrium alloy phase diagrams.

4.5.1 Free Energies

Let us start, in this subsection, by describing and justifying a semi-empirical expression for the molar Gibbs free energy of an epitaxial pseudobinary alloy grown coherently on a thick substrate. The three main components of the molar Gibbs free energy that we need to account for are (1) the enthalpy of mixing, due to the internal distortional energies of the various tetrahedra, (2) the entropy of mixing, and (3) the coherency energy, due to the external distortional energies of the various tetrahedra.

We describe the internal distortions, as discussed in Section 4.1, by an enthalpy of mixing of the regular solution form

$$h_{\text{int}} = \Omega x_{\text{epi}}(1 - x_{\text{epi}}), \tag{4.95}$$

where the interaction parameter, Ω, is identified with that calculated in Equation 4.13. The description could easily be improved further through the use of a sub-regular solution form, in order to account for composition-dependent elastic constants. The description could also easily be improved by allowing the mixing enthalpy to depend on temperature through a composition- *and* temperature-dependent heat capacity.

We describe the entropy of mixing by the ideal solution form:

$$s = -k\left[x_{\text{epi}} \ln(x_{\text{epi}}) + (1 - x_{\text{epi}}) \ln(1 - x_{\text{epi}})\right]. \tag{4.96}$$

Finally, we describe the external distortions, following the treatment of Section 4.2, with a coherency energy of the form[29]

$$h_{\text{ext}} = C_{\text{eff}}\eta^2(x_{\text{epi}} - x_{\text{sub}})^2 a_{\text{avg}}^3/4, \tag{4.97}$$

where

$$\begin{aligned} C_{\text{eff}} &= (1 - x_{\text{epi}})\left(C_{\text{GaAs},11} + C_{\text{GaAs},12} - \frac{2C_{\text{GaAs},12}^2}{C_{\text{GaAs},11}}\right) \\ &\quad + x_{\text{epi}}\left(C_{\text{InAs},11} + C_{\text{InAs},12} - \frac{2C_{\text{InAs},12}^2}{C_{\text{InAs},11}}\right) \end{aligned} \tag{4.98}$$

is an effective elastic coefficient that varies linearly between that of GaAs and that of InAs[30] and

$$\eta = 2\frac{a_{\text{InAs,o}} - a_{\text{GaAs,o}}}{a_{\text{InAs,o}} + a_{\text{GaAs,o}}} \tag{4.99}$$

[29] J.W. Cahn, "On spinodal decomposition," *Acta Metall.* **9**, 795 (1961).

[30] F.C. Larché, W.C. Johnson, C.S. Chiang, and G. Martin, "Influence of substrate-induced misfit stresses on the miscibility gap in epitaxial layers: application to III-V alloys," *J. Appl. Phys.* **64**, 5251 (1988).

is the coefficient of linear expansion per unit composition change.

We note in passing that this elastic energy term, present only for coherent epitaxy, can be an important determinant of the overall driving force for epitaxy. Coherent alloys whose lattice parameters differ from that of the substrate will have higher molar Gibbs free energies than those that are lattice matched. Condensation of lattice-matched alloys will therefore be favored over condensation of lattice-mismatched alloys, as has been observed for both liquid phase epitaxy[31] as well as MBE.[32]

We emphasize here that Equation 4.97 only applies under special circumstances. In particular, it only applies to the geometry we are considering — namely, a thin epitaxial film whose parallel lattice parameter is constrained to be that of its infinitely thick substrate but whose perpendicular lattice parameter is free to adjust — if all phases present have some physical dimension that is large compared to the film thickness. More general treatments of coherent phase equilibria are complicated immensely by the possibility that the elastic coherency energies depend on the details of the phase morphology, which in turn depend on the relative amounts of the different phases present.[33]

Finally, the total molar Gibbs free energy of the disordered alloy is

$$g^{\text{dis}}(x_{\text{epi}}, x_{\text{sub}}) = h_{\text{int}}(x_{\text{epi}}) - Ts(x_{\text{epi}}) + h_{\text{ext}}(x_{\text{epi}}, x_{\text{sub}}), \qquad (4.100)$$

and depends on the compositions of both the epitaxial film and the substrate. As can be seen from the top panels of Figure 4.15, the semi-empirical expression of Equation 4.100 reproduces surprisingly well the molar Gibbs free energies deduced from the CVM calculation shown in Figure 4.13.

Now, as discussed in Section 4.4, ordered and disordered phases should really be treated on a single footing. Doing so requires, however, a microscopic treatment that is difficult to incorporate into a semi-empirical model. Instead, we treat ordered alloys as if they were distinct "compound" phases which exist only within a narrow range of special compositions, as illustrated in the bottom panels of Figure 4.15. In other words, we write

---

[31] G.B. Stringfellow, "The importance of lattice mismatch in the growth of GaInP epitaxial crystals," *J. Appl. Phys.* **43**, 3455 (1972); and R.E. Nahory, M.A. Pollack, E.D. Beebe, J.C. DeWinter, and M. Ilegems, "The liquid phase epitaxy of AlGaAsSb and the importance of strain effects near the miscibility gap," *J. Electrochem. Soc.* **125**, 1053 (1978).

[32] M. Allovon, J. Primot, Y. Gao, and M. Quillec, "Auto lattice matching effect for AlInAs grown by MBE at high substrate temperature," *J. Electron. Mater.* **18**, 505 (1989).

[33] J.W. Cahn and F.C. Larché, "A simple model for coherent equilibrium," *Acta Metall.* **11**, 1915 (1984); W.C. Johnson and C.S. Chiang, "Phase equilibrium and stability of elastically stressed heteroepitaxial thin films," *J. Appl. Phys.* **64**, 1155 (1988).

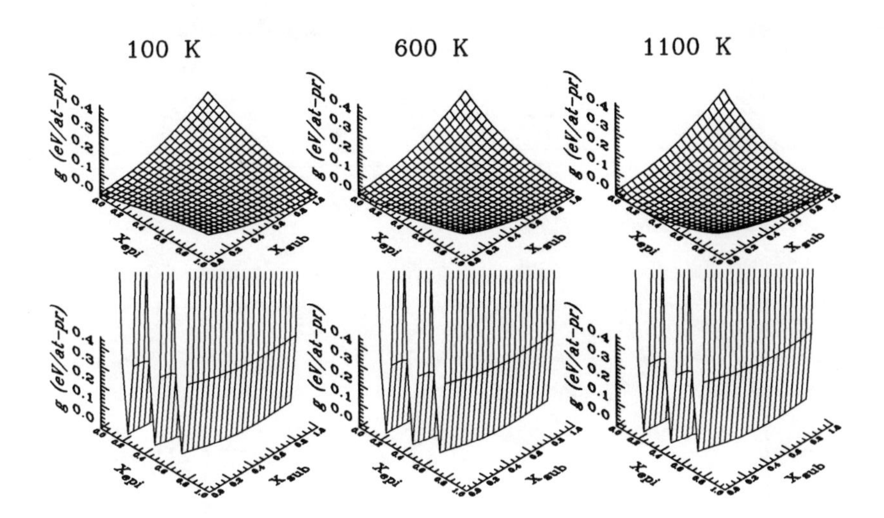

Figure 4.15: Semi-empirical molar Gibbs free energies of disordered (top) and ordered (bottom) $In_{1-x}Ga_xAs$ at composition x_{epi} grown coherently on substrates having the lattice parameters of bulk $In_{1-x}Ga_xAs$ at composition x_{sub}.

their free energies as

$$\begin{aligned} g^{\alpha}(x_{\text{epi}},, x_{\text{sub}}) &= h^{\alpha} + A(x_{\text{epi}} - 1/4)^2 + h_{\text{ext}}(x_{\text{epi}}, x_{\text{sub}}) \\ g^{\beta}(x_{\text{epi}}, x_{\text{sub}}) &= h^{\beta} + A(x_{\text{epi}} - 1/2)^2 + h_{\text{ext}}(x_{\text{epi}}, x_{\text{sub}}) \\ g^{\gamma}(x_{\text{epi}}, x_{\text{sub}}) &= h^{\gamma} + A(x_{\text{epi}} - 3/4)^2 + h_{\text{ext}}(x_{\text{epi}}, x_{\text{sub}}). \quad (4.101) \end{aligned}$$

The first terms in these equations are the enthalpies of the ordered compounds, which for InGaAs we identify with those listed in Table 4.7. The second terms are phenomenological terms reflecting expected sharp dependences of the ordering enthalpies on composition near the special compositions, with A a large constant. The third terms are the energies given by Equation 4.97.

4.5.2 Phase Diagrams

In Subsection 4.5.1, we described a semi-empirical expression for the molar Gibbs free energy of a pseudobinary alloy. In this subsection, we use these free energies and the common tangent prescription described in Chapter 3 to calculate two-dimensional x_{epi}-T cuts through the full three-dimensional

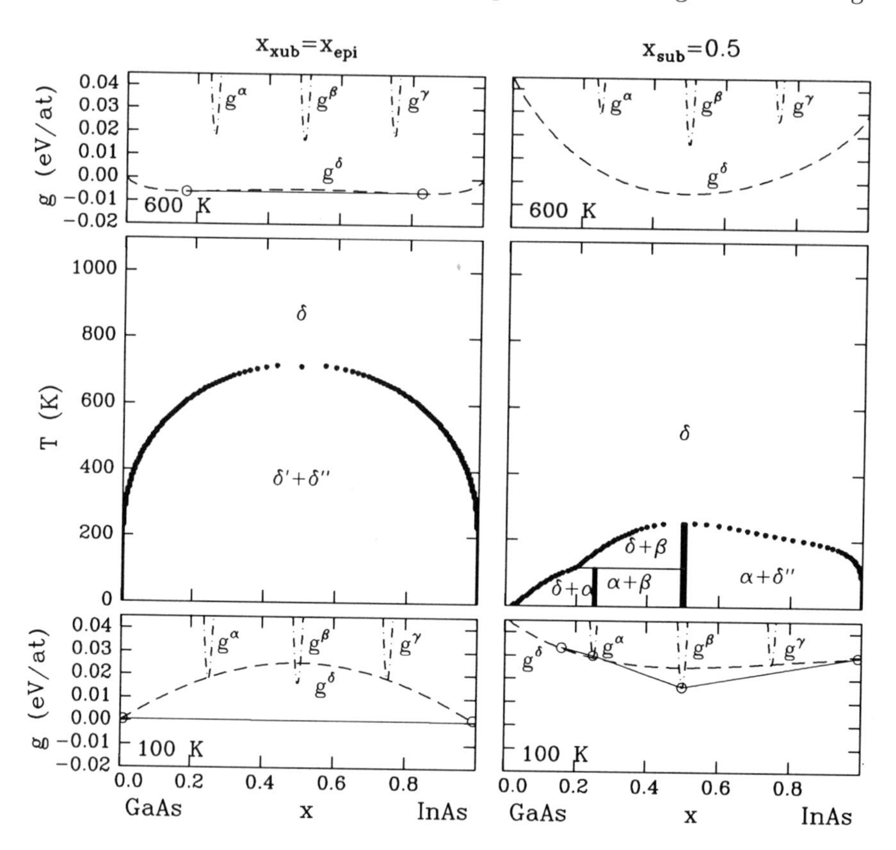

Figure 4.16: $x_{\rm epi}$-T phase diagrams for $In_{1-x}Ga_xAs$ during (right) coherent epitaxy on a substrate of composition $x_{\rm sub}$ and during (left) incoherent epitaxy. Above and below each phase diagram are also shown the molar Gibbs free energies of the various phases at 600 K and 100 K, their common tangents, and the critical compositions (open circles) determined by those common tangents.

$x_{\rm epi}$-$x_{\rm sub}$-T phase diagram, as illustrated in Figure 4.16, or calculate the full $x_{\rm epi}$-$x_{\rm sub}$-T phase diagram itself, as illustrated in Figure 4.17.

In both cases the common tangents may be drawn in two ways. On the one hand, if we constrain the epitaxial film to be coherent, then $x_{\rm sub}$ must be preserved, and so, as illustrated in the right halves of Figures 4.16 and 4.17, we must take horizontal tangents at constant $x_{\rm sub}$. On the other hand, if we do not constrain the epitaxial film to be coherent, then $x_{\rm sub}$ is free to accommodate $x_{\rm epi}$, and so, as illustrated in the left half of Figures 4.16 and 4.17, we must take diagonal tangents for which $x_{\rm sub} = x_{\rm epi}$.

Incoherent Phase Equilibria

Consider first the incoherent case for which $x_{\text{sub}} = x_{\text{epi}}$. Then, Equation 4.100 reduces to

$$g^{\text{dis,inc}}(x_{\text{epi}}) = h_{\text{int}}(x_{\text{epi}}) - Ts(x_{\text{epi}}), \tag{4.102}$$

since h_{ext} vanishes for $x_{\text{epi}} = x_{\text{sub}}$.

At high temperatures, the mixing entropy term in Equation 4.102 causes the molar Gibbs free energy of the disordered phase to be concave up, and to lie below the molar Gibbs free energies of all of the ordered phases. A disordered InGaAs alloy cannot, for any composition, lower its molar Gibbs free energy either by phase-separating into disordered InAs and GaAs rich clusters or into ordered stoichiometric phases.

At low temperatures, the mixing entropy term becomes small, and is no longer sufficient to bring the molar Gibbs free energy of the disordered phase below those of the ordered phases. Therefore, a uniform disordered alloy can decrease its molar Gibbs free energy by phase-separating into either a combination of two ordered phases or a combination of an ordered phase and an InAs or GaAs rich disordered phase.

In addition, however, the (positive) mixing enthalpy term in Equation 4.102 causes the molar Gibbs free energy of the disordered phase to now be concave enough down that it becomes, near its endpoint compositions, lower than those of all of the ordered phases. Therefore, the ordered phases are themselves unstable with respect to phase separation into pure InAs and GaAs disordered phases. A miscibility gap opens up that destabilizes the ordering.

Coherent Phase Equilibria

Consider now the coherent case for which $x_{\text{sub}} = \text{constant}$. Then, the molar Gibbs free energy does not simplify to Equation 4.102, and the full Equation 4.100 must be used.

In this case, at high temperatures, *both* the mixing entropy term and the elastic energy term cause the molar Gibbs free energy of the disordered phase to be concave up. Because of the contribution from both terms, the molar Gibbs free energy remains concave up to lower temperatures, and of course the miscibility gap shifts to lower temperatures.

At low temperatures, the mixing entropy term becomes small, and again is no longer sufficient to bring the molar Gibbs free energy of the disordered phase below those of the ordered phases. Therefore, a uniform disordered alloy can again decrease its molar Gibbs free energy by phase-separating

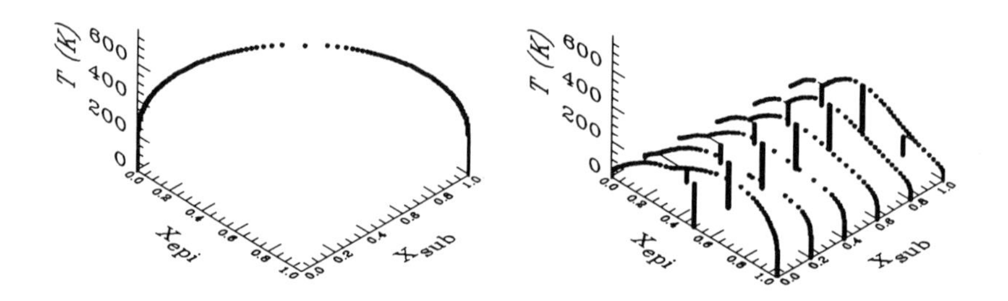

Figure 4.17: Pseudobinary x_{epi}-x_{sub}-T phase diagrams of $In_{1-x}Ga_xAs$. In the usual, "incoherent" diagram on the left, variations in x_{epi} are accompanied by identical variations in the effective composition of the substrate, x_{sub}. In the "coherent" diagram on the right, variations in x_{epi} occur at constant x_{sub}.

into either a combination of two ordered phases or a combination of an ordered phase and an InAs or GaAs rich disordered phase.

In this case, however, the (positive) mixing enthalpy term is countered by the elastic energy term, and the molar Gibbs free energy of the disordered phase remains concave up. Therefore, the ordered phases remain lower in energy than the disordered phase, even near its endpoint compositions. The ordered phases are therefore stable with respect to phase separation into (nearly) pure InAs and (nearly) pure GaAs disordered phases. Coherency suppresses the miscibility gap. Then, if ordered phases are present, as in this example, coherency stabilizes them.[34] If, however, ordered phases are not present, then a uniform disordered alloy will persist to lower temperatures (perhaps even to 0 K) than in the incoherent case.[35]

Suggested Reading

1. W.A. Harrison, *Electronic Structure and the Properties of Solids* (W.H. Freeman, San Francisco, 1980).

2. E.A. Guggenheim, *Thermodynamics* (North-Holland, Amsterdam, 1959).

[34] C.P. Flynn, "Strain-assisted epitaxial growth of new ordered compounds," *Phys. Rev. Lett.* **57**, 599 (1986).

[35] G.B. Stringfellow, "Spinodal decomposition and clustering in III/V alloys," *J. Electron. Mater.* **11**, 903 (1982); and M. Quillec, C. Daguet, J.L. Benchimol, and H. Launois, "InGaAsP alloy stabilization by the InP substrate inside an unstable region in liquid phase epitaxy," *Appl. Phys. Lett.* **40**, 325 (1982).

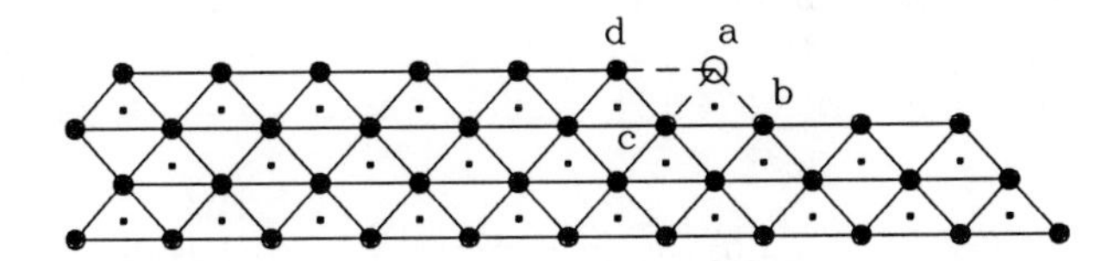

Figure 4.18: Intermediate stage of building of a 2D lattice of face-centered triangles.

3. T. Hill, *Introduction to Statistical Thermodynamics* (Addison Wesley, Reading, MA, 1960).

4. G.B. Stringfellow, *Organometallic Vapor-Phase Epitaxy: Theory and Practice* (Academic Press, Boston, 1989).

5. A. Zunger and D.M. Wood, "Structural phenomena in coherent epitaxial solids," *J. Cryst. Growth* **98**, 1 (1989).

Exercises

1. Calculate the distortion energies of the cluster shown in Figure 4.1 for (a) As atoms forced onto VCA and CRA positions and (b) for an As atom at its actual equilibrium position.

2. From Figure 4.13, it can be seen that there is a greater tendency toward phase decomposition for strained but coherent epitaxy of InGaAs on a GaAs substrate than on an InAs substrate. Why?

3. What is the limiting value of the order parameter in the pair approximation of the CVM [Equation 4.46] when $2u_1 \gg u_0 + u_2$, i.e., when AA and BB pairs are greatly preferred over AB pairs?

4. Derive Equation 4.48 for the temperature dependence of the order parameter in the pair approximation of the CVM.

5. Calculate the entropy of the two-dimensional face-centered triangular lattice shown in Figure 4.18 in the point, pair, and triangle CVM approximations.

6. Construct a ball-and-stick 3D zincblende lattice and identify the tetrahedra, triangles, and pairs associated with adding point a in Figure 4.8.

7. Derive Equation 4.83.

8. Draw envelopes of minimum molar Gibbs free energies for the various 2D cuts shown at the bottom of Figure 4.13, and identify the composition ranges within which various phases or phase mixtures are stable.

9. Construct phase decomposition scenarios for which coherency energies either depend on, or are independent of, the amounts of the different phases present.

10. Starting from Equation 4.100, derive expressions for the chemical potentials of InAs and GaAs in coherent and incoherent epitaxial InGaAs.

11. Using Equations 3.59 and 4.100, derive an expression for how much the vapor pressure of Ga over coherent $In_{1-x}Ga_xAs$ lattice-matched to InP differs from that over incoherent $In_{1-x}Ga_xAs$.

Chapter 5

Coherency and Semi-coherency

In Chapter 4, we described how the thermodynamics of epitaxial alloy films depend on whether those films are coherent or not with their underlying substrate. Films that are coherent often tend to form ordered compounds at certain stoichiometric compositions, while films that are not often tend to separate into their pure-component "endpoint" phases. Coherency with an underlying substrate is thus a crucial determinant of the compositional integrity of alloy films.

Coherency is also a crucial determinant of other properties of alloy films. Consider, e.g., an epitaxial layer whose bulk lattice parameter differs from that of its substrate. On the one hand, if the layer is coherent with its substrate, it will be mechanically strained, and its electronic and opto-electronic properties will be modified through strain-induced changes in electronic band structure.[1] On the other hand, if the layer is not coherent with its substrate, then structural defects must be present, some of which degrade significantly the performance of semiconductor devices.

In this chapter, we discuss the conditions under which coherency between film and substrate can be maintained. In particular, we will focus on the transition from coherency to "semi-coherency." A coherent interface is one that is crystallographically perfect, and that separates epitaxial and substrate atoms in perfect "registry" with each other. If the bulk lattice parameters of the epitaxial layer and the substrate differ, then the epitaxial layer accommodates by developing in-plane strain. A semi-coherent

[1] G.C. Osbourn, "Strained-layer superlattices from lattice mismatched materials," *J. Appl. Phys.* **53**, 1586 (1982).

interface, in contrast, is one for which the registry between epitaxial and substrate atoms is punctuated by occasional localized regions of disregistry, i.e., by dislocations. These localized regions of disregistry compensate for lattice parameter misfit between the epitaxial layer and the substrate, allowing the in-plane strain of the epitaxial layer to relax.

We begin, in Section 5.1, by discussing the energies associated with those two kinds of interfaces. The energy associated with a coherent interface is due solely to "coherency strain" in the epitaxial film, and increases linearly with film thickness. The energy associated with a semi-coherent interface is due partly to coherency strain and partly to "misfit" dislocations at the interface. Much of the energy of the misfit dislocations is due to the disregistered atoms at the dislocation core, and is independent of film thickness. Therefore, thin coherent films will tend to have lower energies than thin semi-coherent films, but thick coherent films will tend to have higher energies than thick semi-coherent films.[2]

As a consequence, in the early stages of film growth, an epitaxial film will usually be coherent with its substrate. Only when the film becomes thick enough will it tend to become semi-coherent with the substrate, and even then, it may not actually become semi-coherent. To become semi-coherent, misfit dislocations must be created at the film/substrate interface, but that creation may be impeded by kinetic barriers. Therefore, in Section 5.2 we discuss the forces, or "excess stresses," acting to create misfit dislocations, and in Section 5.3 we describe how an understanding of those forces can be used to develop semi-empirical macroscopic descriptions of the overall kinetics of strain relaxation.

Note that this chapter deals only with the most common form of heteroepitaxy, in which the film has the same crystal structure as the substrate. Then, provided the lattice parameters of the film and substrate are not too mismatched, epitaxy will occur, and the orientation of the film will mimic that of the substrate. From a practical point of view, we need only be concerned with predicting the conditions under which the film will be coherent or semi-coherent. This chapter also deals only with the simplest form of heteroepitaxy, in which the film grows as layers, rather than as islands (see, e.g., Exercise 2 and Chapter 6).

We emphasize, though, that a deposited film need not have the same crystal structure as the substrate.[3] In such cases, it is not always easy to predict (1) whether epitaxy will even occur at all and (2) even if it does, what the orientation relationship will be between the film and the substrate.

---

[2]F.C. Frank and J.H. van der Merwe, "One-dimensional dislocations. II. Misfitting monolayers and oriented overgrowth," *Proc. R. Soc. London* **A198**, 216 (1949).

[3]E. Grünbaum, "List of epitaxial systems," in *Epitaxial Growth*, J.W. Matthews, Ed. (Academic Press, New York, 1975), pp. 611-673.

These two questions are among the most basic in the science of epitaxy, and have been studied for nearly a century, beginning with the work of Barker[4] and Royer.[5] However, they are also exceedingly difficult questions that are far from being fully answered.

From a purely crystallographic point of view, one anticipates that those orientation relationships will be favored for which the three dimensional film and substrate lattices coincide most closely at the two-dimensional interface.[6] For example, such purely crystallographic considerations are evidently responsible[7] for what are known as the Nishiyama-Wasserman[8] and Kurdjumov-Sachs[9] orientation relationships between fcc and bcc crystals found both in solid-phase precipitation reactions[10] as well as in vapor-phase epitaxy.[11]

However, it will not always be sufficient to consider the crystallography of the known equilibrium bulk phases. Occasionally, it will be possible to epitaxially stabilize crystal phases which are not normally stable in bulk form.[12] Elemental tin, e.g., adopts a metastable diamond structure when deposited epitaxially on the (001) surfaces of InSb and CdTe.[13]

Moreover, the epitaxial film may also be chemically different from the substrate, and so the orientation relationship will depend not just on crystallography, but on bond chemistry as well. For these reasons, an under-

---

[4]T.V. Barker, "Contributions of the theory of isomorphism based on experiments on the regular growths of crystals of one substance on those of another," *J. Chem. Soc. Trans.* **89**, 1120 (1906).

[5]L. Royer, "Recherches expérimentales sur l'épitaxie ou orientation mutuelle de cristaux d'espèces différentes," *Bull. Soc. Franc. Mineral* **51**, 7 (1928).

[6]R.W. Balluffi, A. Brokman, and A.H. King, "CSL/DSC lattice model for general crystal-crystal boundaries and their line defects," *Acta. metall.* **30**, 1453 (1982); and A. Zur and T.C. McGill, "Lattice match: an application to heteroepitaxy," *J. Appl. Phys.* **55**, 378 (1984).

[7]R. Ramirez, A. Rahman, and I.K. Schuller, "Epitaxy and superlattice growth," *Phys. Rev.* **B30**, 6208 (1984).

[8]Z. Nishiyama, "X-ray investigation of the mechanism of the transformation from face-centred cubic lattice to body-centred cubic," *Sci. Rep. Tohoku Univ.* **23**, 638 (1934); and G. Wasserman, *Arch. Eisenhuettenwes.* **126**, 647 (1933).

[9]G. Kurdjumov and G. Sachs, "Über den Mechanismus der Stahlhärtung," *Z. Phys.* **64**, 325 (1930).

[10]U. Dahmen, "Orientation relationships in precipitation systems," *Acta Metall.* **30**, 63 (1982).

[11]L.A. Bruce and H. Jaeger, "Geometric factors in f.c.c. and b.c.c. metal-on-metal epitaxy III. The alignments of (111) f.c.c.–(110) b.c.c. epitaxed metal pairs," *Phil. Mag.* **A38**, 223 (1978).

[12]R. Bruinsma and A. Zangwill, "Structural transitions in epitaxial overlayers," *J. Physique* **47**, 2055 (1986).

[13]R.F.C. Farrow, D.S. Robertson, G.M. Williams, A.G. Cullis, G.R. Jones, I.M. Young, and P.N.J. Dennis, "The growth of metastable, heteroepitaxial films of α-Sn by metal beam epitaxy," *J. Cryst. Growth* **54**, 507 (1981).

standing of orientation relationships in epitaxy is an enormously complicated ongoing area of research that will not be treated here.

5.1 Energetics of Misfit Accommodation

Let us begin, in this section, by discussing the *energetics* of epitaxial films attached through coherent or semi-coherent interfaces to substrates with different lattice parameters. We discuss first, in Subsection 5.1.1, the coherency strains and energies associated with the epitaxial films. Then, we discuss, in Subsection 5.1.2, the strain fields and energies associated with misfit dislocations at the interfaces between the epitaxial films and their substrates, with particular emphasis on face-centered cubic (fcc) and diamond lattices. Finally, we discuss, in Subsection 5.1.3, the dependence of both kinds of energies on misfit dislocation density. Minimizing the sum of the two energies with respect to misfit dislocation density determines how the overall misfit is partitioned, in equilibrium, between coherency strain and misfit dislocation density. We will find that for thin, low misfit films, energy is minimized when the misfit dislocation density is zero. For thick, high misfit films, however, energy is minimized when the misfit dislocation density is nonzero.[14]

5.1.1 Coherency Strain

Let us start, in this subsection, by discussing the strain energy associated with epitaxial films that are coherent with their substrates. In particular, consider the simplest strained heterostructure: a single, thin, planar layer of one material and a thick substrate of a different material. As illustrated in Figure 5.1, in the absence of a connection between the two materials, each is unstrained and will adopt its own bulk lattice parameter — either $a_{\mathrm{epi,o}}$ or a_{sub}. Note that we neglect changes in the lattice parameter of a free-standing film due to surface stresses, changes that may be important for very thin films.[15]

Suppose we exert a compressive in-plane force on the epitaxial layer and an equal but opposing tensile in-plane force on the substrate. Then, the in-plane lattice parameter of the epitaxial film will shrink and that of the substrate will grow. If the bulk lattice parameter of the epitaxial layer were larger than that of the substrate, as is the case in Figure 5.1, then the

[14] J.H. van der Merwe, "Crystal interfaces. Part II. Finite overgrowths," *J. Appl. Phys.* **34**, 123 (1963).

[15] R.C. Cammarata and K. Sieradzki, "Surface stress effects on the critical film thickness for epitaxy," *Appl. Phys. Lett.* **55**, 1197 (1989).

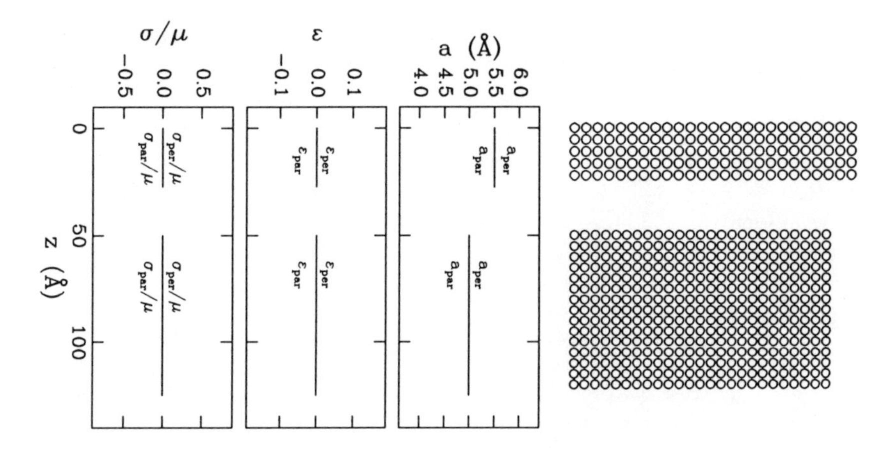

Figure 5.1: Hypothetical simple cubic epitaxial layer and substrate with bulk lattice parameters 5.0 and 5.5 Å, respectively. The epitaxial layer is imagined to be disconnected from the substrate, and so is free to adopt its bulk lattice parameter. As a consequence, it is both unstrained and unstressed.

two in-plane lattice parameters can eventually be made to match, and the epitaxial layer can be joined coherently to the substrate.

Note that if the substrate is much thicker than the epitaxial layer, then it will experience a much lower average in-plane stress than will the epitaxial layer, and its lattice parameter will change much less. Therefore, we make the usual approximation that *all* of the lattice parameter misfit is accommodated by strain in the epitaxial layer, rather than in the substrate. In the general case, though, partitioning of lattice parameter misfits between film and substrate, and even between layers within a multilayered film, must be taken into account.[16]

Note also that the Hooke's law energies associated with straining the epitaxial layer and substrate are each proportional to thickness and to the square of the change in lattice parameter. Since the changes in lattice parameter are proportional to applied stress, which is inversely proportional to thickness, the Hooke's law energies are themselves inversely proportional to thickness. Therefore, we can also make the approximation that all of the strain energy associated with coherently joining the epitaxial layer to the substrate is in the epitaxial layer, rather than in the substrate. In other words, just as we saw in Section 4.1.2, most of the energy associated with

[16] Z.C. Feng and H.D. Liu, "Generalized formula for curvature radius and layer stresses caused by thermal strain in semiconductor multilayer structures," *J. Appl. Phys.* **54**, 83 (1983).

coupled spring systems is stored in the weaker and more deformed spring.

To calculate the actual strain energy in a fully coherent epitaxial layer, we follow the discussion in Section 4.2, in which a generalized Hooke's law was written in terms of the elastic coefficients C_{ij}. That law is also commonly written, for cubic materials, in terms of Poisson's ratio, ν (defined as the negative of the ratio between lateral and longitudinal strains under uniaxial longitudinal stress), and the shear modulus, μ (defined as the ratio between applied shear stress and shear strain under pure shear):

$$\begin{pmatrix} \epsilon_x \\ \epsilon_y \\ \epsilon_z \end{pmatrix} = \frac{1}{2\mu(1+\nu)} \begin{pmatrix} 1 & -\nu & -\nu \\ -\nu & 1 & -\nu \\ -\nu & -\nu & 1 \end{pmatrix} \begin{pmatrix} \sigma_x \\ \sigma_y \\ \sigma_z \end{pmatrix}. \tag{5.1}$$

The relationships between the C_{ij}, ν, and μ are

$$\begin{aligned} C_{11} &= 2\mu\left(\frac{1-\nu}{1-2\nu}\right) \\ C_{12} &= 2\mu\left(\frac{\nu}{1-2\nu}\right). \end{aligned} \tag{5.2}$$

The shear modulus itself is related to the modulus of elasticity, E, by $2\mu = E/(1+\nu)$.

For concreteness, let us assume, as is common, that the epitaxial film and its substrate are not only cubic, but are oriented along one of the $\langle 100 \rangle$ cubic symmetry directions.[17] Then, the in-plane strains are symmetric and can be taken to be along the x and y axes. If we denote in-plane quantities as "parallel," and out-of-plane quantities as "perpendicular," then we can write

$$\begin{pmatrix} \epsilon_\parallel \\ \epsilon_\perp \end{pmatrix} = \frac{1}{2\mu(1+\nu)} \begin{pmatrix} 1-\nu & -\nu \\ -2\nu & 1 \end{pmatrix} \begin{pmatrix} \sigma_\parallel \\ \sigma_\perp \end{pmatrix}, \tag{5.3}$$

which is just the inverse of Equation 4.20.

Equation 5.3 contains two known and two unknown quantities. The first known quantity is the parallel strain, $\epsilon_\parallel$, which is determined by the lattice mismatch. The second known quantity is the perpendicular stress, $\sigma_\perp$, which, since the epitaxial layer is free to expand vertically, vanishes. Therefore, Equation 5.3 determines the two unknown quantities — the parallel stress, $\sigma_\parallel$, and perpendicular strain, $\epsilon_\perp$ — in terms of $\epsilon_\parallel$ only:

$$\sigma_\parallel = 2\mu\left(\frac{1+\nu}{1-\nu}\right)\epsilon_\parallel \tag{5.4}$$

[17] Otherwise, more general expressions are required. See, e.g., J.P. Hirth, "On dislocation injection into coherently strained multilayer structures," *S. Afr. J. Phys.* **9**, 72 (1986).

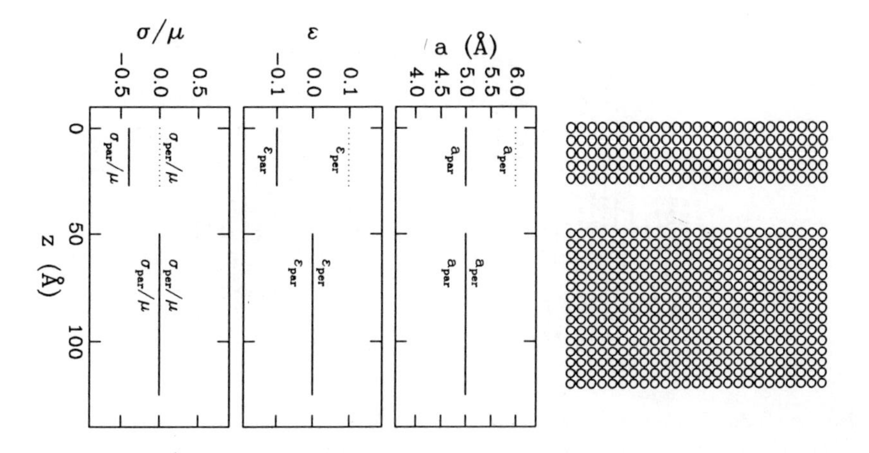

Figure 5.2: Hypothetical simple cubic epitaxial layer and substrate with bulk lattice parameters 5.5 and 5.0 Å, respectively. The epitaxial layer is still imagined disconnected from the substrate, but has been strained in a direction parallel to the interface so that its parallel lattice parameter matches that of the substrate. As a consequence, it develops both a parallel (in-plane) stress and a perpendicular (out-of-plane) strain.

$$\epsilon_\perp = \frac{-2\nu}{1-\nu}\epsilon_\parallel. \tag{5.5}$$

As illustrated in Figure 5.2, if the epitaxial layer is strained in a direction parallel to the interface so that its parallel lattice parameter matches that of the substrate, then it must develop a parallel stress. It also develops a perpendicular strain, in the same direction as that which would preserve unit-cell volume. In fact, if $\epsilon_\perp$ were exactly $-2\epsilon_\parallel$, or if $2\nu/(1-\nu)$ were exactly 2, then unit-cell volume would be exactly preserved. Poisson's ratio, however, lies in the range 0.25–0.35 for most materials, so that $2\nu/(1-\nu)$ is actually approximately 1, and unit-cell volume is only approximately conserved.

The "coherency" energy associated with strain in the epitaxial layer can now be calculated, per unit area, to be

$$u_{\rm coh} = \frac{1}{2}h\left(2\sigma_\parallel\epsilon_\parallel + \sigma_\perp\epsilon_\perp\right) = 2\mu\left(\frac{1+\nu}{1-\nu}\right)h\epsilon_\parallel^2, \tag{5.6}$$

where h is the thickness of the film.

In an epitaxial film composed of multilayers each with a different lattice parameter, the multilayer coherency energy will just be a sum of (or integral

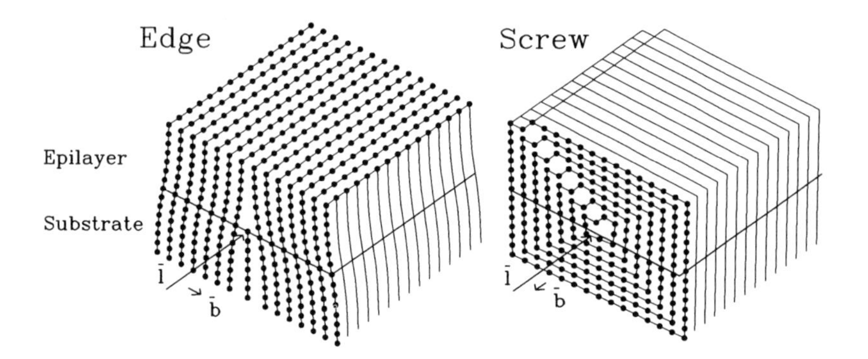

Figure 5.3: Pure edge (left) and pure screw (right) dislocations lying in an interface separating an epitaxial film from its substrate.

over) expressions such as Equation 5.6 for each layer:

$$u_{\text{coh}} = 2\mu \left(\frac{1+\nu}{1-\nu} \right) \sum_i h_i \epsilon_{i,\|}^2, \tag{5.7}$$

where h_i and $\epsilon_{i,\|}$ are the thicknesses and parallel strains of the ith layer.

5.1.2 Misfit Dislocations

In Subsection 5.1.1, we discussed the strain energy associated with epitaxial films that are coherent with their substrates. In this subsection, we discuss the energy associated with epitaxial films that are semi-coherent with their substrates. In particular, consider a single-layer heterostructure in which the perfect coherent registry between the epitaxial film and substrate is broken by a localized region of "disregistry." In the simplest case, as illustrated in the left half of Figure 5.3, the disregistry might consist of a half plane missing from the epitaxial film. Physically, we might imagine that the half plane had been "squeezed" upward out of the epitaxial film by a compressive coherency stress, thereby relieving some (or all) of that coherency stress.

Geometrically, the disregistry can be thought of as formed by making what is known as a "Volterra" cut in the epitaxial film perpendicular to the interface, removing a plane of atoms, and then rejoining the remaining crystal by inwardly collapsing atoms in the adjacent planes. The disregistry can then be seen to be equivalent to a negative edge dislocation along the line labeled $\bar{l}$ in the left half of Figure 5.3, with Burgers vector along the line labeled $\bar{b}$. In a sense, such interface dislocations act to relieve the coherency strain in the epitaxial layer by concentrating the lattice misfit into localized regions of disregistry confined mainly to the interface.

Note that pure edge dislocations such as this, whose Burgers vectors are both perpendicular to the dislocation line *and* in the plane of the interface, are the most efficient means for relieving coherency strain. Screw dislocations such as that illustrated in the right half of Figure 5.3, whose Burgers vectors are parallel to the dislocation line, do not relieve coherency strain. Likewise, edge dislocations whose Burgers vectors are perpendicular both to the dislocation line and to the plane of the interface, do not relieve coherency strain. Therefore, in the general case of "mixed" dislocations, having both edge and screw character, only that component of the Burgers vector that is both "edgelike" *and* in the plane of the interface acts to relieve coherency strain. In particular, if, as illustrated in Figure 5.23 on page 195, λ is the angle between (a) the Burgers vector and (b) the direction that is both normal to the dislocation line and that lies within the plane of the interface, then only the component,

$$b_{\mathrm{edg},\|} \equiv b \cos \lambda, \tag{5.8}$$

acts to relieve lattice misfit.

Also note that dislocations with partial or full edge character move most easily by gliding *within* the plane containing both the dislocation line and its Burgers vector. The pure edge dislocation illustrated in Figure 5.3, e.g., will move most easily within the interface between the epilayer and the substrate. Therefore, such a dislocation, if created at the free surface, would be unable to glide to the interface between the epilayer and the substrate. Instead, it would be constrained to glide parallel to that interface.

To be *practically* effective at relieving misfit strain, then, dislocations must usually have some component of their Burgers vector out of the interface. Otherwise, they must move by "climbing" *out* of the plane containing both the dislocation line and its Burgers vector. Such motion requires the creation or annihilation of vacancies at the dislocation core, and hence a diffusive flux of vacancies either away from or toward the dislocation core. For example, to move the edge dislocation illustrated in the left half of Figure 5.3 down from the interface by one lattice spacing, a row of vacancies must be removed from the dislocation core. Such vacancy removal ultimately requires diffusion away from the core, which usually only becomes significant at fairly high temperatures.[18]

[18] E.A. Fitzgerald, P.D. Kirchner, R.E. Proano, G.D. Pettit, J.M. Woodall and D.G. Ast, "Totally relaxed Ge_xSi_{1-x} layers with low threading dislocation densities grown on Si substrates," *Appl. Phys. Lett.* **59**, 811 (1991).

While acting to relieve misfit strain, interface dislocations also cost energy, due to the disruption in bonding associated with the disregistered atoms at their core and to the long-range elastic stress and strain fields away from their core. For pure edge and screw dislocations, the energies per unit length associated with the elastic stresses and strains in a cylindrical ring surrounding a long straight dislocation core can be shown, in a continuum model, to be approximately

$$\begin{aligned} U_{\text{edg}} &= \frac{\mu b_{\text{edg}}^2}{4\pi(1-\nu)} \ln(R/r_o) \\ U_{\text{scr}} &= \frac{\mu b_{\text{scr}}^2}{4\pi} \ln(R/r_o), \end{aligned} \tag{5.9}$$

where r_o and R are the inner and outer radii of the cylinder. The energies associated with the disregistered core atoms, however, are difficult to determine. Instead, it is common to adjust the inner "cutoff" radius r_o so that the core energies are included in Equations 5.9. In practice, a value of $r_o = b/4$ for covalent semiconductors is often used.

For dislocations having mixed character, the energy is the sum of Equations 5.9, with the edge and screw components of the Burgers vectors used accordingly. If, as illustrated in Figure 5.23 on page 195, β is the angle between the Burgers vector and the dislocation line, then the edge component is $b \sin\beta$ and the screw component is $b\cos\beta$. Therefore,

$$\begin{aligned} U_{\text{dis}} &= \frac{\mu b^2}{4\pi}\left(\frac{\sin^2\beta}{1-\nu} + \cos^2\beta\right) \ln(4R/b) \\ &= \frac{\mu b^2}{4\pi}\left(\frac{1-\nu\cos^2\beta}{1-\nu}\right) \ln(4R/b). \end{aligned} \tag{5.10}$$

Note that, because of the long-range nature of the elastic stresses and strains, the dislocation energy diverges logarithmically with the radius of the outer radius of the cylinder. Therefore, a dislocation embedded in an infinite crystal has infinite energy. In fact, the long-range elastic stresses and strains are always disrupted (and bounded) either by free surfaces or by the stress and strain fields of neighboring dislocations.

For example, if a free surface at $z = 0$ is placed a distance h away from a dislocation at $z = h$, the normal and shear stress components acting on the surface must vanish, because the surface is free to expand outward or contract inward. The effect of the surface can be accounted for approximately by placing an "imaginary" dislocation of the opposite sign at $z = -h$, thereby largely cancelling the long-range stress field at distances much greater than h from the dislocation core. The energy associated with

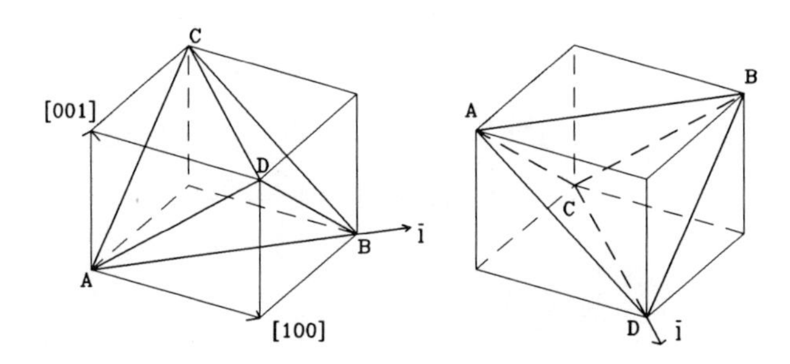

Figure 5.4: Dislocation lines lying along [110] (left) and [1$\bar{1}$0] (right) directions. The edges of the tetrahedra inscribed within each unit cube represent the possible directions of the Burgers vectors for each of those dislocations.

a dislocation a distance h from a free surface is therefore approximately

$$U_{\text{dis}} = \frac{\mu b^2}{4\pi}\left(\frac{1-\nu\cos^2\beta}{1-\nu}\right)\ln(4h/b). \tag{5.11}$$

Note that the energy is proportional to b^2, because the strains around the dislocations are proportional to b, and the energy is proportional to the square of the strains. Therefore, dislocations with shorter Burgers vectors will be more common than those with longer Burgers vectors. For this reason, the most common Burgers vectors in fcc-based diamond and zincblende lattices are of the $\frac{1}{2}\langle 110\rangle$ type, since these are the shortest lattice vectors in these crystals.[19] Since there are six possible $\langle 110\rangle$ directions, there are six possible directions for the Burgers vectors. These six directions are the edges of the tetrahedra shown in Figure 5.4.[20]

Consider, for example, misfit dislocations lying along either [110] or [1$\bar{1}$0] directions, as shown in the left and right halves, respectively, of Figure 5.4. For (001) oriented fcc-lattice-based epitaxial films, these two dislocation line directions are the most common, as they lie both in the (001) interface plane as well as in one of the close-packed {111} slip planes within which dislocations move most readily. Dislocations having these line directions

[19] Dissociation into "partial dislocations" having shorter Burgers vectors separated by stacking faults is also possible.

[20] N. Thompson, "Dislocation nodes in face-centred cubic lattices," *Proc. Phys. Soc.* **B66**, 481 (1955).

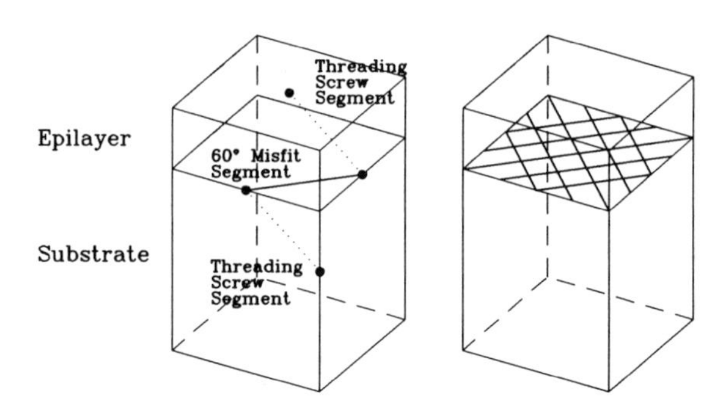

Figure 5.5: Left: $[\bar{1}01]$ threading screw dislocation segments and $[110]$ 60° misfit dislocation segment with Burgers vector along $[\bar{1}01]$ direction. Right: crossed grid of two arrays of misfit dislocations along the $[\bar{1}10]$ and $[110]$ directions.

can be one of three types, depending on the directions of their Burgers vectors. If the Burgers vector of the $\bar{l} = [110]$ dislocation illustrated in the left half of Figure 5.4 lies along the line A–B, parallel to $\bar{l}$, then it is screw in character. If its Burgers vector lies along the line C–D, perpendicular to $\bar{l}$, then it is edge in character. If its Burgers vector lies along any of the four other lines, A–C, A–D, B–C or B–D, at 60° to $\bar{l}$, then it is a "mixed" 60° dislocation. Likewise, the $\bar{l} = [1\bar{1}0]$ dislocation illustrated in the right half of Figure 5.4 will either be screw, edge, or 60° mixed, depending on the direction of its Burgers vector.

An example of a commonly observed dislocation configuration is illustrated in the left half of Figure 5.5. An $\bar{l} = [\bar{1}01]$ screw dislocation with $\bar{b} = [\bar{1}01]/2$ is shown threading up diagonally from the substrate into the epilayer. Just at the epilayer/substrate interface, the dislocation has bent over to form a misfit dislocation segment with $\bar{l} = [110]$. Since Burgers vectors must be preserved along the length of any particular dislocation, the misfit dislocation segment is a 60° dislocation with $\cos\beta = \hat{l}\cdot\hat{b} = \cos 60° = 1/2$. A plan-view transmission electron micrograph of a crossed grid of such misfit dislocations is shown in Figure 5.6.

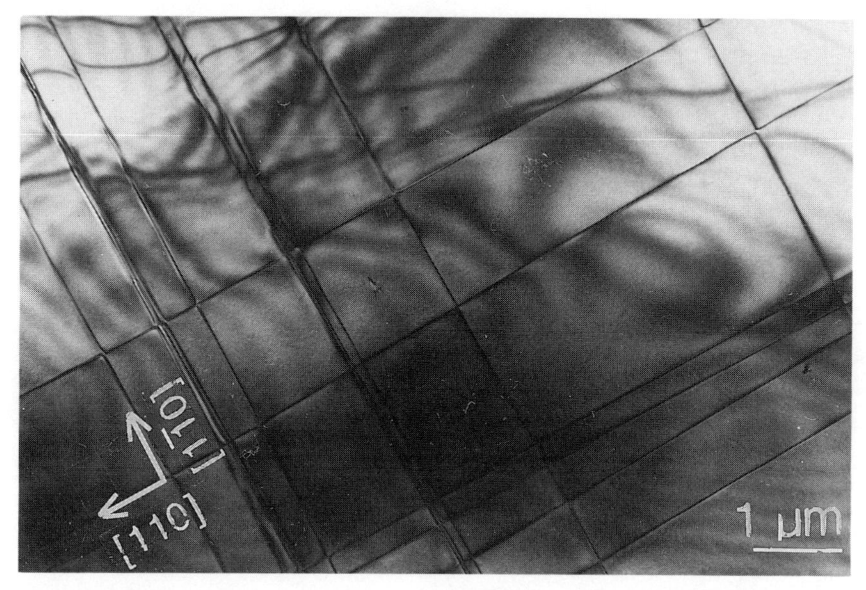

Figure 5.6: Bright-field plan-view transmission electron micrograph of the interface between a 200-nm $Si_{0.9}Ge_{0.1}$ layer grown on a Si (001) substrate.[a] The misfit dislocations are arranged in a crossed grid running along the $\langle 110 \rangle$ directions within the (001) interface.

[a]Y. Fukuda, Y. Kohama, M. Seki, and Y. Ohmachi, "Misfit dislocation structures at MBE-grown $Si_{1-x}Ge_x$/Si interfaces, *Jpn. J. Appl. Phys.* **27**, 1593 (1988).

5.1.3 Equilibrium Strains and Dislocation Densities

In Subsection 5.1.1 we discussed the strain energy cost associated with a perfectly coherent interface, and in Subsection 5.1.2 we discussed the dislocation energy cost associated with a semi-coherent interface. In this subsection, we ask which of the two interfaces costs the least energy, and hence will be thermodynamically preferred. To answer this question, let us calculate how the two kinds of energies depend on misfit dislocation density. For concreteness, we assume that the semi-coherent interface is composed, as illustrated in the right half of Figure 5.5, of a crossed-grid of two identical arrays of dislocations, each having a linear density of ρ_{md}.

For a fully coherent interface, for which $\rho_{\mathrm{md}} = 0$, there is only the coherency strain energy, which we have already calculated to be $2\mu[(1+\nu)/(1-\nu)]hf^2$, where f is the lattice parameter misfit between the epitaxial layer and the substrate. For a semi-coherent interface, for which $\rho_{\mathrm{md}} > 0$, the misfit will be partially taken up by localized regions of disregistry, thereby decreasing the coherency strain energy, but increasing the misfit dislocation energy.

To see how the coherency strain energy will decrease with misfit dislocation density, we note that in one dimension, the misfit taken up by dislocations, f_{dis}, is the lattice displacement parallel to the interface per dislocation, $b_{\text{edg},\|}$, divided by the spacing between dislocations, $1/\rho_{\text{md}}$. In other words,

$$f_{\text{dis}} = \rho_{\text{md}} b_{\text{edg},\|}. \tag{5.12}$$

Therefore, the dislocation density that would relieve all the misfit in one dimension would be $\rho_{\text{md}} = f/b_{\text{edg},\|}$. If, on average, the strain in the epitaxial film decreases linearly with the dislocation density, then

$$\epsilon_{\|} \approx f - f_{\text{dis}} = f - \rho_{\text{md}} b_{\text{edg},\|}. \tag{5.13}$$

The dependence of the coherency strain energy on misfit dislocation density can then be written as

$$u_{\text{coh}} = 2\mu \left(\frac{1+\nu}{1-\nu} \right) h \left(f - \rho_{\text{md}} b_{\text{edg},\|} \right)^2. \tag{5.14}$$

As indicated by the dashed lines in the left and right halves of Figure 5.7, the coherency strain energy depends parabolically on dislocation density, and vanishes when $\rho_{\text{md}} = f/b_{\text{edg},\|}$.

At the same time, the energy associated with the dislocations themselves will increase as the misfit dislocation density increases. For most applications, it is sufficient to approximate the energy associated with each of the two dislocation arrays to be the dislocation density times the energy of an isolated dislocation, or

$$u_{\text{dis}} \approx \rho_{\text{md}} \frac{\mu b^2}{4\pi} \left(\frac{1-\nu\cos^2\beta}{1-\nu} \right) \ln\left(\frac{4h}{b} \right). \tag{5.15}$$

This linear dependence of the dislocation array energy on dislocation density is shown as the dotted lines in the left and right halves of Figure 5.7.

For more precise calculations, however, we note that interactions between dislocation should be taken into account. The reason is that when the dislocation spacing is less than the film thickness, the stress fields of individual dislocations are not fully screened from each other by the free surface, and mediate an "interaction" between them.[21]

---

[21] See, e.g., J.P. Hirth and X. Feng, "Critical layer thickness for misfit dislocation stability in multilayer structures," *J. Appl Phys.* **67**, 3343 (1990); and J.R. Willis, S.C. Jain, and R. Bullough, "The energy of an array of dislocations: implications for strain relaxation in semiconductor heterostructures," *Philos. Mag.* **A62**, 115 (1990).

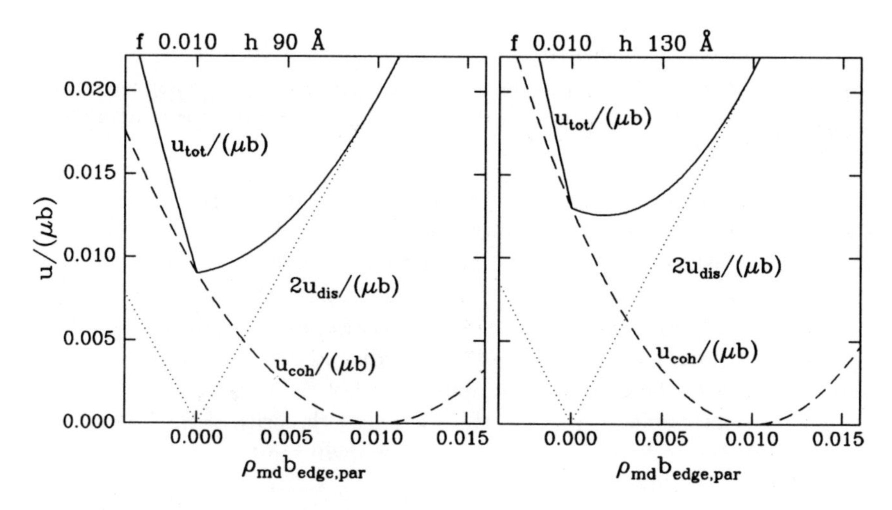

Figure 5.7: Areal energy densities, normalized by the product of the shear modulus and the Burgers vector, as a function of the misfit taken up by dislocations, $\rho_{md} b_{edg,\|}$. The dotted lines represent the energies of two dislocation arrays; the dashed lines represent the coherency strain energies; the full lines represent the sum. The film on the left is thin enough that it is stable when it is coherent with the substrate; the film on the right is thicker and is stable when it is semi-coherent with the substrate.

The total areal energy density is the sum of the areal energy densities associated with the coherency strain and *both* of the dislocation arrays, or

$$\begin{aligned} u_{tot} &= u_{coh} + 2u_{dis} \\ &= 2\mu \left(\frac{1+\nu}{1-\nu}\right) h \left(f - \rho_{md} b_{edg,\|}\right)^2 \\ &\quad + \rho_{md} \frac{\mu b^2}{2\pi} \left(\frac{1 - \nu\cos^2\beta}{1-\nu}\right) \ln\left(\frac{4h}{b}\right). \end{aligned} \tag{5.16}$$

This dependence of u_{tot} on ρ_{md} displays two distinct kinds of behavior, as illustrated in Figure 5.7. For thin, low-misfit films, the total energy is minimum at $\rho_{md} = 0$. Misfit dislocations cost more energy than is regained by release of coherency strain. For thick, high-misfit films, however, the total energy is minimum at $\rho_{md} > 0$. The introduction of *some* misfit dislocations costs less energy than is regained by release of coherency strain.

Mathematically, these two kinds of behaviors arise according to whether the energy associated with either of the dislocation arrays, $u_{tot}/2$, increases or decreases for an incremental increase in ρ_{md} from $\rho_{md} = 0$. In other words, according to whether

$$\left[\frac{\partial u_{\mathrm{tot}}}{2\partial \rho_{\mathrm{md}}}\right]_{\rho_{\mathrm{md}}=0} = -2\mu\left(\frac{1+\nu}{1-\nu}\right)hb_{\mathrm{edg},\|}f + \frac{\mu b^2}{4\pi}\left(\frac{1-\nu\cos^2\beta}{1-\nu}\right)\ln(4h/b) \tag{5.17}$$

is greater than or less than zero. On the one hand, if it is greater than zero, then the change in energy upon introducing the first few misfit dislocations is positive. Misfit dislocations will *not* tend to form, and the fully strained, coherent epilayer will be thermodynamically stable. On the other hand, if it is less than zero, then the change in energy upon introducing the first few misfit dislocations is negative. Misfit dislocations *will* tend to form, and the strain in the epilayer will tend to "relax."

The critical misfit for a given thickness and the critical thickness for a given misfit, beyond which misfit dislocations will tend to form, are determined by the condition $(1/2)[\partial u_{\mathrm{tot}}/\partial \rho_{\mathrm{md}}]_{\rho_{\mathrm{md}}=0} = 0$, or

$$\begin{aligned} f_{\mathrm{c}} &= \frac{b}{8\pi h\cos\lambda}\left(\frac{1-\nu\cos^2\beta}{1+\nu}\right)\ln(4h/b) \\ h_{\mathrm{c}} &= \frac{b}{8\pi f\cos\lambda}\left(\frac{1-\nu\cos^2\beta}{1+\nu}\right)\ln(4h_{\mathrm{c}}/b) \end{aligned} \tag{5.18}$$

where we have used Equation 5.8, $b_{\mathrm{edg},\|} = b\cos\lambda$. These expressions reproduce exactly those derived originally by Matthews and Blakeslee[22] and more recently by Ball and van der Merwe.[23] They are illustrated in Figure 5.8 for $\lambda = \beta = 60°$, which is often the case for fcc-lattice-based diamond and zincblende crystals. Films having thickness/misfit combinations below the curves are stable against the introduction of misfit dislocations; films having thickness/misfit combinations above the curves are not.

Also shown in Figure 5.8 are experimental data points corresponding to $In_{1-x}Ga_xAs$ films grown on GaAs substrates and $Si_{1-x}Ge_x$ films grown on Si substrates. As can be seen, the boundary separating the coherent from the semi-coherent films is given very closely by Equation 5.18. That equation is also believed to describe the thermodynamic boundary dividing coherent from semi-coherent epitaxy of metal films.[24]

Above the critical layer thickness, the energy decreases at first upon the introduction of the first few misfit dislocations, but eventually increases

[22] J.W. Matthews and A.E. Blakeslee, "Defects in epitaxial multilayers I. Misfit dislocations," *J. Cryst. Growth* **27**, 118 (1974).

[23] C.A.B. Ball and J.H. van der Merwe, "The growth of dislocation-free layers," in *Dislocations in Solids*, F.R.N. Nabarro, Ed. (North-Holland, Amsterdam, 1983), Chap. 27.

[24] Y. Kuk, L.C. Feldman, and P.J. Silvermann, "Transition from the pseudomorphic state to the nonregistered state in epitaxial growth of Au on Pd (111)," *Phys. Rev. Lett.* **50**, 511 (1983).

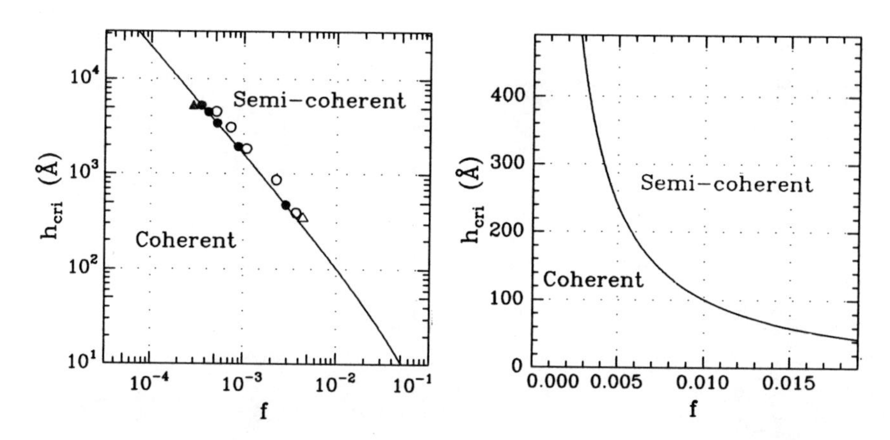

Figure 5.8: Logarithmic (left) and linear (right) plots of critical layer thicknesses for a given misfit (or, alternatively, critical layer misfits for a given thickness), as deduced from Equation 5.18. The triangles represent experiments in which $In_{1-x}Ga_xAs$ layers were grown on GaAs substrates and then annealed.[a] The circles represent experiments in which $Si_{1-x}Ge_x$ layers were grown on Si substrates and then annealed.[b] The filled data points correspond to structures that maintained coherency; the open data points correspond to structures that became semi-coherent.

[a]P.S. Peercy, B.W. Dodson, J.Y. Tsao, E.D. Jones, D.R. Myers, T.E. Zipperian, L.R. Dawson, R.M. Biefeld, J.F. Klem and C.R. Hills, "Stability of strained quantum-well field effect transistors," *IEEE Electron Dev. Lett.* **9**, 621 (1988).

[b]D.C. Houghton, C.J. Gibbings, C.G. Tuppen, M.H. Lyons, and M.A.G. Halliwell, "Equilibrium critical thickness for $Si_{1-x}Ge_x$ strained layers on (100) Si," *Appl. Phys. Lett.* **56**, 460 (1990).

again. The dislocation density that minimizes u_{tot} can be found by solving for that ρ_{md} for which the derivative

$$\begin{aligned}\frac{\partial u_{\text{tot}}}{2\partial\rho_{\text{md}}} &= -2\mu\left(\frac{1+\nu}{1-\nu}\right)hb_{\text{edg},\|}\left(f-\rho_{\text{md}}b_{\text{edg},\|}\right)\\ &\quad+\frac{\mu b^2}{4\pi}\left(\frac{1-\nu\cos^2\beta}{1-\nu}\right)\ln\left(\frac{4h}{b}\right)\end{aligned} \tag{5.19}$$

vanishes. In other words, the *equilibrium* dislocation density is given by

$$\begin{aligned}\rho_{\text{md,equ}} &= \frac{f}{b\cos\lambda}\\ &\quad-\frac{b}{8\pi h\cos^2\lambda}\left(\frac{1-\nu\cos^2\beta}{1+\nu}\right)\ln\left(\frac{4h}{b}\right)\end{aligned}$$

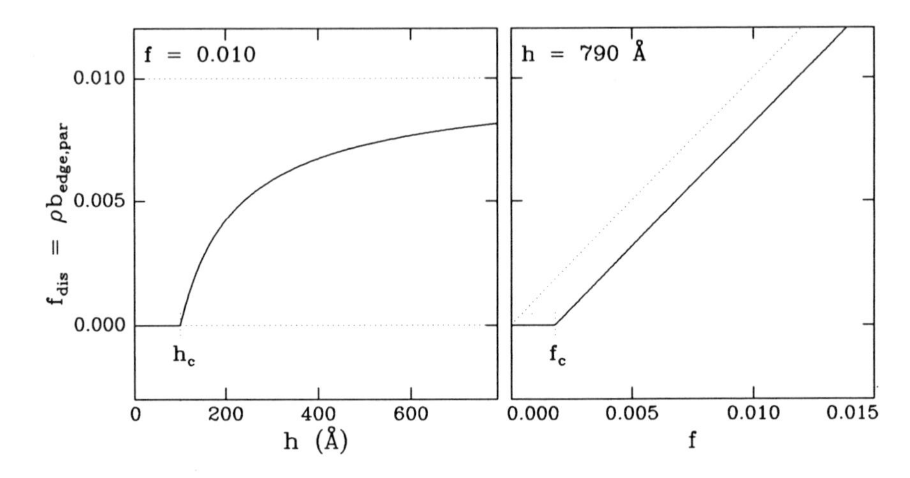

Figure 5.9: Equilibrium misfit strain ($f_{\rm dis}$) taken up by dislocations as a function of thickness at constant misfit (left) or as a function of misfit at constant thickness (right). Below either the critical thickness ($h_{\rm c}$) or critical misfit ($f_{\rm c}$) the strain taken up by dislocations is zero; above the critical thickness or critical misfit it becomes an increasingly large fraction of the total misfit (f).

$$= \frac{f}{b\cos\lambda}\left[1 - \frac{h_{\rm c}}{h}\right]. \tag{5.20}$$

This equation determines $\rho_{\rm md}$ self-consistently in terms of $\rho_{\rm md}$ itself, and can be solved iteratively; a reasonable initial guess is $\rho_{\rm md} = f/(2b\cos\lambda)$. Once determined, the equilibrium $\rho_{\rm md}$ then determines that portion of the total misfit, $f_{\rm dis}$, that is taken up by dislocations through Equation 5.12.

For a given misfit, $f_{\rm dis}$ is zero for thicknesses less than the critical layer thickness, but increases sharply for thicknesses greater than the critical layer thickness. This dependence is shown in the left half of Figure 5.9. Note that even for $h > h_{\rm c}$, the equilibrium dislocation density is less than that which would eliminate all coherency strain. In other words, even above the critical layer thickness the coherency strain in the epilayer is only partially, and not fully, relaxed.

Likewise, for a given thickness, $\rho_{\rm md}$ is zero for misfits less than the critical layer misfit, but increases logarithmically for misfits greater than the critical layer misfit. This dependence is shown in the right half of Figure 5.9. Again, even for $f > f_{\rm c}$, the equilibrium dislocation density is always less than that which would eliminate all coherency strain.

5.2 Forces on Dislocations

In Section 5.1, we found that the transition between coherency and semi-coherency corresponded to the thicknesses and misfits at which introduction of the first few misfit dislocations became *energetically* favorable. Ultimately, though, the transition between coherency and semi-coherency also requires the *motion* of dislocations to (or near to) the epilayer/substrate interface. There must then be forces acting on the dislocations to cause them to move. In this section, we describe the forces acting on dislocations in strained heterostructures. We will find that the thermodynamic transition between coherency and semi-coherency also corresponds exactly to the thicknesses and misfits at which the forces acting to elongate existing misfit dislocations are positive or negative. We will also find that the average force acting to elongate existing misfit dislocations, the "excess stress" of the structure, is a natural measure of the driving force for strain relaxation by misfit dislocation creation.

We will begin, in Subsection 5.2.1, by describing excess stress in a simple structure: a single strained surface layer grown on a very thick substrate. Then, in Subsection 5.2.2, we describe excess stress in a more complicated structure: a single strained layer buried within a very thick substrate. Finally, in Subsection 5.2.3, we generalize the concept of excess stress to even very complicated heterostructures, for which the excess stress depends on depth within the structure.

5.2.1 Strained Surface Layers

We start, in this subsection, by describing excess stress in a simple structure: a single strained surface layer grown on a very thick substrate. Consider a dislocation "threading" upward through the epilayer/substrate interface and into the epilayer itself, as illustrated in the left half of Figure 5.10. If the dislocation bends over, then new length of misfit dislocation will be created at the epilayer/substrate interface. If, in steady state, the shape of the threading segment as it moves from A–C to B–D does not change, then the net change in energy is due solely to the new misfit segment C–D, which we may imagine has moved downward from AB.

The energy gained by moving unit length of that segment a distance h downward (or, equivalently, by bending the threading dislocation unit length to the right) is $h\hat{z}$ dotted into what is known as the Peach-Koehler force, $d\bar{F} = (\bar{b} \cdot \bar{\bar{\sigma}}) \times \hat{l}$, which describes the force acting on unit length of dislocation in an external stress field. In particular, the bending force due

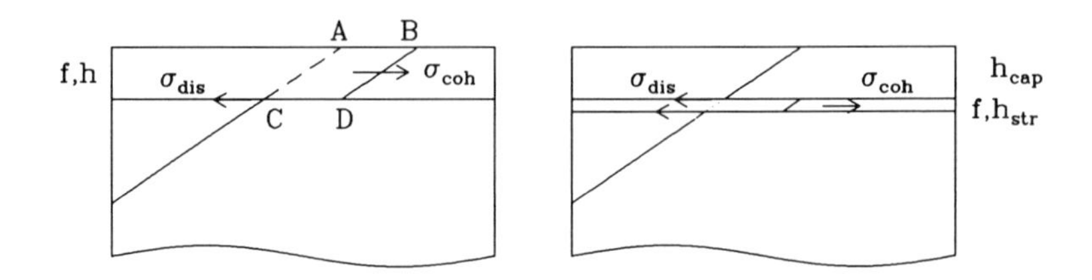

Figure 5.10: Strain relaxation by single (left) and double (right) kinking of a threading dislocation to form misfit dislocations.

to the coherency stress in the film is

$$\bar{F}_{\text{coh}} = h\hat{z} \cdot \{(\bar{b} \cdot \bar{\bar{\sigma}}) \times \hat{l}\}, \tag{5.21}$$

where

$$\bar{\bar{\sigma}} = \begin{pmatrix} \sigma_{\text{coh}} & 0 & 0 \\ 0 & \sigma_{\text{coh}} & 0 \\ 0 & 0 & 0 \end{pmatrix} \tag{5.22}$$

is the stress tensor in the epilayer film, and, using Equation 5.4,

$$\sigma_{\text{coh}} = \sigma_{||} = 2\mu \left(\frac{1+\nu}{1-\nu} \right) \epsilon_{||} \tag{5.23}$$

is the in-plane stress acting in the epilayer.

For the 60° misfit dislocation segment illustrated in Figure 5.5, whose Burgers vector is $\bar{b} = [\bar{1}01]b/\sqrt{2}$ and whose line direction within the interface is $\hat{l} = [110]/\sqrt{2}$, the force is

$$\bar{F}_{\text{coh}} = \left[\begin{array}{ccc} 0 & 0 & h \end{array} \right] \cdot \begin{pmatrix} 0 \\ 0 \\ -b\sigma_{\text{coh}}/2 \end{pmatrix}, \tag{5.24}$$

with a magnitude of

$$F_{\text{coh}} = b\sigma_{\text{coh}} h/2. \tag{5.25}$$

Since for this geometry

$$b_{\text{edg},||} = b \cos \lambda = b \cos 60^\circ = b/2, \tag{5.26}$$

the coherency force can also be written as

$$F_{\text{coh}} = b_{\text{edg},||} \sigma_{\text{coh}} h, \tag{5.27}$$

which can be shown to be generally true for arbitrary geometries (see Exercise 4 at the end of this chapter).

Opposing this force is a line tension associated with the energy required to create the new misfit dislocation segment C–D. From Equation 5.11, this force, or the energy per unit length, is

$$F_{\text{dis}} = \frac{\mu b^2}{4\pi}\left(\frac{1-\nu\cos^2\beta}{1-\nu}\right)\ln(4h/b), \tag{5.28}$$

where we have assumed noninteracting dislocations. If we recast this equation into the form $F_{\text{dis}} = b_{\text{edg},\|}\sigma_{\text{dis}}h$, then we can write

$$\sigma_{\text{dis}} = \frac{\mu b}{4\pi h\cos\lambda}\left(\frac{1-\nu\cos^2\beta}{1-\nu}\right)\ln(4h/b). \tag{5.29}$$

which is an effective stress, σ_{dis}, associated with the dislocation line tension that opposes the coherency stress, σ_{coh}.

The net, or "excess" stress driving the bending of threading dislocations to form single-kink misfit segments is therefore

$$\begin{aligned}\sigma_{\text{exc}}^{\text{SK}} &= \sigma_{\text{coh}} - \sigma_{\text{dis}}\\ &= 2\mu\left(\frac{1+\nu}{1-\nu}\right)\left(f-\rho_{\text{md}}b_{\text{edg},\|}\right)\\ &\quad - \frac{\mu b}{4\pi h\cos\lambda}\left(\frac{1-\nu\cos^2\beta}{1-\nu}\right)\ln(4h/b).\end{aligned} \tag{5.30}$$

When $\sigma_{\text{exc}}^{\text{SK}} > 0$, threading dislocations will tend to bend over to form strain-relaxing misfit segments. When $\sigma_{\text{exc}}^{\text{SK}} < 0$, threading dislocations that have bent over to form strain-relaxing misfit segments will tend to straighten. When $\sigma_{\text{exc}}^{\text{SK}} = 0$, threading dislocations will have neither tendency.[25]

Note that this excess stress reproduces exactly the variation of energy with dislocation density found in Equation 5.19, assuming noninteracting dislocations. Therefore, the condition $[\sigma_{\text{exc}}^{\text{SK}}]_{\rho_{\text{md}}=0} = 0$ is equivalent to the condition $(1/2)[\partial u_{\text{tot}}/\partial\rho_{\text{md}}]_{\rho_{\text{md}}=0} = 0$ for the thickness/misfit boundary between coherent and semi-coherent films, and the condition $\sigma_{\text{exc}}^{\text{SK}} = 0$ is equivalent to the condition $(1/2)\partial u_{\text{tot}}/\partial\rho_{\text{md}} = 0$ for the equilibrium misfit dislocation density beyond the critical layer thickness. Physically, the force required to form new misfit dislocations by bending of existing threading dislocations is equivalent to that required to increase the density of a dislocation array by "squeezing" laterally on the dislocation array.

[25] L.B. Freund, "The driving force for glide of a threading dislocation in a strained epitaxial layer on a substrate," *J. Mech. Phys. Solids* **38**, 657 (1990).

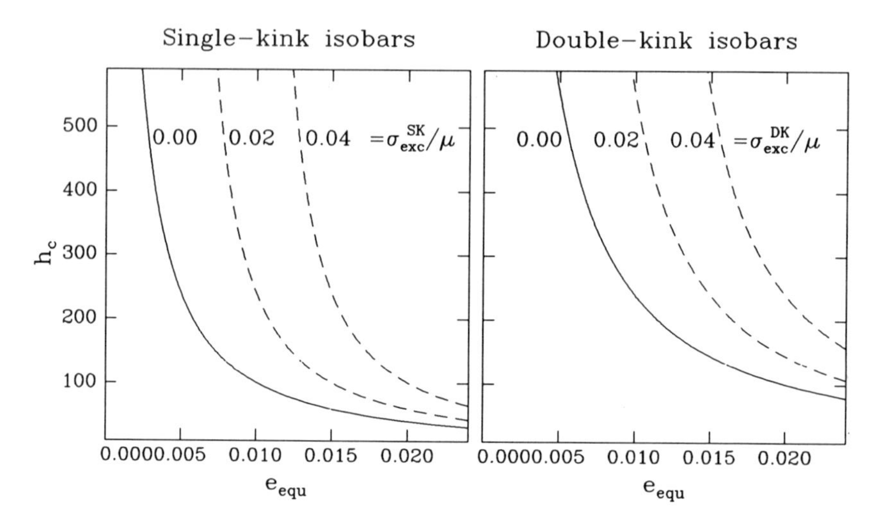

Figure 5.11: Contours of constant single-kink (left) and double-kink (right) excess stresses on a thickness/equivalent-strain diagram.

As we will discuss later, it is often possible, for kinetic reasons, to grow coherent epilayers to thicknesses well beyond those for which they should become semi-coherent. In these cases, the excess stress evaluated at $\rho_{md} = 0$ is a useful measure for the degree of metastability of the structure. The critical thickness/misfit relationship for *a given degree of metastability*, or a given value of σ_{exc}^{SK}, is then found, from Equation 5.30, to be

$$f_c(\sigma_{exc}^{SK}, h) = \frac{1}{2}\left(\frac{1-\nu}{1+\nu}\right)\frac{\sigma_{exc}^{SK}}{\mu} + \frac{b}{8\pi h\cos\lambda}\left(\frac{1-\nu\cos^2\beta}{1+\nu}\right)\ln(4h/b)\,. \tag{5.31}$$

These metastable critical thicknesses and misfits are illustrated in the left half of Figure 5.11 for various values of σ_{exc}^{SK}/μ. For thin ($\approx$ 150 Å) "quantum-well" type structures, which are capped immediately by unstrained material, it is often possible to grow coherent metastable structures up to values of $\sigma_{exc}^{SK}/\mu = 0.04$. For thicker structures, the maximum σ_{exc}^{SK}/μ values decrease considerably. In the limit of very thick structures, grown for very long times, the maximum σ_{exc}^{SK}/μ values are zero, and the equilibrium critical layer thickness boundary holds.

5.2.2 Strained Buried Layers

In Subsection 5.2.1, we described excess stress in a simple structure consisting of a strained surface layer. In this subsection, we describe excess stress in a more complicated structure consisting of a strained *buried* layer. Consider a dislocation threading upward through this buried strained layer, as illustrated in the right half of Figure 5.10. The buried strained layer has thickness $h_{\rm str}$, and is capped by an unstrained layer of thickness $h_{\rm cap}$. For this structure, it is possible for the dislocation to bend twice, thereby relieving strain only within the buried layer.[26] We can apply all the arguments of the previous subsection to calculate the net bending force, except that the line tension force must now be taken twice. The excess stress driving misfit dislocation formation by this macroscopic "double-kinking" of a threading dislocation is therefore

$$\begin{aligned} \sigma_{\rm exc}^{\rm DK} &= \sigma_{\rm coh} - \sigma_{\rm dis,1} - \sigma_{\rm dis,2} \\ &= 2\mu\left(\frac{1+\nu}{1-\nu}\right)(f - \rho_{\rm md} b_{\rm edg,\parallel}) \\ &\quad - \frac{\mu b}{4\pi R_1 \cos\lambda}\left(\frac{1-\nu\cos^2\beta}{1-\nu}\right)\ln(4R_1/b) \\ &\quad - \frac{\mu b}{4\pi R_2 \cos\lambda}\left(\frac{1-\nu\cos^2\beta}{1-\nu}\right)\ln(4R_2/b), \end{aligned} \tag{5.32}$$

where $\rho_{\rm md}$ is now the density of dislocations at *each* interface surrounding the buried strained layer.

Note that the stresses associated with the dislocation line tensions are different for the two dislocations, because they may have different cutoff "screening" distances for their elastic energies. The cutoff distance for the dislocation farthest from the free surface will be approximately the distance to the dislocation closest to the free surface, or h. However, the cutoff distance for the dislocation closest to the free surface will be the smaller of the distances to the free surface, $h_{\rm cap}$, or to the adjacent dislocation, or approximately $h_{\rm eff} = h_{\rm str}h_{\rm cap}/(h_{\rm str}+h_{\rm cap})$. Therefore,

$$\begin{aligned} \sigma_{\rm exc}^{\rm DK} &= 2\mu\left(\frac{1+\nu}{1-\nu}\right)(f - \rho_{\rm md} b_{\rm edg,\parallel}) \\ &\quad - \frac{\mu b}{4\pi h_{\rm str} \cos\lambda}\left(\frac{1-\nu\cos^2\beta}{1-\nu}\right)\ln\left(\frac{4h_{\rm str}}{b}\right) \\ &\quad - \frac{\mu b}{4\pi h_{\rm eff} \cos\lambda}\left(\frac{1-\nu\cos^2\beta}{1-\nu}\right)\ln\left(\frac{4h_{\rm eff}}{b}\right) \end{aligned} \tag{5.33}$$

In the limit $h_{\rm cap} \rightarrow 0$ (or, to avoid singularities, $h_{\rm cap} \rightarrow b/4$), the

[26] W.D. Nix, D.B. Noble, and J.F. Turlo, "Mechanisms and kinetics of misfit dislocation formation in heteroepitaxial thin films," *Mat. Res. Soc. Symp. Proc.* **188**, 315 (1990).

double-kink excess stress becomes

$$\sigma_{\rm exc}^{\rm DK} = 2\mu\left(\frac{1+\nu}{1-\nu}\right)(f-\rho_{\rm md}b_{{\rm edg},\|}) - \frac{\mu b}{4\pi h_{\rm str}\cos\lambda}\left(\frac{1-\nu\cos^2\beta}{1-\nu}\right)\ln\left(\frac{4h_{\rm str}}{b}\right), \tag{5.34}$$

and is equivalent to the single-kink excess stress. Effectively, there is no cap, and the energy of the dislocation elongating along the surface is zero.

In the opposite limit $h_{\rm cap}\to\infty$, the double-kink excess stress becomes

$$\sigma_{\rm exc}^{\rm DK} = 2\mu\left(\frac{1+\nu}{1-\nu}\right)(f-\rho_{\rm md}b_{{\rm edg},\|}) - \frac{\mu b}{2\pi h_{\rm str}\cos\lambda}\left(\frac{1-\nu\cos^2\beta}{1-\nu}\right)\ln\left(\frac{4h_{\rm str}}{b}\right). \tag{5.35}$$

In this limit, the critical thickness/misfit relationship for a given degree of metastability, or a given value of $\sigma_{\rm exc}^{\rm DK}$, can be calculated to be

$$f_{\rm c}(\sigma_{\rm exc}^{\rm DK},h) = \frac{1}{2}\left(\frac{1-\nu}{1+\nu}\right)\frac{\sigma_{\rm exc}^{\rm DK}}{\mu} + \frac{b}{4\pi h\cos\lambda}\left(\frac{1-\nu\cos^2\beta}{1+\nu}\right)\ln\left(\frac{4h_{\rm str}}{b}\right). \tag{5.36}$$

These metastable critical layer thicknesses are illustrated in the right half of Figure 5.11. The equilibrium critical layer thickness boundary is determined by $\sigma_{\rm exc}^{\rm DK}=0$, and is seen to be shifted to the right from the single-kink curves. Because the line tension enters in twice, for strained layers having the same thickness and misfit, this double-kink mechanism for strain relaxation is usually less likely than the single-kink mechanism discussed earlier.

5.2.3 Generalized Excess Stress

In Subsections 5.2.1 and 5.2.2, we described the excess stresses associated with two strained layer structures, one in which the layer is at the surface, another in which the layer is buried. In this subsection, we describe excess stress in general structures composed of multilayers of different misfits and thicknesses. Such structures are susceptible to either single-kink or double-kink relaxation at different depths within the structure. In other words, dislocations may bend anywhere within a given structure. Then, it is useful to generalize the driving force for that bending to include a dependence on depth.[27]

---

[27] J.Y. Tsao and B.W. Dodson, "Excess stress and the stability of strained heterostructures," *Appl. Phys. Lett.* **53**, 848 (1988).

In doing so, we note that for many applications it is only necessary to calculate the excess stress in unrelaxed, fully coherent structures. Therefore, we restrict ourselves to the simplest case of unrelaxed ($\rho_{\rm md} = 0$) structures. Structures that are partially relaxed ($\rho_{\rm md} \neq 0$) are considerably more difficult to treat.

If we make the approximation that the elastic moduli of the different layers are equal, then for single-kink relaxation, the depth-dependent excess stress can be written as

$$\sigma_{\rm exc}^{\rm SK}(z) = 2\mu\left(\frac{1+\nu}{1-\nu}\right)\epsilon_{\rm equ}^{\rm SK}(z) - \frac{\mu b}{4\pi z\cos\lambda}\left(\frac{1-\nu\cos^2\beta}{1-\nu}\right)\ln\left(\frac{4z}{b}\right), \quad (5.37)$$

where z is the depth from the free surface, and

$$\epsilon_{\rm equ}^{\rm SK}(z) = \int_0^z e(z')\frac{dz'}{z}, \quad (5.38)$$

the equivalent strain, is the average parallel strain associated with the structure from the surface to that depth.

Physically, the coherency stress acting to bend a dislocation at a depth z is an integral of the strains over the length of the dislocation above that depth. The dislocation line tension stress acting to straighten a dislocation at a depth z is the energy associated with creating a dislocation at that depth. If $\sigma_{\rm exc}^{\rm SK}(z) < 0$ at a particular depth z, misfit dislocation formation at that depth leads to an increase in energy. If $\sigma_{\rm exc}^{\rm SK}(z) > 0$ at a particular depth z, misfit dislocation formation at that depth leads to a decrease in energy.

Note that even if $\sigma_{\rm exc}^{\rm SK}(z) > 0$ at a particular depth, a threading dislocation will not necessarily bend there. Kinetic limitations may prevent such bending, and there may be other depths in the structure at which the excess stress is even higher, and which will be even more favored for misfit dislocation creation.

To illustrate this concept of a depth-dependent excess stress, in Figures 5.12 and 5.13 we show $\sigma_{\rm exc}^{\rm SK}(z)$ for two double quantum-well heterostructures. In the unstrained caps, $\sigma_{\rm exc}^{\rm SK}(z)$ increases gradually from the surface, as the line tension stress associated with misfit dislocation creation decreases. In the strained layers themselves, $\sigma_{\rm exc}^{\rm SK}(z)$ increases more quickly, as the coherency stresses increase. In the unstrained buffers beneath the strained layers $\sigma_{\rm exc}^{\rm SK}(z)$ decreases, as the coherency stress associated with the strained layer is "diluted," so that the average coherency stress above a depth z decreases.

In these examples, the single-kink excess stress is maximum at the rear of the deepest buried strained layers. The rear of that strained layer is

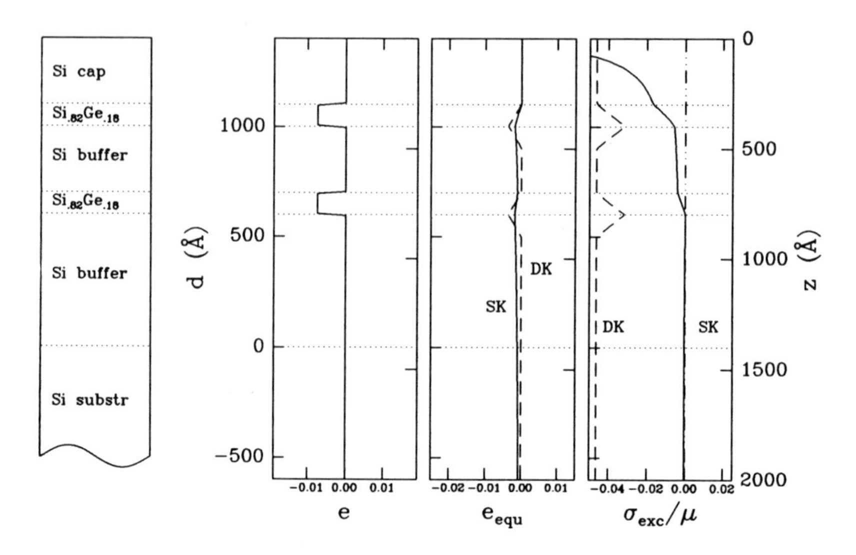

Figure 5.12: Strains, equivalent strains, and excess stresses in a double buried quantum-well heterostructure. Both the single- and double-kink excess stresses maximize at a depth of 800 Å, but only the single-kink excess stress exceeds zero, and even then just barely. Therefore, this structure is stable with respect to double-kink strain relaxation, but slightly unstable with respect to single-kink strain relaxation.

therefore the weakest point in the structure, where misfit dislocations are most likely to form. It is important to emphasize, though, that the rear of the shallowest buried strained layer is also a weak point, at which misfit dislocations may form.

The double-kink excess stress can be generalized in a similar way:

$$\sigma_{\text{exc}}^{\text{DK}}(z) = 2\mu\left(\frac{1+\nu}{1-\nu}\right)\epsilon_{\text{equ}}^{\text{DK}}(z) - \frac{\mu b}{2\pi h\cos\lambda}\left(\frac{1-\nu\cos^2\beta}{1-\nu}\right)\ln\left(\frac{4h}{b}\right), \tag{5.39}$$

where z is the depth of the lower kink from the free surface, h is the thickness of material between the kinks, and

$$\epsilon^{\text{DK}}(z) = \int_z^{z+h} \epsilon(z')\frac{dz'}{h} \tag{5.40}$$

is the equivalent strain associated with the material between the kinks.

Note that the double-kink excess stress depends not just on depth, but on the thickness of material between the kinks. It will be maximum when it is matched to the thicknesses of the buried strained layers. For example,

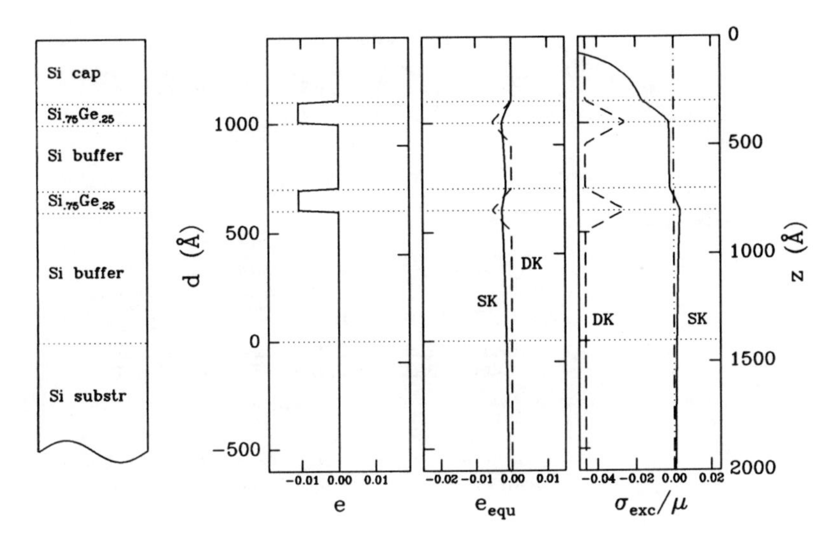

Figure 5.13: Strains, equivalent strains, and excess stresses in a double buried quantum well heterostructure. Both the single- and double-kink excess stresses maximize at a depth of 800 Å, but only the single-kink excess stress exceeds zero. Therefore, this structure is stable with respect to double-kink strain relaxation, but unstable with respect to single-kink strain relaxation.

Figures 5.12 and 5.13 show the depth-dependent double-kink excess stress for the double buried quantum-well heterostructures evaluated at h equal to the actual thicknesses, h_{str}, of the buried quantum wells. The excess stresses can be seen to be small everywhere except in the quantum wells themselves.

Also note that even at their maxima, at the rear of the buried strained layers, the double-kink excess stresses are less than the single-kink excess stresses. These particular buried structures will therefore be more likely to relax by generation of single-kink rather than double-kink misfit dislocations.

If we require a structure to be absolutely stable, both with respect to single-kink and double-kink relaxation, then we require $\sigma_{\text{exc}}^{\text{SK}}(z) < 0$ and $\sigma_{\text{exc}}^{\text{DK}}(z) < 0$ for all z. For the single buried strained layer structure shown in Figure 5.14, whose weakest point is at a depth $z = h_{\text{cap}} + h_{\text{str}}$, where h_{cap} is the thickness of the unstrained capping layer, we must then satisfy both

$$f_{\text{equ}}^{\text{SK}} \leq \frac{b}{8\pi h_{\text{equ}}^{\text{SK}} \cos\lambda}\left(\frac{1-\nu\cos^2\beta}{1+\nu}\right)\ln\left(\frac{4h_{\text{equ}}^{\text{SK}}}{b}\right) \qquad (5.41)$$

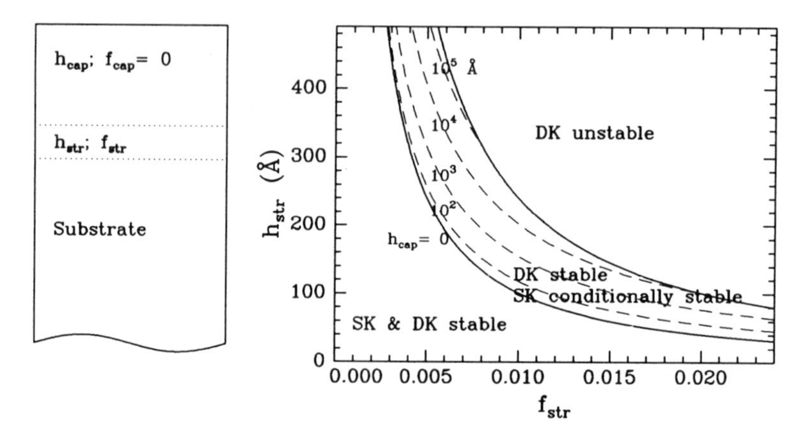

Figure 5.14: Stability curves for structure consisting of a strained layer of thickness $h_{\rm str}$ and misfit $f_{\rm str}$, buried underneath a capping layer of thickness $h_{\rm cap}$ and misfit $f_{\rm cap}$.

$$f_{\rm equ}^{\rm DK} \leq \frac{b}{4\pi h_{\rm equ}^{\rm DK}\cos\lambda}\left(\frac{1-\nu\cos^2\beta}{1+\nu}\right)\ln\left(\frac{4h_{\rm equ}^{\rm DK}}{b}\right), \tag{5.42}$$

where

$$\begin{aligned} \epsilon_{\rm equ}^{\rm SK} &= \frac{f_{\rm str}h_{\rm str}}{h_{\rm cap}+h_{\rm str}} \\ h_{\rm equ}^{\rm SK} &= h_{\rm cap}+h_{\rm str} \end{aligned} \tag{5.43}$$

$$\begin{aligned} \epsilon_{\rm equ}^{\rm DK} &= f_{\rm str} \\ h_{\rm equ}^{\rm DK} &= h_{\rm str} \end{aligned} \tag{5.44}$$

are the single- and double-kink equivalent strains and thicknesses. Note that we have made the approximation $h_{\rm equ}^{\rm DK} = h_{\rm str}h_{\rm cap}/(h_{\rm str}+h_{\rm cap}) = h_{\rm str}$ for the double-kink equivalent thicknesses.

Substituting Equations 5.43 into Equation 5.41, we find that for a given thickness cap, $h_{\rm cap}$, and a given thickness buried strained layer, $h_{\rm str}$, the critical strained layer misfit below which the structure will be stable with respect to single-kink relaxation is

$$f_{\rm str} = \frac{h_{\rm cap}+h_{\rm str}}{h_{\rm str}}\frac{b}{8\pi(h_{\rm cap}+h_{\rm str})\cos\lambda}\left(\frac{1-\nu\cos^2\beta}{1+\nu}\right)\ln\left[\frac{4(h_{\rm cap}+h_{\rm str})}{b}\right]. \tag{5.45}$$

The resulting critical layer misfits for a given thickness (or, equivalently, the critical layer thicknesses for a given misfit) are shown as the dashed curves in Figure 5.14 for various capping layer thicknesses. Buried strained layer thickness and misfit combinations that lie to the left of the curves are stable with respect to single-kink relaxation.

If the cap thickness is zero, then the critical thickness curve (leftmost solid curve) for simple surface strained layers is obtained. As the cap becomes thicker, the critical thickness curves shift to the right, as the unstrained cap "dilutes" the strain of the buried layer and stabilizes it. Ultimately, for infinitely thick caps, the structure becomes more unstable with respect to double-kink relaxation than to single-kink relaxation. In this limit, the critical strained layer misfit below which the structure will be stable is found by substituting Equations 5.44 into Equation 5.42, or

$$f_{\text{str}} = \frac{b}{4\pi h_{\text{str}} \cos\lambda} \left(\frac{1 - \nu \cos^2 \beta}{1 + \nu} \right) \ln \left(\frac{4 h_{\text{str}}}{b} \right). \tag{5.46}$$

The resulting critical layer misfit for a given thickness (or, equivalently, the critical layer thickness for a given misfit) is shown as the rightmost solid curve in Figure 5.14. Buried strained layer thickness and misfit combinations that lie to the right of the curve are not stable with respect to double-kink relaxation.

To the left of the leftmost solid curve, then, structures are absolutely stable with respect to both single- and double-kink relaxation, regardless of cap thickness. Within the window between the solid curves, structures are absolutely stable with respect to double-kink relaxation, but require stabilization with respect to single-kink relaxation by a finite-thickness cap layer. To the right of the rightmost solid curve, structures are not stable with respect to double-kink relaxation, even if they have been stabilized against single-kink relaxation by an infinitely thick cap layer.

5.3 Relaxation of Strain

In Sections 5.1 and 5.2, we described the *thermodynamics* of the creation of misfit dislocations. In particular, we described the driving force, or "excess stress," acting to bend vertical dislocation segments into misfit dislocations lying in the interface between the strained layer and the substrate. Basically, the sign of σ_{exc} determines whether or not misfit dislocations will have a tendency to form, while the magnitude of σ_{exc} determines the driving force for them to form. Even if $\sigma_{\text{exc}} > 0$, however, misfit dislocations will not form instantly during growth.[28] Instead, they will form at a finite

[28] A.T. Fiory, J.C. Bean, R. Hull, and S. Nakahara, "Thermal relaxation of metastable strained-layer Ge_xSi_{1-x}/Si epitaxy," *Phys. Rev.* **B31**, 4063 (1984); and E. Kasper, "Growth and properties of Si/SiGe superlattices," *Surface Sci.* **174**, 630 (1986).

rate determined, to first order, by the magnitude of σ_{exc}.

In this section, we discuss the *kinetics* of the creation of misfit dislocations. We begin, in Subsection 5.3.1, with a brief introduction to the dynamics of dislocations in bulk materials, as summarized in what are known as deformation-mechanism maps. Then, in Subsection 5.3.2, we describe a simple qualitative model for the dynamics of dislocations in epitaxial thin films, and use it to simulate, in an approximate way, the evolution of strain and misfit dislocation density during actual growth and processing of strained heterostructures. Finally, in Subsection 5.3.3, we discuss the construction of stability diagrams, which describe the stress-temperature-time regimes within which strained heterostructures will be metastable to various amounts of relaxation.

5.3.1 Deformation Mechanism Maps

Let us begin, in this subsection, with a brief introduction to the dynamics of dislocations in bulk materials. At the outset, we note that the mechanisms underlying the introduction, motion, and multiplication of dislocations in bulk materials are exceedingly complex. The mechanisms are many, and each may be important only under certain conditions. To illustrate this, let us first consider some of the ways in which *bulk* materials deform plastically under the application of externally imposed stresses. The classic way of representing the plastic deformation of bulk materials is through the use of deformation-mechanism maps.[29] These maps are stress-temperature diagrams on which are indicated regimes within which various mechanisms for plastic deformation are dominant.

Consider, for example, the deformation-mechanism maps illustrated in Figures 5.15 and 5.16 for Si and Ge, respectively. At relatively low temperatures and high stresses, deformation is dominated by "low-temperature plasticity," in which dislocations move mainly by conservative motion, or glide, within the plane containing both the dislocation line and its Burgers vector. At relatively high temperatures and moderate stresses, deformation is dominated by "power-law creep," in which dislocations are increasingly able to move by the nonconservative motion, or climb, of dislocations out of the plane containing both the dislocation line and its Burgers vector. At the lowest stresses, deformation of polycrystalline materials is dominated by "diffusional flow," in which, even in the absence of dislocations, grain boundaries move and change shape via diffusion of matter through the grains or along the grain boundaries themselves.

[29] H.J. Frost and M.F. Ashby, *Deformation-Mechanism Maps* (Pergamon, Oxford, 1982).

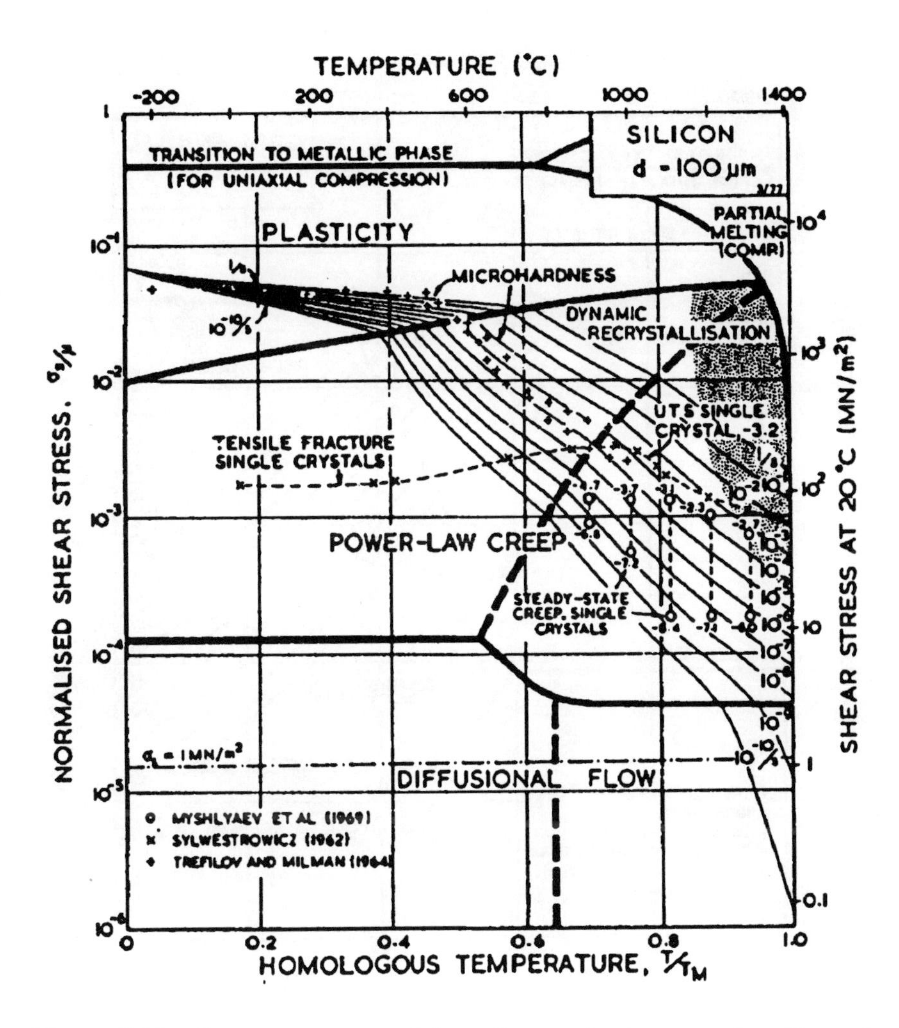

Figure 5.15: Stress–temperature deformation mechanism map for silicon of grain size 100 μm.[a] Iso-strain rate contours are drawn from 1/s to 10^{-10}/s.

[a]Reprinted from H.J. Frost and M.F. Ashby, *Deformation-Mechanism Maps* (Pergamon, Oxford, 1982), p. 71.

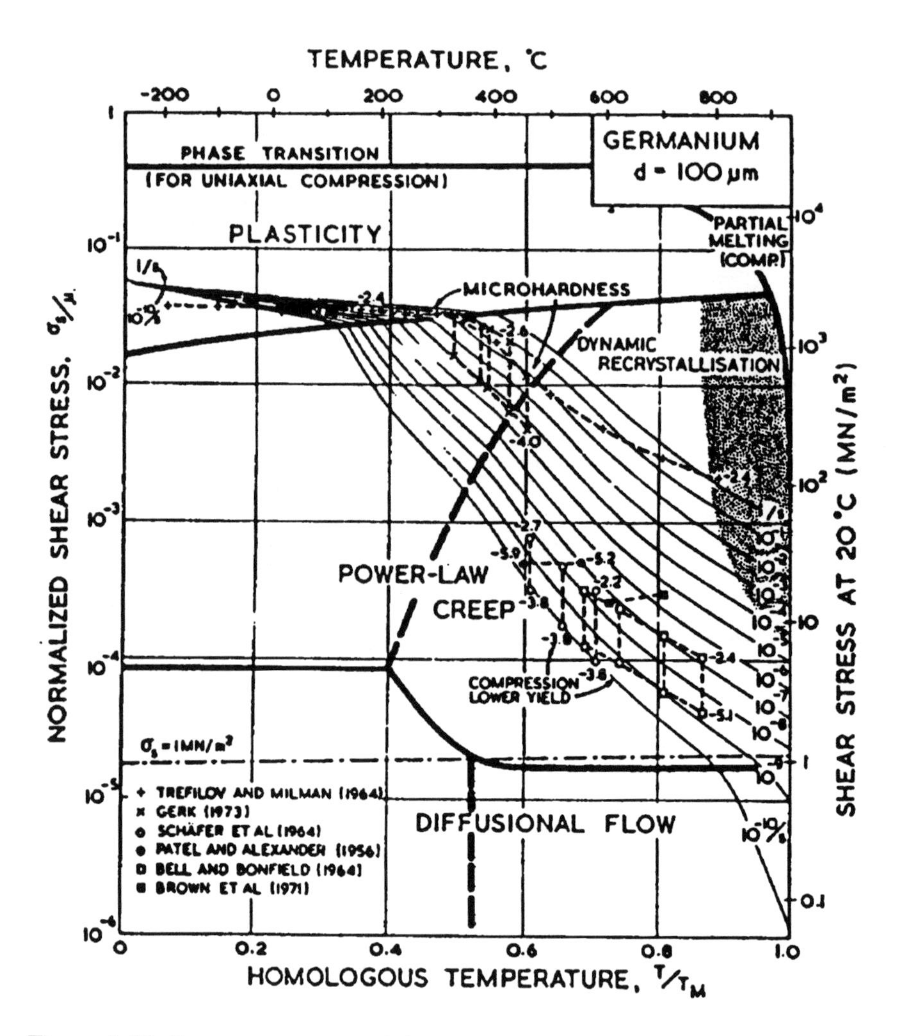

Figure 5.16: Stress–temperature deformation mechanism map for germanium of grain size 100 μm.[a] Iso-strain rate contours are drawn from 1/s to 10^{-10}/s.

[a]Reprinted from H.J. Frost and M.F. Ashby, *Deformation-Mechanism Maps* (Pergamon, Oxford, 1982), p. 73.

Superimposed on these diagrams are iso-strain rate contours, which indicate the temperature-dependent stresses required to cause a given strain rate. Generally, lower stresses are required to cause a given strain rate at higher temperatures, due to increased dislocation mobilities. This increased dislocation mobility may occur for a number of reasons. For example, the nucleation rate of microscopic double kinks, by which dislocations glide laterally on an atomic scale, may increase. The rate at which dislocations pass through obstacles may increase. The rate at which vacancies diffuse to and from dislocations may also increase, thereby increasing the rate at which dislocations climb.

We emphasize that deformation-mechanism maps represent an enormous simplification of a number of complex mechanisms, and can only be a rough guide to deformation behavior. In particular, their construction requires the assumption of a particular microstructure, e.g., dislocation density and, in polycrystalline materials, grain size. As materials deform, however, their microstructure will change; if the change is severe, the corresponding change in the deformation-mechanism map may also be quite severe. In other words, a complete picture of plastic deformation must include the time evolution of dislocation densities and other aspects of microstructure, and how that evolving microstructure in turn influences the further evolution of dislocation densities.

5.3.2 A Simple Phenomenological Model

In Subsection 5.3.1, we discussed briefly plastic deformation in bulk materials. The geometry of thin film single-crystal heterostructures is much simpler than that of a bulk polycrystalline material, and so in principle should be correspondingly easier to treat. However, this has not yet proven so. A general treatment of the plastic deformation of thin film strained heterostructures must itself include a number of complex microscopic mechanisms. In this subsection, we briefly discuss these microscopic mechanisms, and then discuss a simple phenomenological model based on these mechanisms.

Consider the microscopic mechanisms illustrated in Figure 5.17. First, because the initial threading dislocation densities in electronic-grade semiconductor substrates are exceedingly low and cannot by themselves account for the amounts of strain relaxation commonly observed, nucleation of new dislocation loops must be included.[30] These loops are most likely "half-

[30] P.M.J. Marée, J.C. Barbour, J.F. van der Veen, K.L. Kavanagh, C.W.T. Bulle-Lieuwma, and M.P.A. Viegers, "Generation of misfit dislocations in semiconductors," *J. Appl. Phys.* **62**, 4413 (1987); and R. People and J.C. Bean, "Calculation of critical layer thickness versus lattice mismatch for Ge_xSi_{1-x}/Si strained-layer heterostructures,"

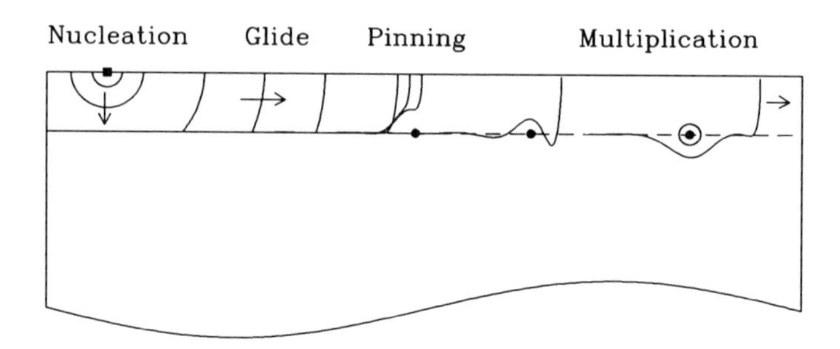

Figure 5.17: Schematic illustrations of possible microscopic deformation mechanisms operative during strain relaxation.

loops" nucleated at the free surface, perhaps catalyzed by defects or other stress concentrators.[31] In compound semiconductors, the situation is even more complicated, due to possible dependences of the nucleation rate on surface conditions and chemistry.

Second, the outward "bowing" motion of these dislocation half-loops to form misfit dislocation segments at the epilayer/substrate interface must be included.[32] This motion will be dominated by glide at low temperatures, but will increasingly have a climb component at higher temperatures. It may also be mediated by nucleation of microscopic single kinks at the free surface in very thin films, or by nucleation of microscopic double kinks in thicker films.[33]

Third, the slowing and occasional pinning of these dislocation half-loops as they move and encounter other dislocation segments must be included.[34] Such pinning has been observed during *in situ* transmission

Appl. Phys. Lett. **47**, 322 (1985) and **49**, 229 (1986).

[31]B.W. Dodson, "Nature of misfit dislocation sources in strained-layer semiconductor structures," *Appl. Phys. Lett.* **53**, 394 (1988); C.J. Gibbings, C.G. Tuppen, and M. Hockly, "Dislocation nucleation and propagation in $Si_{0.95}Ge_{0.05}$ layers on silicon," *Appl. Phys. Lett.* **54**, 148 (1989); and D.J. Eaglesham, E.P. Kvam, D.M. Maher, C.J. Humphreys, and J.C. Bean, "Dislocation nucleation near the critical thickness in GeSi/Si strained layers" *Phil. Mag.* **A59**, 1059 (1989).

[32]J.W. Matthews, S. Mader, and T.B. Light, "Accommodation of misfit across the interface between crystals of semiconducting elements or compounds," *J. Appl. Phys.* **41**, 3800 (1970).

[33]R. Hull, J.C. Bean, D. Bahnck, L.J. Peticolas, Jr., K.T. Short, and F.C. Unterwald, "Interpretation of dislocation propagation velocities in strained Ge_xSi_{1-x}/Si(100) heterostructures by the diffusive kink pair model," *J. Appl. Phys,* to be published.

[34]L.B. Freund, "A criterion for arrest of a threading dislocation in a strained epitaxial layer due to an interface misfit dislocation in its path" *J. Appl. Phys.* **68**, 2073 (1990).

electron microscopy,[35] and is likely to be extremely important in the later stages of strain relaxation,[36] when both crossed grids of dislocations have become quite dense.

Fourth, the unpinning and possible multiplication[37] of dislocation half-loops as they "bow through" obstacles such as other dislocations must be included. These processes have not been directly observed, but may be important in the later stages of strain relaxation.

Finally, the way in which all of these microscopic processes depend on depth within the structure must be included. Treating depth-dependent strain relaxation would represent a nontrivial extension of existing theories, but would be particularly important for compositional graded strained heterostructures, or for strained heterostructures composed of multiple layers, each having its own misfit.

As a consequence, all current models treat only some of these processes, and even then only in simplified ways. For concreteness, let us consider here one model,[38] based on the phenomenology of deformation in bulk diamond-structure materials.[39] The model is not the most complete,[40] but is simple and predicts at least qualitatively much of what is known about strain relaxation.

The model assumes that dislocations multiply at a rate proportional to (a) the velocity at which they move, (b) the number of dislocations present, and (c) the excess stress. If the number of dislocations is itself proportional to the amount of strain relaxation, γ, and if the dislocation glide and climb velocities are both thermally activated and proportional to the excess stress, then we can write

$$\frac{d\gamma}{dt} = \frac{\sigma_{\text{exc}}^2(\gamma)}{\mu^2} \left(\Gamma_g e^{-Q_g/kT} + \Gamma_c e^{-Q_c/kT} \right) (\gamma + \gamma_o) , \qquad (5.47)$$

[35] R. Hull and J.C. Bean, "Variation in misfit dislocation behavior as a function of strain in the GeSi/Si system" *Appl. Phys. Lett.* **54**, 925 (1989).

[36] B.W. Dodson, "Work hardening and strain relaxation in strained-layer buffers," *Appl. Phys. Let.* **53**, 37 (1988).

[37] W. Hagen and H. Strunk, "A new type of source generating misfit dislocations," *Appl. Phys.* **17**, 85 (1978).

[38] B.W. Dodson and J.Y. Tsao, "Relaxation of strained-layer semiconductor structures via plastic flow," *Appl. Phys. Lett.* **51**, 1325-1327 (1987); B.W. Dodson and J.Y. Tsao, "Erratum: Relaxation of strained-layer semiconductor structures via plastic flow," *Appl. Phys. Lett.* **52**, 852 (1988); and R. People, "Comment on 'Relaxation of strained-layer semiconductor structures via plastic flow'," *Appl. Phys. Lett.* **53**, 1127 (1988).

[39] H. Alexander and P. Haasen, "Dislocations and plastic flow in the diamond structure," in *Solid State Physics* Vol. 22, F. Seitz and D. Turnbull, Eds. (Academic Press, New York, 1968), pp. 27-158.

[40] See, e.g., D.C. Houghton, "Strain relaxation kinetics in $Si_{1-x}Ge_x$/Si heterostructures," *J. Appl. Phys.* **70**, 2136 (1991), and Exercise 6 at the end of this chapter.

or

$$\frac{d\ln(\gamma+\gamma_o)}{dt} = \frac{\sigma_{\text{exc}}^2(\gamma)}{\mu^2}\left(\Gamma_{\text{g}}e^{-Q_{\text{g}}/kT}+\Gamma_{\text{c}}e^{-Q_{\text{c}}/kT}\right). \qquad (5.48)$$

In these equations, Γ_{g} and Γ_{c} are glide and climb rate prefactors, Q_{g} and Q_{c} are glide and climb activation energies, and γ_o represents a constant "source" term.

Note that the form of Equation 5.48 is general, but the actual values of the kinetic parameters depend on the orientation of the slip planes with respect to the epilayer/substrate interface, and on the direction of slip within those planes. More general treatments can be formulated by replacing the excess in-plane stress with the excess stress resolved on the slip plane and acting in the direction of slip within that plane.[41] For (001) oriented films in the $Si_{1-x}Ge_x$ system, approximate fits to relaxation data give[42] rate prefactors of $\Gamma_{\text{g}} = 2\times10^{10}\ \text{s}^{-1}$ and $\Gamma_{\text{c}} = 3\times10^{21}\ \text{s}^{-1}$, a stress-dependent glide activation energy of $Q_{\text{g}} = Q_{\text{g},o}[1-\sigma_{\text{exc}}(\gamma)/\sigma_o]$, where $Q_{\text{g},o} = 16kT_{\text{m}}$ and $\sigma_o \approx 0.1\mu$, a stress-independent climb activation energy of $Q_{\text{c}} = 30kT_{\text{m}}$, and a "source" term of magnitude $\gamma_o \approx 10^{-4}$. Here, T_{m} is the melting temperature of the $Si_{1-x}Ge_x$ alloy.

Note that the excess stress in Equation 5.48 depends nonlinearly on the actual equivalent strain, ϵ_{equ} which in turn depends on the degree of relaxation, γ:

$$\epsilon_{\text{equ}} = f_{\text{equ}} - \gamma. \qquad (5.49)$$

Therefore, Equation 5.48 is a highly nonlinear differential equation whose full solution requires numerical techniques. However, for practical device heterostructures which are adversely affected by dislocations, small amounts of relaxation ($\gamma < 10^{-3}$) are often of greatest interest. Since these relaxations are less than the unrelaxed equivalent strains in typical structures, the excess stresss may be considered independent of the amount of relaxation. Then, it is straightforward to integrate Equation 5.48 numerically to deduce the time-dependent strain relaxation, γ, and by differentiation to deduce the time-dependent strain relaxation rate, $\dot{\gamma}$.

5.3.3 Time, Temperature and Excess Stress

In Subsection 5.3.2, we described a simple phenomenological model for the relaxation of excess stress and strain. In this subsection, we illustrate the

[41] L.B. Freund, "The driving force for glide of a threading dislocation in a strained epitaxial layer on a substrate," *J. Mech. Phys. Solids* **38**, 657 (1990).

[42] R. Hull, J.C. Bean, D.J. Werder, and R.E. Leibenguth, "*In situ* observations of misfit dislocation propagation in Ge_xSi_{1-x}/Si(100) heterostructures," *Appl. Phys. Lett.* **52**, 1605 (1988); and B.W. Dodson and J.Y. Tsao, "Non-Newtonian strain relaxation in highly strained SiGe heterostructures," *Appl. Phys. Lett.* **53**, 2498 (1988).

time evolution of excess stress and strain[43] in the two simple structures shown in Figures 5.12 and 5.13. We imagine that, as the two structures are grown, they are subjected to the hypothetical (but realistic) temperature cycles shown in Figures 5.18 and 5.19. In each case, growth of a 600-Å Si buffer layer at 750°C commences at $t = 0$. Toward the end of growth of this buffer, the temperature is ramped down to 550°C for growth of the two 100-Å-thick buried $Si_{1-x}Ge_x$ strained layers. During growth of the strained layers, the single-kink excess stresses (evaluated at the rear of the deepest strained layer) increase, but during growth of the unstrained spacer and capping layers they decrease. At the end of growth, the temperature is ramped down to room temperature (25°C). Finally, we have included the possibility of a 45-s, 900°C post-growth rapid thermal anneal for dopant activation or oxide growth.

For the weakly strained structure in Figure 5.18, the final structure has a single-kink excess stress that is barely positive, and so is fairly resistant to strain relief by plastic flow. Even the 45-s rapid thermal anneal at 900°C causes a strain relaxation less than 10^{-7}. Note that this amount of relaxation may be considered nearly unobservable, even by dislocation counting, since it corresponds to on the order of one dislocation per centimeter. Note also that just after growth of the final strained layer the structure passes through an intermediate structure for which the single-kink excess stress is greatest. However, because the growth temperature is low, negligible strain relaxation occurs.

For the moderately strained structure shown in Figure 5.19, the final structure has a single-kink excess stress that is larger, and hence is less resistant to strain relief by plastic flow. Indeed, during the 45-s rapid thermal anneal at 900°C, strain relaxation is significant. Again, note that just after growth of the final strained layer the structure passes through an intermediate structure for which the single-kink excess stress is greatest. However, because the growth temperature is low, negligible strain relaxation occurs.

5.3.4 Stability Diagrams

From the discussions in Subsections 5.3.1–5.3.3, it is clear that the major determinant of the stability of coherent strained heterostructures is its excess stress, convolved with the time-temperature cycle that it experiences during growth and processing. If the excess stress everywhere in the structure is at all times less than zero, then the coherent structure is absolutely stable. If, during some time interval, the excess stress anywhere in the

[43] J.Y. Tsao and B.W. Dodson, "Time, temperature and excess stress: relaxation in strained heterostructures," *Surf. Sci.* **228**, 260 (1990).

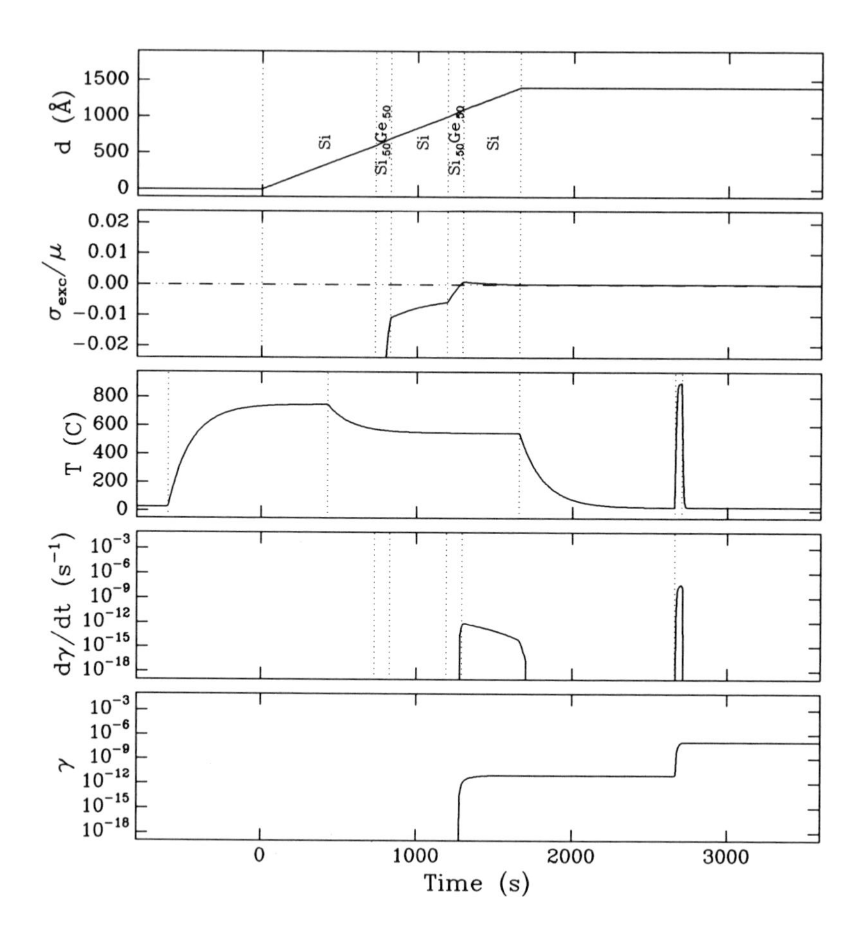

Figure 5.18: Thickness, excess stress (evaluated at the rear of the deepest strained layer), temperature, strain rate, and strain relaxation in a slightly metastable double buried quantum-well heterostructure.

structure rises above zero, then some strain relaxation will occur. However, the *amount* of strain relaxation may be small if the temperature during that time interval is low, or if the duration of the time interval is short. In other words, it is the *time at temperature while the excess stress is highest* that determines whether significant strain relaxation will occur.

For a given time duration, then, the two parameters that most directly determine the amount of strain relaxation that will occur are the excess

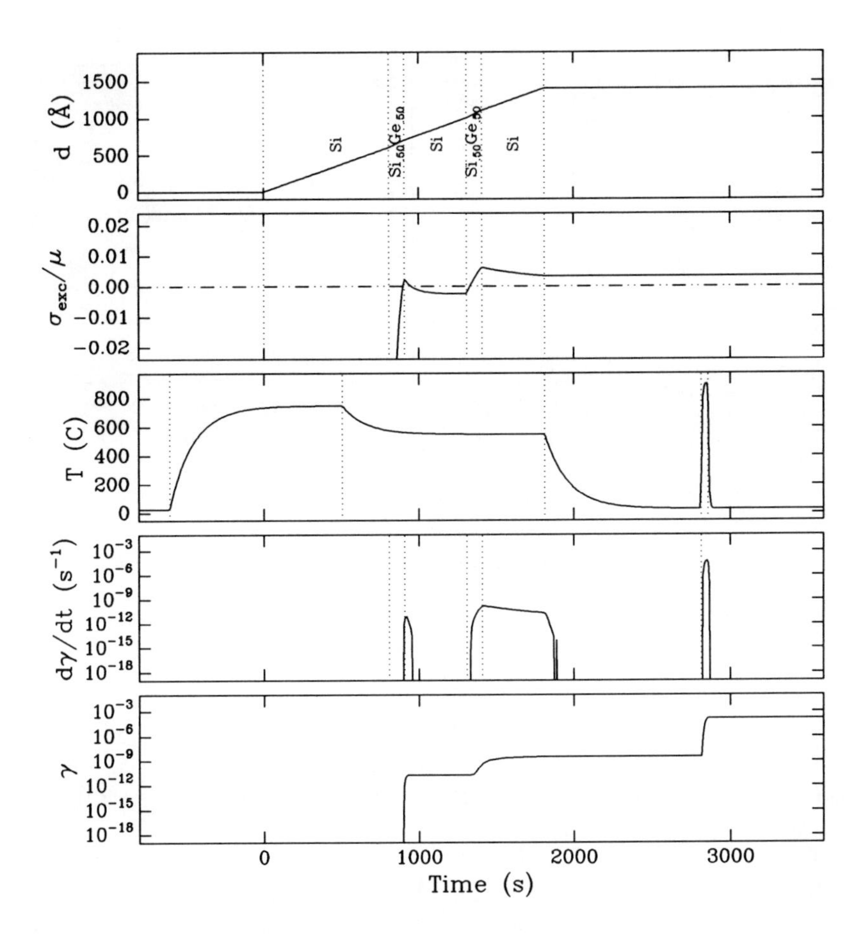

Figure 5.19: Thickness, excess stress (evaluated at the rear of the deepest strained layer), temperature, strain rate, and strain relaxation in a highly metastable double buried quantum-well heterostructure.

stress and temperature.[44] In this subsection, we describe the use of excess stress versus temperature stability diagrams for depicting various regimes of strain relaxation. To illustrate, we show, in Figure 5.20, contours of constant strain relaxation plotted on a stress–temperature diagram. The contours were calculated according to the simple phenomenological model described earlier by Equation 5.48, and so should only be taken as qual-

[44] J.Y. Tsao, B.W. Dodson, S.T. Picraux, and D.M. Cornelison, "Critical stresses for $Si_{1-x}Ge_x$ strained-layer plasticity," *Phys. Rev. Lett.* **59**, 2455 (1987).

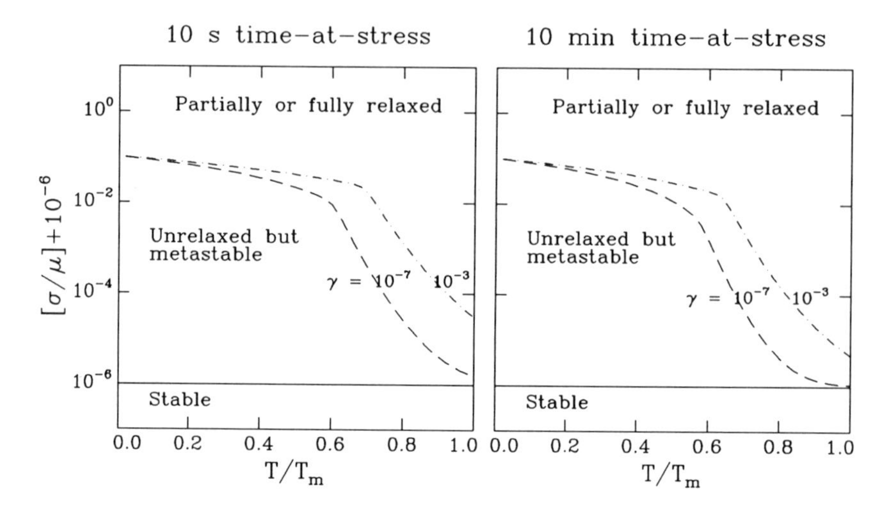

Figure 5.20: Stress–temperature stability diagrams for strained $Si_{1-x}Ge_x$ heterostructures, assuming times-at-stress of 10 s (left) and 10 min (right). Structures lying below the solid lines are absolutely stable. Structures lying below the dashed lines will have relaxed by less than $\approx 10^{-7}$. Structures lying below the dot-dashed lines will have relaxed by less than $\approx 10^{-3}$.

itative guides. Nevertheless, they illustrate how such kinetic models can be used to construct these "stability" diagrams. Such diagrams are *practical* guides to the degree of relaxation that can be expected for a given structure.

The diagram on the left in Figure 5.20 was calculated assuming a "time-at-stress" of 10 s. Such a diagram would be appropriate for the growth of a buried strained quantum well, in which the excess stress of the structure reaches its maximum just after the buried strained layer has been grown, but diminishes quickly thereafter upon initiation of growth of the unstrained capping layer.

The diagram on the right in Figure 5.20 was calculated assuming a time-at-stress of 10 min. Such a diagram would be appropriate for the growth of a thick surface strained layer, in which the excess stress of the structure reaches its maximum gradually during growth, and persists during the cool-down after growth has terminated.

In both diagrams, structures lying below the solid lines, whose excess stresses are less than zero, are absolutely stable. Structures lying below the dashed lines will have relaxed by less than $\approx 10^{-7}$. This amount of strain relaxation is essentially negligible, because it corresponds to on the order of one misfit dislocation per centimeter. Structures lying below the dot-dashed lines will have relaxed by less than $\approx 10^{-3}$. This amount of strain relaxation is not negligible, because it corresponds to on the order of one misfit dislocation per micrometer.

Note that the definition of the stress-temperature boundary at which strain relaxation just becomes observable depends on the sensitivity of the technique used to measure the relaxation.[45] On the one hand, if the measurement technique is sensitive to isolated dislocations in a large field of view, as x-ray topography or etch-pit delineation might be, then at high temperatures the critical stresses approach zero. On the other hand, if the measurement technique is less sensitive (e.g., x-ray diffraction or ion-beam channeling), then the critical stresses may differ significantly from zero, and various degrees of metastability will be observed.

Suggested Reading

1. D. Hull and D.J. Bacon, *Introduction to Dislocations*, 3rd Ed., International Series on Materials Science and Technology, Vol. 37 (Pergamon Press, Oxford, 1984).

2. J.P. Hirth and J. Lothe, *Theory of Dislocations*, 2nd Ed. (John Wiley and Sons, New York, 1982).

3. T.P. Pearsall, Volume Ed., *Strained-Layer Superlattices: Materials Science and Technology*, Semiconductors and Semimetals Vol. 33, R.K. Willardson and A.C. Beer, Series Eds. (Academic Press, Boston, 1991).

4. H.J. Frost and M.F. Ashby, *Deformation-Mechanism Maps* (Pergamon, Oxford, 1982).

Exercises

1. An alternative route to misfit accommodation involves tilting of the epitaxial layer with respect to the substrate.[46] The interface con-

[45] I.J. Fritz, "Role of experimental resolution in measurements of critical layer thickness for strained-layer epitaxy," *Appl. Phys. Lett.* **51**, 1080 (1987).

[46] See, e.g., H. Brooks, "Theory of internal boundaries," in *Metal Interfaces* (American Society of Metals, 1952), pp. 20-64; W.A. Jesser, "On the extension of Frank's formula to crystals with different lattice parameters," *Phys. Stat. Sol.* **A20**, 63 (1973); G.H. Olsen and R.T. Smith, "Misorientation and tetragonal distortion in heteroepitaxial vapor-grown III-V structures," *Phys. Stat. Sol.* **A31**, 739 (1975); R. Du and C.P. Flynn, "Asymmetric coherent tilt boundaries formed by molecular beam epitaxy," *J. Phys.* **C2**, 1335 (1990); and J.E. Ayers, S.K. Ghandhi, and L.J. Schowalter, "Crystallographic

Figure 5.21: Untilted (left) and tilted (right) epitaxy.

tains, instead of "misfit" dislocations with Burgers vectors parallel to the interface, "tilting" dislocations with Burgers vectors *perpendicular* to the interface, and forms what is known as an asymmetric tilt boundary.[47] Consider the one-dimensional boundary shown in Figure 5.21, containing a linear array of such tilting dislocations. Show that the parallel strain in the epitaxial layer decreases with tilt angle, θ, according to

$$\epsilon = 1 - \frac{1-f}{\cos\theta}, \tag{5.50}$$

where f is the misfit between the epitaxial layer and the substrate. Then show that the dislocation density increases with tilt angle according to

$$\rho \approx \theta / b_{\mathrm{edg},\perp}, \tag{5.51}$$

where $b_{\mathrm{edg},\perp}$ is the magnitude of the edge component of the Burgers vector perpendicular to the interface.

Finally, calculate and compare the dislocation density dependence of the coherency strain and dislocation array energies, and deduce the "critical layer thickness" associated with strain relaxation by tilting dislocations. Is the critical layer thickness greater than or less than that associated with strain relaxation by misfit dislocations? Does the total energy increase or decrease at first for small tilts? All other things equal, which is more likely — strain relaxation by tilting or misfit dislocations?

2. A second alternative route to misfit accommodation is through the introduction of islanding or surface roughness.[48] Consider the two

---

tilting of heteroepitaxial layers," submitted to *J. Cryst. Growth.*

[47] S. Amelinckx and W. Dekeyser, "The structure and properties of grain boundaries," in *Solid State Physics*, Vol. 8, F. Seitz and D. Turnbull, Eds. (Academic Press, New York, 1959), pp. 325–499.

[48] D.J. Eaglesham and M. Cerullo, "Dislocation-free Stranski-Krastanow growth of Ge

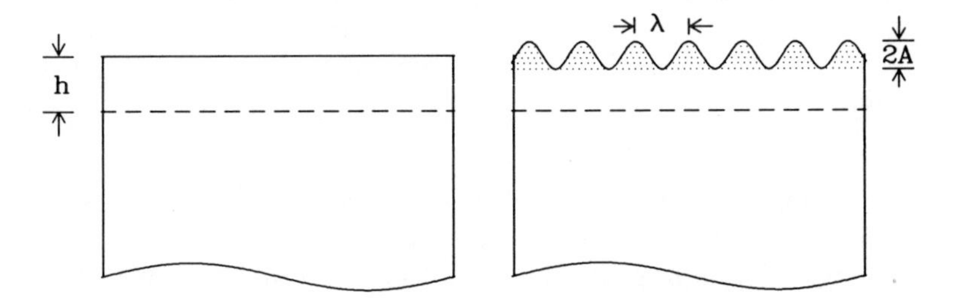

Figure 5.22: Uncorrugated (left) and corrugated (right) epitaxial strained layers.

structures in Figure 5.22, the left composed of a planar strained layer of thickness h, the right composed of a strained layer which has developed a one-dimensional sinusoidal corrugation $A \sin(2\pi x/\lambda)$.

Suppose, due to lateral relaxation, that the strain in the corrugated (dotted) part of the strained layer is reduced to zero.[49] What is the total coherency energy, u_{coh}, associated with the entire strained layer? Suppose that the surface energy per unit area of the strained layer is γ. What is the surface energy, u_{surf}, due to the corrugation? How does the total energy, $u_{\text{coh}} + u_{\text{surf}}$, depend on A and λ? For what values of λ does it decrease with increasing A, and hence for what wavelengths is the surface unstable to roughening.[50] Are corrugations more or less likely in high or low surface energy systems? How might a surface that lowers the surface energy make corrugations *less* likely?[51]

3. Even in relaxed films there may be a thermodynamic driving force for surface roughening. Qualitatively, how does the *equilibrium* energy

---

on Si (100)," *Phys. Rev. Lett.* **64**, 1943 (1990); S. Guha, A. Madhukar, and K.C. Rajkumar, "Onset of incoherency and defect introduction in the initial stages of molecular beam epitaxical growth of highly strained $In_{1-x}Ga_xAs$ on GaAs (100)," *Appl. Phys. Lett.* **57**, 2110 (1990); and K. Sakamoto, T. Sakamoto, S. Nagao, G. Hashiguchi, K. Kuniyoshi, and Y. Bando, "Reflection high-energy electron diffraction intensity oscillations during Ge_xSi_{1-x} MBE growth on Si (001) substrates," *Jpn. J. Appl. Phys.* **26**, 666 (1987).

[49] This is a very crude assumption; for better assumptions see, e.g., S. Luryi and E. Suhir, "New approach to the high quality epitaxial growth of lattice-mismatched materials," *Appl. Phys. Lett.* **49**, 140 (1986).

[50] D.J. Srolovitz, "On the stability of surfaces of stressed solids," *Acta Metall.* **37**, 621 (1989); and C.W. Snyder, B.G. Orr, D. Kessler, and L.M. Sander, "Effect of strain on surface morphology in highly strained InGaAs films," *Phys. Rev. Lett.* **66**, 3032 (1991).

[51] M. Coppel, M.C. Reuter, E. Kaxiras, and R.M. Tromp, "Surfactants in epitaxial growth," *Phys. Rev. Lett.* **63**, 632 (1989).

per unit volume, u_{tot}/h, of a misfitting layer depend on the height of the film, both below and above the critical layer thickness? Just at the critical layer thickness, can the film reduce its energy by decomposing into some regions infinitesimally thicker, and other regions infinitesimally thinner?

4. Consider a misfit dislocation lying along the y-axis, as illustrated in Figure 5.23. Its Burgers vector $\bar{b}$ can be defined either by the pair of angles λ and δ, or by the pair of angles α and β. Show that the Burgers vector of the dislocation is

$$\bar{b} = b \begin{pmatrix} \cos\lambda \\ \sin\lambda\cos\delta \\ \sin\lambda\sin\delta \end{pmatrix} = b \begin{pmatrix} \sin\alpha\sin\beta \\ \cos\beta \\ \cos\alpha\sin\beta \end{pmatrix}, \quad (5.52)$$

and that the Peach-Koehler coherency force acting to create unit length of the dislocation is

$$\begin{aligned} F_{coh} &= h\hat{z} \cdot \left[(\bar{b} \cdot \bar{\bar{\sigma}}) \times \hat{l} \right] \\ &= b\sigma_{coh} h \cos\lambda = b\sigma_{coh} h \sin\alpha\sin\beta \\ &= b_{edg,||}\sigma_{coh} h. \end{aligned} \quad (5.53)$$

For a $\beta = 60°$ dislocation with, as illustrated in Figures 5.4 and 5.5, $\bar{l} = [110]$ and $\bar{b} = [\bar{1}0\bar{1}]/2$, what are the angles α, λ, γ, and δ?

5. Consider the double quantum well structure shown in Figures 5.24, in which two strained quantum wells of thicknesses h_{str} and strains f_{str} are spaced apart by an unstrained layer of thickness h_{spa}, and capped by an unstrained layer of thickness h_{cap}. Given h_{str} and f_{str}, what must h_{cap} be in order for the structure to be stable with respect to misfit dislocation formation at a depth $z = z_1$? What must h_{spa} be in order for the structure to be stable with respect to misfit dislocation formation at a depth $z = z_2$? Derive expressions for the time-evolution of the single-kink excess stresses at z_1 and z_2 during growth of the structure.

6. Suppose one considers two types of dislocations, misfit dislocations lying in the epilayer/substrate interface, with density ρ_{md}, and dislocation segments threading upward to the epilayer surface, with density n_{td}. The units of ρ_{md} and n_{td} are cm^{-1} and cm^{-2}, respectively. Suppose that misfit dislocations are created exclusively by lateral bending of threading segments at velocity v; that threading segments are created exclusively by half-loop nucleation at the free surface at a rate

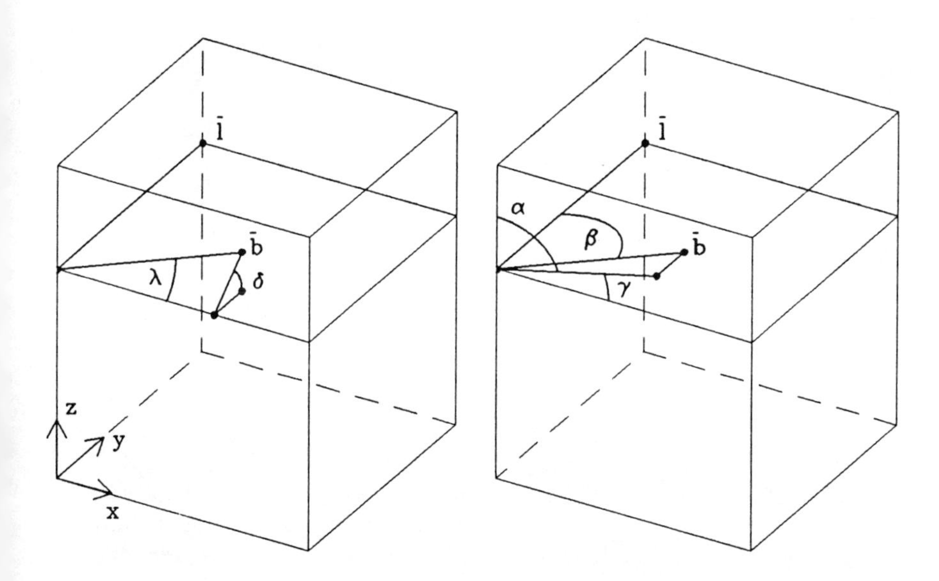

Figure 5.23: Angles commonly used to define misfit dislocations and their Burgers vectors. For convenience, the dislocation line, $\bar{l}$, is taken to be oriented along the y-axis. Left: λ is the angle between (a) the Burgers vector and (b) the direction that is both normal to the dislocation line and within the plane of the interface; δ is the angle between (a) the dislocation line and (b) the projection of the Burgers vector onto the plane containing the dislocation that is perpendicular to the plane of the interface. Right: β is the angle between (a) the Burgers vector and (b) the dislocation line; α is the angle between (a) the slip plane containing both $\bar{b}$ and $\bar{l}$ and (b) the perpendicular to the plane of the interface; and γ is the angle between (a) the slip plane containing both $\bar{b}$ and $\bar{l}$ and (b) the plane of the interface.

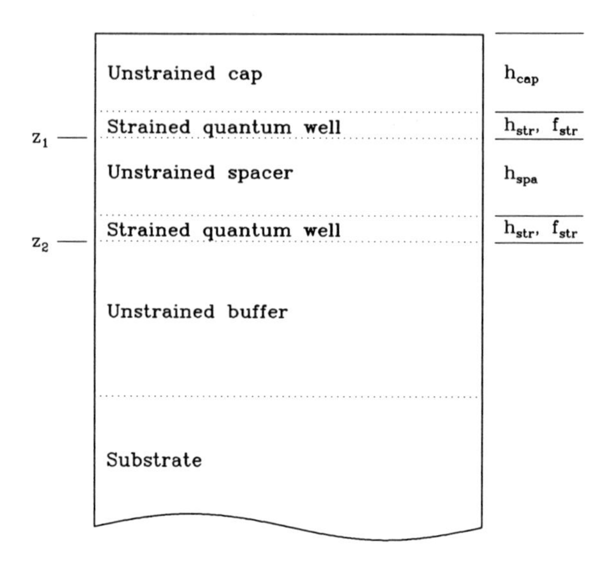

Figure 5.24: A generic strained double quantum-well heterostructure.

j; that threading segments are pinned with probability η by interactions with misfit dislocations; and that multiplication of threading segments by interactions with misfit dislocations is negligible. Show that

$$\begin{aligned} \dot{\rho}_{\rm md} &= v n_{\rm td} \\ \dot{n}_{\rm td} &= j - \eta v n_{\rm td} \rho_{\rm md} \end{aligned} \tag{5.54}$$

are a set of coupled first-order differential equations for the time evolution of the two kinds of dislocation densities.[52]

7. Suppose again that the nucleation rate of dislocation half-loops is j

[52]R. Hull, J.C. Bean, and C. Buescher, "A phenomenological description of strain relaxation in $Ge_xSi_{1-x}/Si(100)$ heterostructures," *J. Appl. Phys.* **66**, 5837 (1989).

and that the velocity at which they propagate to form misfit dislocations is v. Suppose the maximum dislocation propagation length is l, due to pinning by lithographically fabricated boundaries.[53] Show that the misfit dislocation creation rate in the low velocity limit is $\dot{\rho_{md}} = vjt$. What is the misfit dislocation creation rate in the high velocity limit? Does this rate increase or decrease with a decrease in the spatial scale of the lithographic patterning?

[53] E.A. Fitzgerald, G.P. Watson, R.E. Proano, D.G. Ast, P.D. Kirchner, G.D. Pettit, and J.M. Woodall, "Nucleation mechanisms and the elimination of misfit dislocations at mismatched interfaces by reduction in growth area," *J. Appl. Phys.* **65**, 2220 (1989).

Part III

Surface Morphology and Composition

In Part I, we described phase transformations from vapors to bulk crystals, and in Part II, we described phase transformations from vapors to thin epitaxial films. In both parts, we were primarily interested in the properties of starting and ending states, so as to understand the *thermodynamic* competition between different possible transformations. Ultimately, though, all these transformations are mediated by *kinetic* processes, many of which occur on the surface. Therefore, the paths along which various transformations occur can depend crucially on the properties of the surface.

In this part, we discuss the equilibrium and nonequilibrium properties of surfaces. Two of the properties most important to MBE are the morphology and composition of surfaces. In principle, these two aspects of surfaces may be coupled in a complex, interdependent way. In this book, however, we neglect these interdependencies. We start, in Chapter 6, by treating surface morphology assuming that surface composition is unimportant. Then, in Chapter 7, we treat surface composition assuming that surface morphology is unimportant.

Chapter 6

Surface Morphology

We start, in this chapter, by treating surface morphology. At the outset, it is important to distinguish between two aspects of surface morphology: *structure*, the crystallography of defect-free surfaces, and *microstructure*, the distribution of point and line defects that interrupt that perfect crystallography. In this chapter we will be mainly concerned with microstructure and, to a much lesser extent, structure.

It is also important to distinguish between two kinds of surfaces: high-symmetry *singular* surfaces, at whose orientations surface free energies are cusped and have discontinuous first derivatives; and *vicinal* surfaces miscut slightly from singular orientations, composed of singular terraces separated by steps. In this chapter we will be concerned with both of these kinds of surfaces.

We first ask, in Section 6.1: what are the statistics of defects on singular and vicinal surfaces in equilibrium with their vapor, i.e., in the absence of net growth? We will find, not surprisingly, that those statistics depend both on temperature as well as on the average orientation of the surface. Moreover, those statistics are themselves a major determinant of the orientation dependences of surface free energies.

We then ask, in Section 6.2: *given* full knowledge of surface free energies, what is the equilibrium morphology of crystals, surfaces, and thin films? On the one hand, in one-material systems, e.g., "homoepitaxial" films of one material on substrates of the same material, morphology is determined by the orientation dependence of the surface free energy. On the other hand, in two-material systems, e.g., for "heteroepitaxial" films of one material on substrates of a different material, morphology is also determined by interface and volume free energies.

We finally ask, in Section 6.3: what is the defect microstructure of

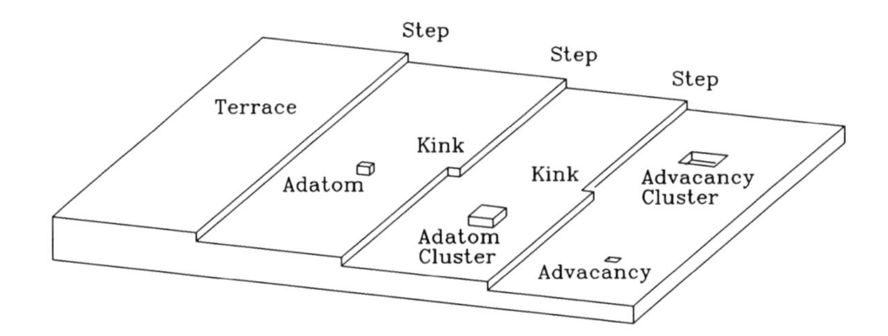

Figure 6.1: Terraces, ledges, kinks, adatoms, adatom clusters, advacancies, and advacancy clusters on a Kossel crystal.

surfaces *not* in equilibrium with their vapor, i.e., in the presence of net growth? We will find that growth is essentially a competition between surface defects of various kinds for adatoms arriving from the vapor. This competition results in a rich and often oscillatory time evolution to the overall microstructure of the surface.

6.1 Statistics of Adatoms, Kinks, and Steps

Let us start, in this section, by discussing defects on surfaces. Consider the idealized (001) surface of a cubic elemental crystal. For simplicity, we suppose it to be "unreconstructed," in that bonds dangling into free space do not rehybridize into pairs or higher order atom arrangements. The important microstructural features of the idealized surface of such a "Kossel" crystal[1] are illustrated in Figure 6.1.

At low to medium temperatures, the dominant microstructural features are terraces, steps and kinks. The terraces can be considered planar defects in a bulk three-dimensional crystal. Separating terraces of different heights are ledges, or steps, which are line defects on a two-dimensional surface. Finally, along these steps there may also be kinks, which are point defects on one-dimensional steps.

[1]W. Kossel, *Nachr. Ges. Wiss. Gottingen*, p. 135 (1927); I.N. Stranski, *Z. Phys. Chem.* **136**, 259 (1928).

At higher temperatures, or away from equilibrium, microstructural features such as adatoms and advacancies, either isolated or clustered into two-dimensional islands, become important. We will begin, in Subsection 6.1.1, by treating adatoms on singular surfaces. Then, in Subsection 6.1.2, we treat kinks in isolated steps. Finally, in Subsection 6.1.3, we treat interacting steps on vicinal surfaces.

6.1.1 Adatoms on Singular Surfaces

Let us start, in this subsection, by considering adatoms, which we imagine adding one by one to a flat, singular surface. There are two extreme ways in which the adatoms can be distributed on this surface. First, they can cluster together predominantly into a half sheet, as illustrated at the bottom of Figure 6.2, so as to maximize the number of lateral in-plane bonds and hence minimize energy. Second, they can distribute randomly, as illustrated at the top of Figure 6.2, so as to maximize configurational entropy.

To describe qualitatively the competition between these two kinds of distributions,[2] consider the number of bonds formed as a new adatom arrives on the surface. The new adatom has four dangling lateral bonds and one dangling vertical bond, but has also "annihilated" the dangling vertical bond of the atom underneath it. Therefore, the adatom has associated with it four "missing" bonds. If each bond has an energy w, then the adatom has associated with it an energy $4w$.

Note, though, that as the adatom coverage, θ, on the surface builds up, adatoms will occasionally find themselves next to other adatoms. If the adatoms are distributed randomly, then the sites adjacent to a given adatom have a probability θ of being occupied. Since there are four such sites, the energy associated with that adatom decreases by $4w\theta$. The energy per adatom is therefore $4w - 4w\theta$, or $4w(1-\theta)$. Altogether, the energy per surface site is the adatom coverage times the energy per adatom, or

$$u_{\text{adat}} = 4w\theta(1-\theta). \tag{6.1}$$

This energy is exactly that (see Table 3.1 on page 50) associated with a two-component strictly regular solution in which the two components are considered to be adatoms and "missing" adatoms. Viewed in this way, the

[2]See, e.g., K.A. Jackson, "Theory of crystal growth," in *Treatise on Solid State Chemistry, Vol. 5*, N.B. Hannay, Ed. (Plenum Press, New York, 1975), pp. 233-282; and D.E. Temkin, "O molekulyarnoi sherokhovatosti granitsy kristall-rasplav (On molecular roughness of the crystal-melt interface)," in *Mekhanizm i kinetika kristallizatsii (Mechanism and Kinetics of Crystallization)*, N.N. Sirota, Ed. (Nauka i Tekhnika, Minsk, 1964), p. 86.

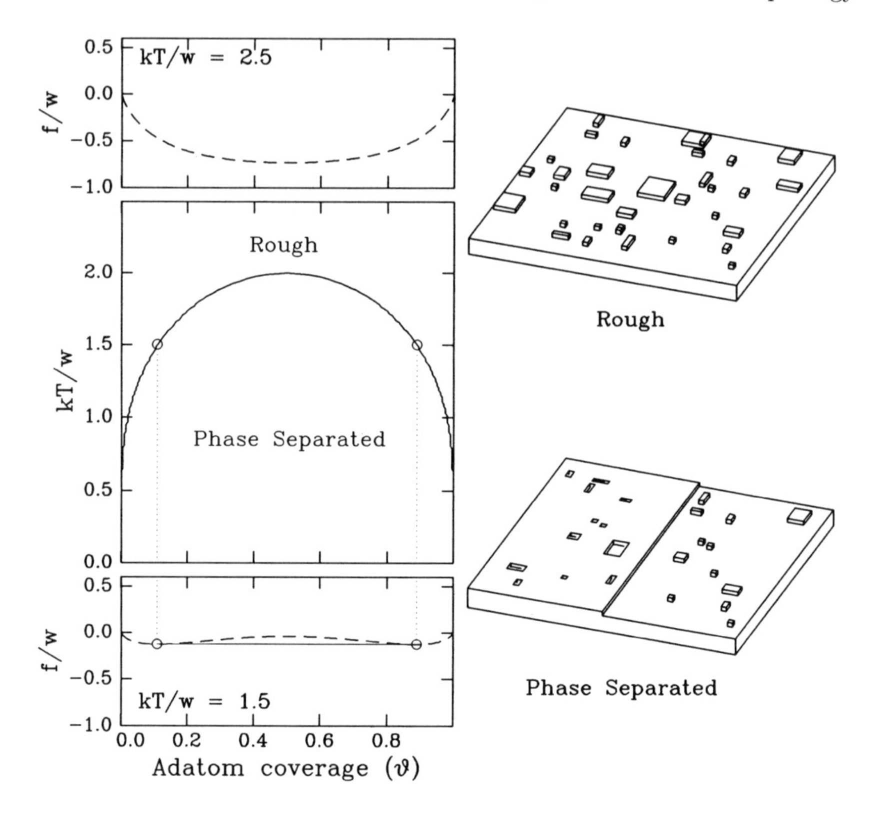

Figure 6.2: x-T phase diagrams for surface roughening. Above and below each phase diagram are also shown the normalized molar free energies of the adlayer phases at $kT/w = 2.5$ and $kT/w = 1.5$, their common tangents, and the critical compositions (open circles) determined by those common tangents.

ideal configurational entropy of mixing per surface site is, by analogy to Equation 3.24,

$$-\frac{s_{\text{adat}}}{k} = \theta \ln \theta + (1-\theta)\ln(1-\theta). \tag{6.2}$$

The free energy per surface site, normalized to the bond strength, is then

$$\begin{aligned} \frac{f_{\text{adat}}}{w} &= \frac{u_{\text{adat}} - Ts_{\text{adat}}}{w} \\ &= 4\theta(1-\theta) + \frac{kT}{w}\left[\theta\ln\theta + (1-\theta)\ln(1-\theta)\right]. \end{aligned} \tag{6.3}$$

This normalized free energy is shown in Figure 6.2 for two different normalized temperatures.

At low temperatures, the bond energy contribution dominates, and the free energy curve is basically concave down. Hence, an adlayer having an average coverage of 1/2 can minimize its free energy by "phase-separating" into regions having near-zero coverage and other regions having near-unity coverage. Note that, just as in the discussion of Section 3.1, the $\theta = 0$ and $\theta = 1$ intercepts of the tangents to the free energy curve are the chemical potentials of the missing adatoms and adatoms, respectively. Therefore, the two phases can only be in equilibrium on the surface if the chemical potentials of their two components are equal. In other words, again following the discussion of Section 3.1, the "compositions" of the two phases are determined by the familiar common tangent construction. Physically, the adlayer minimizes its free energy if most of the adatoms condense into a smooth sheet having a large number of lateral in-plane bonds, with a few stray adatoms to increase configurational entropy.

At high temperatures, the entropy contribution dominates, and the free energy is everywhere concave up. Then, adlayers of *any* composition are stable against phase separation into clusters of adatoms and clusters of missing adatoms. The adatoms are distributed randomly and the surface appears microscopically "rough."

The critical temperature separating smooth, phase-separated adlayers from microscopically rough adlayers is the so-called roughening temperature. It is essentially the critical temperature above which the miscibility gap in this two-component solution vanishes. Since the miscibility gap vanishes when the free energy curve at $\theta = 0.5$ just becomes concave up, the critical temperature is that temperature at which $[\partial^2 f/\partial\theta^2]_{\theta=0.5} = 0$, or $T_{\mathrm{r,adat}} = 2w/k$. Note that the enthalpy of sublimation for this Kossel crystal is the bond energy (w) times the number of bonds per atom (6), divided by the number of atoms per bond (2). Therefore, $\Delta h_{\mathrm{sub}} = 3w$, and we have

$$T_{\mathrm{r,adat}} \approx \frac{2}{3}\Delta h_{\mathrm{sub}}. \tag{6.4}$$

We emphasize that this equation can only give a crude indication of the actual roughening temperature of real crystal surfaces. Its derivation neglected, among other things, multilayer roughness, next-nearest-neighbor and longer-range adatom-adatom interactions, and possible dependences of adatom energies on cluster sizes due to surface reconstruction effects, all of which will tend to decrease $T_{\mathrm{r,adat}}$. Nevertheless, the main idea is that a critical temperature exists above which the equilibrium surface is rough. In some cases, though, this temperature may be above the melting temperature of the crystal, and hence will be unobservable.

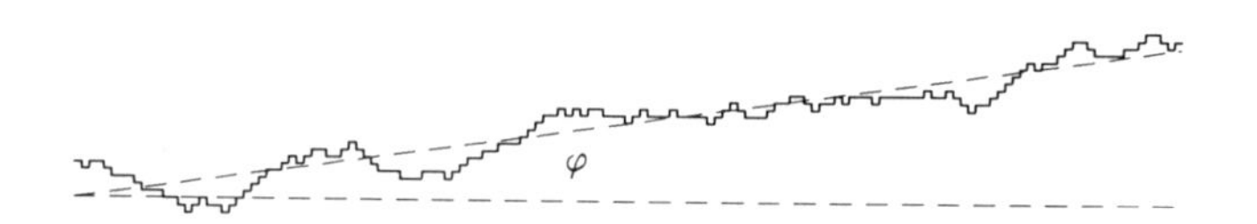

Figure 6.3: A step wandering on the surface of a Kossel crystal due to randomly distributed kinks.

6.1.2 Kinks in Isolated Steps

In Subsection 6.1.1, we considered adatom "excitations" on a singular surface. In practice, real surfaces nearly always contain steps. If the steps are far enough apart not to interact, then their energetics are determined by kink "excitations" along their length. In this subsection, we consider such kink excitations in isolated steps.

Consider the isolated step shown in Figure 6.3. Along this step there may be positive or negative kinks that cause the step to wander randomly.[3] On the one hand, this kink-induced step wandering is favorable, in that it increases the entropy of the step. On the other hand, the kinks themselves are unfavorable, because they cost energy. Indeed, for a simple Kossel crystal, the energy of a single-kink can be calculated, as shown in Fig. 6.4, to be $\epsilon_{\text{kink}} = w/2$, where w is the bond strength. For real crystals, however, the energy of a single kink may be considerably different, due to the reconstructed bonds on the surface.

To quantify the statistics of kinks in steps, let us suppose, for simplicity, that kinks that move steps laterally one lattice unit are much more numerous than those which move steps laterally more than one lattice unit. Note, though, that this approximation breaks down when kink energies are low relative to kT (see e.g., Figure 6.5).

If we nevertheless make this approximation, then we are interested in the probabilities, p_+, p_- and p_o, that an arbitrarily chosen position along a step contains either plus or minus single kinks, or no kink, respectively.[4] Since we have excluded all other possibilities, these must sum to unity:

$$p_+ + p_- + p_o = 0. \tag{6.5}$$

[3] J. Frenkel, "On the surface motion of particles in crystals and the natural roughness of crystalline faces," *J. Phys. U.S.S.R.* **9**, 392 (1945).

[4] W.K. Burton, N. Cabrera, and F.C. Frank, "The growth of crystals and the equilibrium structure of their surfaces," *Philos. Trans. R. Soc. London Ser. A* **243**, 299 (1951).

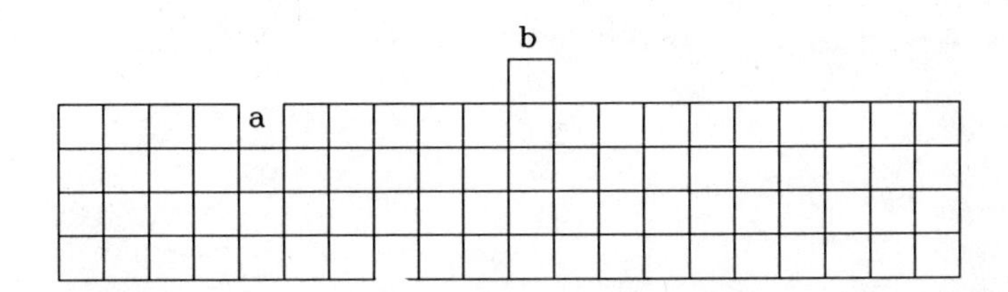

Figure 6.4: In moving an atom from position a to position b, two net bonds are broken, but four kinks are formed. If the bond energy is w, then the energy per kink is $\epsilon_{\text{kink}} = 2w/4 = w/2$.

If we also allow for the step to make a nonzero average angle, ϕ, with the underlying lattice, then the difference between the plus and minus kink probabilities is determined by

$$p_+ - p_- = \tan\phi \equiv p_{\text{ext}}. \tag{6.6}$$

In a sense, p_{ext} defines an "extrinsic" kink probability imposed by the miscut of the step. Then, $p_{\text{int}} = 2p_-$ may be thought of as an "intrinsic" kink probability. Their sum, $p_{\text{int}} + p_{\text{ext}} = p_+ + p_-$, is the total kink probability. If we also assume that the kinks do not interact with each other, then the additional energy of the step due to kinks, per lattice unit along the step, is the total kink probability, times the kink energy, ϵ_{kink}. Hence, the total energy of the step is

$$u_{\text{step}} = \epsilon_{\text{kink}}(p_+ + p_-) + \epsilon_{\text{step}}, \tag{6.7}$$

where ϵ_{step} is the energy per lattice unit of a straight step without kinks.

Since we have assumed the kinks to be independent of each other, the configurational entropy associated with the kinks is determined by the number of ways they may be distinguishably distributed along the length of the step. Following a simple extension of Equation 3.24 to a three-component alloy, the ideal entropy of mixing is

$$-\frac{s_{\text{step}}}{k} = p_+ \ln p_+ + p_- \ln p_- + p_o \ln p_o. \tag{6.8}$$

Altogether, the step free energy is $f_{\text{step}} = u_{\text{step}} - Ts_{\text{step}}$, which we can rewrite in terms of the extrinsic and intrinsic kink probabilities as

$$\begin{aligned} f_{\text{step}} &= \epsilon_{\text{step}} + \epsilon_{\text{kink}}(p_+ + p_-) \\ &\quad + kT[(p_{\text{ext}} + p_{\text{int}}/2)\ln(p_{\text{ext}} + p_{\text{int}}/2) + (p_{\text{int}}/2)\ln(p_{\text{int}}/2) \\ &\quad + (1 - p_{\text{ext}} - p_{\text{int}})\ln(1 - p_{\text{ext}} - p_{\text{int}})]. \end{aligned} \tag{6.9}$$

Figure 6.5: Scanning tunneling micrograph of a Si surface misoriented 0.5° from (001) toward [110]. The surface height decreases from upper left to lower right.[a] On this surface, alternate single-height steps are referred to as type SA and SB, and are smooth and rough, respectively, reflecting the relative energies of kink formation.

[a]B.S. Swartzentruber, *Steps on Si(001): Energetics and Statistical Mechanics* (Ph.D Thesis, U. Wisconsin-Madison, 1992).

For a given extrinsic kink probability, the equilibrium intrinsic kink probability is that which minimizes f_{step}, or

$$\begin{aligned} \frac{\partial f_{\text{step}}}{\partial p_{\text{int}}} &= \epsilon_{\text{kink}} + kT[(1/2)\ln(p_{\text{ext}} + p_{\text{int}}/2) + (1/2)\ln(p_{\text{int}}/2) \\ &\quad - (1/2)\ln(1 - p_{\text{ext}} - p_{\text{int}})] \\ &= 0. \end{aligned} \tag{6.10}$$

Rewriting this in terms of the positive, negative, and missing kink probabilities then gives

$$p_+ p_- = p_o^2 e^{-2\epsilon_{\text{kink}}/kT}. \tag{6.11}$$

Note that this equation reproduces exactly the "quasi-chemical" expression of Equation 4.44. The reason is that equilibrium between kinks on a step can be thought of as a balance between forward and backward chemical reactions, with positive and negative kinks annihilating to form missing kinks, and missing kinks thermally unbinding to form positive and negative kinks.

Equations 6.5, 6.6, and 6.11 are sufficient to determine the three equilibrium kink probabilities, and give, after some algebra,

$$\begin{aligned} p_o^{\text{equ}} &= \frac{1-\sqrt{1-(1-4e^{-2\epsilon_{\text{kink}}/kT})(1-\tan^2\phi)}}{1-4e^{-\epsilon_{\text{kink}}/kT}} \\ p_+^{\text{equ}} &= \frac{\tan\phi}{2}+\sqrt{p_o^2 e^{-2\epsilon_{\text{kink}}/kT}+\frac{1}{4}\tan^2\phi} \\ p_-^{\text{equ}} &= -\frac{\tan\phi}{2}+\sqrt{p_o^2 e^{-2\epsilon_{\text{kink}}/kT}+\frac{1}{4}\tan^2\phi}. \end{aligned} \tag{6.12}$$

For the special case of perfectly cut step for which $\tan\phi = 0$, the energy, entropy, and free energy of an isolated step simplify to

$$\begin{aligned} u_{\text{step}} &= \epsilon_{\text{step}} + 2\epsilon_{\text{kink}} p_- \\ -\frac{s_{\text{step}}}{k} &= 2p_- \ln p_- + (1-2p_-)\ln(1-2p_-) \\ f_{\text{step}} &= u_{\text{step}} - T s_{\text{step}}. \end{aligned} \tag{6.13}$$

These energies, entropies, and free energies, normalized to the energy of a straight step, are plotted in Figure 6.6 as a function of $p_- = p_+$. For concreteness, we have assumed that kinks add an additional energy equal to the energy of the step itself, $\epsilon_{\text{kink}} = \epsilon_{\text{step}} = w/s$, as they would in a Kossel crystal. At all nonzero temperatures, the free energy initially decreases with increasing p_-, due to entropy, and then increases, due to energy. The kink probabilities at which the free energies minimize are given by Equations 6.12, which, in the limit $\tan\phi = 0$, simplify to

$$\begin{aligned} p_o^{\text{equ}} &= \frac{1}{1+2e^{-\epsilon_{\text{kink}}/kT}} \\ p_+^{\text{equ}} = p_-^{\text{equ}} &= \frac{e^{-\epsilon_{\text{kink}}/kT}}{1+2e^{-\epsilon_{\text{kink}}/kT}}. \end{aligned} \tag{6.14}$$

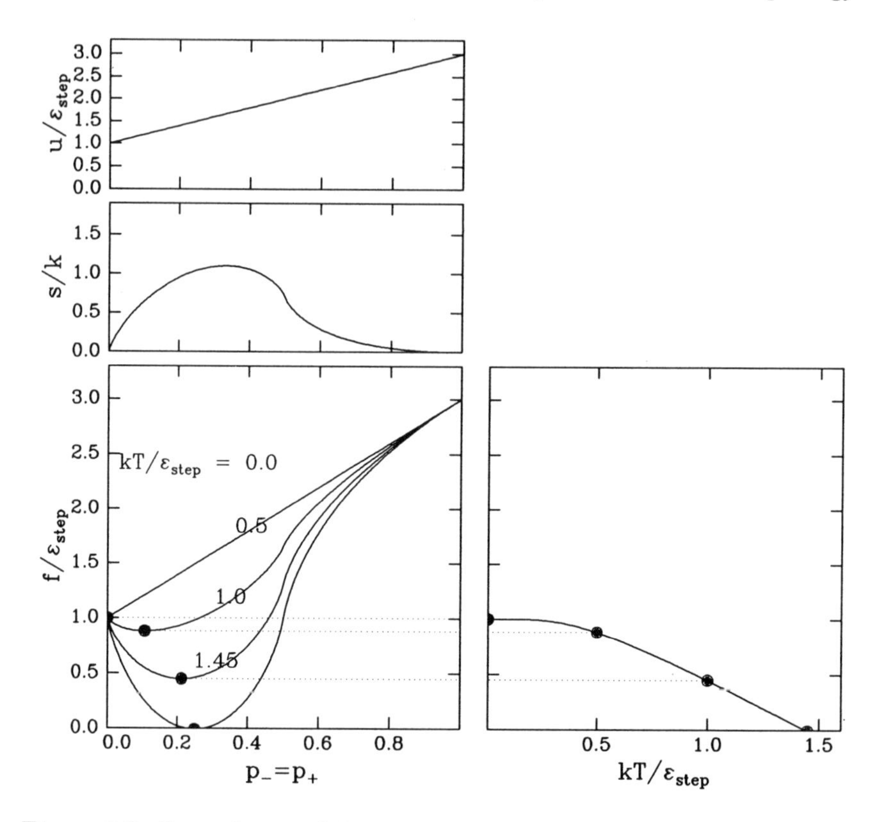

Figure 6.6: Dependences of the energy, entropy, and free energy of a step on the probability of intrinsic kinks. The step is assumed to have no extrinsic kinks, and a kink is assumed to add an additional energy equal to half the energy of a straight step. The energies and free energies are normalized to the energy of a straight step. As the temperature increases, the intrinsic kink probability that minimizes the free energy of the step increases, and the minimum free energy decreases.

At a critical temperature, $T_{r,step}$, the step free energy vanishes at its minimum. Above this temperature, steps will form spontaneously on the surface, and the surface is said to be above its roughening temperature.[5] For a Kossel crystal, this critical temperature is $T_{r,step} \approx \Delta h_{sub}/4$.

Note that this temperature is considerably below that given in the previous subsection by Equation 6.4. Physically, the reason is that, per lattice

[5] An alternative way of calculating the roughening temperature is to calculate the temperature at which the free energies of closed step loops vanishes; see, e.g., A. Zangwill, *Physics at Surfaces* (Cambridge University Press, Cambridge, 1988), pp. 16-17.

site, step excitations on terraces are energetically less costly than adatom and missing adatom excitations. Note also that even this temperature is only a crude indication of the actual roughening temperature of real crystal surfaces. Its derivation neglected, among other things, the possibility that kinks may move steps laterally more than one lattice unit, and the solid-on-solid constraint that prevents steps from crossing each other.

More advanced treatments take both of these effects into account, and are based on an analogy between noncrossing wandering steps on a surface and 1D spinless fermion gases,[6] where the Pauli principle automatically prohibits crossing.[7] These treatments also borrow heavily from studies of domain walls in 2D commensurate adsorbate phases,[8] which are also analogous to 1D spinless fermion gases.[9] The result is that step free energies approach zero at $T_{r,step}$ according to[10]

$$f_{step} \sim e^{-c/\sqrt{|T-T_r|}}, \tag{6.15}$$

and represent a second-order phase transition from smooth to rough.

6.1.3 Steps on Vicinal Surfaces

In Subsection 6.1.2, we calculated the free energy of an isolated step wandering on a surface. The free energy was decreased below that of a perfectly straight step due to the configurational entropy associated with the mixing of positive, negative, and missing kinks. In the absence of step-step interactions, the free energy of a surface depends only on the free energy of the terraces plus those of the steps. For a surface miscut by an angle θ away from the orientation of a singular surface, and hence having a step density per lattice site of $s \equiv \tan\theta$, the free energy, per lattice site, would then be

$$f_{surf} = f_{terr} + f_{step} \tan\theta, \tag{6.16}$$

where f_{terr} is the free energy of the singular, unstepped surface, and f_{step} is given by Equation 6.13.

In this subsection, we consider the possibility that the steps interact, and that those interactions give rise to nonlinear dependences of the surface

[6] C. Jayaprakash, C. Rottman and W.F. Saam, "Simple model for crystal shapes: step-step interactions and facet edges," *Phys. Rev.* **B30**, 6549 (1984).

[7] P.G. de Gennes, "Soluble model for fibrous structures with steric constraints," *J. Chem. Phys.* **48**, 2257 (1968).

[8] J.M. Kosterlitz and D.J. Thouless, "Ordering, metastability and phase transitions in two-dimensional systems," *J. Phys.* **C6**, 1181 (1973).

[9] H.J. Schulz, B.I. Halperin, and C.L. Henley, "Dislocation interaction in an adsorbate solid near the commensurate-incommensurate transition," *Phys. Rev.* **B26**, 3797 (1982).

[10] H.J. Schulz, "Equilibrium shape of crystals," *J. Physique* **46**, 257 (1985).

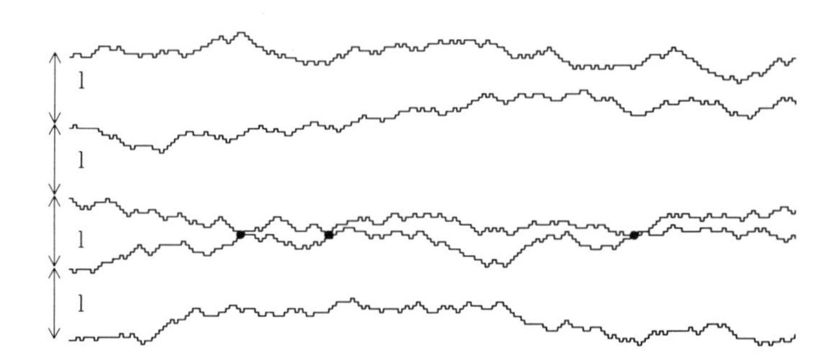

Figure 6.7: Five wandering steps of average spacing l. Two of the steps intersect, by chance, three times.

free energy on $\tan\theta$. Consider the array of steps illustrated in Figure 6.7. On average, they are parallel to each other, but as they wander they occassionally touch. If we do not allow "overhangs" on the surface, then, as mentioned above, the steps are not free to cross each other. Each step is confined by the random wanderings of its immediate neighbors, and its entropy is reduced.[11]

To quantify this entropy reduction, consider again the step intersections shown in Figure 6.7. On the one hand, if the steps were truly independent, then each step intersection point would have two equally likely interpretations: either the steps actually cross, or they bounce back from each other. On the other hand, if the steps cannot cross, then each step intersection can have only the second interpretation. Each step intersection, therefore, has associated with it an entropy decrease of $k\ln 2$.

How often, on average, do the steps intersect? Let $b^2 \equiv p_+a^2 + p_-a^2$ be the mean square lateral displacement of the step per lattice unit. Then, after n lattice units, the step will have wandered laterally on the average $\sqrt{n}b$ lattice units. Therefore, we expect a collision whenever $\sqrt{n}b$ exceeds the mean spacing between steps, l, or every $n = l^2/b^2$ lattice units.[12]

Altogether, the entropy decrease, per lattice unit, is roughly $(1/n)\ln 2 = (b^2/l^2)\ln 2$. More precise calculations, taking into account the simultaneous

[11]E.E. Gruber and W.W. Mullins, "On the theory of anisotropy of crystal surface tension," *J. Phys. Chem. Solids* **28**, 875 (1967); and G.H. Gilmer and J.D. Weeks, "Statistical properties of steps on crystal surfaces," *J. Chem. Phys.* **68**, 950 (1978).

[12]M.E. Fisher and D.S. Fisher, "Wall wandering and the dimensionality dependence of the commensurate-incommensurate transition," *Phys. Rev.* **B25**, 3192 (1982).

wandering of all the steps, give an entropy decrease of[13]

$$\frac{\Delta s_{\text{step}}}{k} = \frac{\pi^2}{6}\frac{b^2/2}{l^2} = \frac{\pi^2}{12}\frac{b^2}{a^2}\tan^2\phi. \tag{6.17}$$

The surface free energy is therefore

$$f_{\text{surf}} = f_{\text{terr}} + f_{\text{step}}\tan\phi + kT\frac{\pi^2}{12}\frac{b^2}{a^2}\tan^3\phi, \tag{6.18}$$

and contains a cubic dependence on step density.

Suppose, now, that in addition to a short-range repulsion preventing step-step crossings, there is also longer range repulsion.[14] Such a repulsion might be generated, e.g., by strain fields in the substrate surrounding each step.[15] For simplicity, suppose the repulsion takes the quadratic form

$$\Delta u_{\text{step}} = A\left[\frac{1}{(l+x)^2} + \frac{1}{(l-x)^2}\right] \approx \frac{2A}{l^2}\left(1 + x^2/l^2\right), \tag{6.19}$$

where $x = \pm l$ are, as illustrated in Figure 6.8, the positions of rigid steps surrounding (and confining) a center, wandering step.

In the presence of this repulsion, the potential energy of the step decreases the less it wanders away from $x = 0$. However, the entropy of the step also decreases, by $\Delta s_{\text{step}} \approx (\pi^2/12)(b^2/x^2)$. The actual wandering will be determined by a balance between the two, or

$$\frac{d(\Delta u_{\text{step}} - T\Delta s_{\text{step}})}{dx} = \frac{d}{dx}\left[\frac{2A}{l^2}\left(1 + \frac{x^2}{l^2}\right) + kT\frac{\pi^2}{12}\frac{b^2}{x^2}\right] = 0. \tag{6.20}$$

Solving Equation 6.20 then gives the equilibrium alley width within which the step will wander:

$$d = \left(\frac{\pi^2}{24}\frac{kTb^2}{A}\right)^{1/4} l. \tag{6.21}$$

[13] C. Jayaprakash, C. Rottman, and W.F. Saam, "Simple model for crystal shapes: step-step interactions and facet edges," *Phys. Rev.* **B30**, 6549 (1984); and V.V. Voronkov, "Free energy of a stepped surface," in *Growth of Crystals*, Vol. 15, E.I. Givargizov and S.A. Grinberg, Eds. (Consultants Bureau, New York, 1988).

[14] Our treatment follows closely that of N.C. Bartelt, T.L. Einstein, and E.D. Williams, "The influence of step-step interactions on step wandering," *Surf. Sci. Lett.* **240**, L591 (1990).

[15] J.M. Blakely and R.L. Schwoebel, "Capillarity and step interactions on solid surfaces," *Surf. Sci.* **26**, 321 (1971); V.I. Marchenko and A. Ya. Parshin, "Elastic properties of crystal surfaces," *Sov. Phys. JETP* **52**, 129 (1980); F.K. Men, W.E. Packard, and M.B. Webb, "Si (100) surface under an externally applied stress," *Phys. Rev. Lett.* **61**, 2469 (1988); and O.L. Alerhand, A.N. Berker, J.D. Joannopoulos, D. Vanderbilt, R.J. Hamers, and J.E. Demuth, "Finite-temperature phase diagram of vicinal Si(100) surfaces," *Phys. Rev. Lett.* **64**, 2406 (1990).

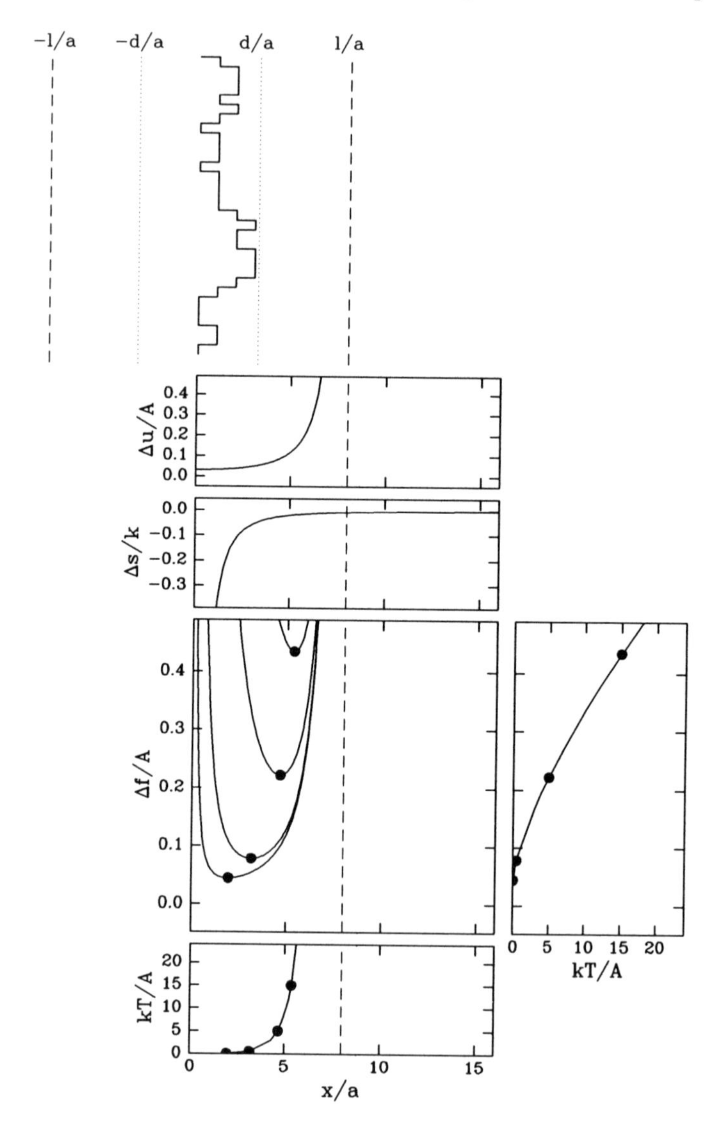

Figure 6.8: Changes in energy, entropy, and free energy as the width of the alley (x/a, in units of lattice constants) within which a step is allowed to wander increases. Both the energy and entropy increase as the width of the alley increases and approaches the mean step spacing (l/a, in units of lattice constants). The equilibrium alley width (d/a, in units of lattice constants) is that which minimizes the free energy. As the ratio between the temperature and the strength of the interaction between steps (kT/A) increases, both the equilibrium alley width and the free energy at that equilibrium alley width increase.

As illustrated in Figure 6.8, this balance will depend on temperature, though only weakly, because of the 1/4 power.

Altogether, the change in free energy, per step, due to step-step interactions, is

$$\Delta f_{\text{step}} = \frac{2A}{l^2}\left(1+\sqrt{\frac{\pi^2 kTb^2}{6A}}\right). \tag{6.22}$$

As expected, the free energy change increases with the step-step interaction strength, A. It also increases with the root-mean-square kink amplitude, b, since the larger b is, the more "difficult" it is to confine the step.

Finally, we can write the surface free energy, per lattice site, as

$$\begin{aligned} f_{\text{surf}} &= f_{\text{terr}} + (f_{\text{step}} + \Delta f_{\text{step}})\tan\theta \\ &= f_{\text{terr}} + f_{\text{step}}\tan\theta + \frac{2A}{a^2}\left(1+\sqrt{\frac{\pi^2 kTb^2}{6A}}\right)\tan^3\theta. \end{aligned} \tag{6.23}$$

Note that the first nonlinear term is cubic, rather than quadratic, in step density. This has consequences, as will be seen in the next section, on the shape of the equilibrium crystal near the $\tan\theta = 0$ orientation.

6.2 Equilibrium Morphology

In Section 6.1, we discussed the statistics of adatoms, kinks, and steps. These statistics are the primary determinants of the orientation dependence of the free energies of vicinal surfaces. In this section, we suppose that we have been *given* complete knowledge of surface free energies, and ask: how do those free energies determine equilibrium morphologies? For macroscopic crystals of constant volume, we will find, in Subsection 6.2.1, that the equilibrium shape is determined by the orientation dependence of the surface free energy through what is known as the Wulff construction. For "planar" surfaces of constant average orientation, we will find, in Subsection 6.2.2, that the equilibrium morphology can be deduced from the orientation dependence of the surface free energy using a common tangent construction. Some average orientations will be stable, while others will tend to break up into combinations of other orientations. Finally, for thin heteroepitaxial films of one material on substrates of another material, we will find, in Subsection 6.2.3, that equilibrium morphologies are determined not only by surface free energies, but by interface and volume free energies as well.

6.2.1 Shapes of Crystals: Wulff's Theorem

We start, in this subsection, by considering a macroscopic crystal of a single material whose overall volume is specified. What shape will this crystal have in equilibrium? We discuss, in turn, three related constructions for equilibrium crystal shapes. The most basic is known as Wulff's construction; from Wulff's construction may be derived what is known as Herring's construction; and from Herring's construction in turn may be derived what is known as Andreev's construction.

Wulff's Construction

Intuitively, we expect the equilibrium shape of a crystal of constant volume to be such that those surfaces whose orientations have less energy will have greater area, while those whose orientations have greater energy will have lesser area.

For example, consider the rectangular prism illustrated in Figure 6.9, bounded by rectangular faces of specific surface free energies γ_x, γ_y, and γ_z. If the distances of each face from the crystal center are h_x, h_y, and h_z, then the face areas are $h_y h_z$, $h_x h_z$, and $h_x h_t$, and the total surface free energy is

$$E = 2\gamma_x h_y h_z + 2\gamma_y h_x h_z + 2\gamma_z h_x h_y. \tag{6.24}$$

If we require the volume, $V = 8h_x h_y h_z$, to be constant, then we can write

$$E = \frac{\gamma_x V}{4h_x} + \frac{\gamma_y V}{4h_y} + 2\gamma_z h_x h_y. \tag{6.25}$$

To find the distances h_x and h_y that minimize the energy, we set $\partial E/\partial h_x = \partial E/\partial h_y = 0$, giving

$$\gamma_x h_y h_z = \gamma_y h_x h_z = \gamma_z h_x h_y = (\gamma_x \gamma_y \gamma_z)^{1/3} V^{2/3} = \text{constant}. \tag{6.26}$$

In other words, the free energies of all the faces of the equilibrium crystal are equal.

Note also that the areas of the faces are inversely proportional to their distances from the center of the crystal [e.g., $h_x h_y = V/(8h_z)$]. Therefore, those distances are in turn proportional to the specific surface free energies:

$$\frac{\gamma_x}{h_x} = \frac{\gamma_y}{h_y} = \frac{\gamma_z}{h_z} = \left(\frac{\gamma_x \gamma_y \gamma_z}{V}\right)^{1/3}. \tag{6.27}$$

In other words, faces of high specific surface free energy lie farther from the center of the crystal than those of low specific surface free energy, and therefore have lower relative surface areas.[16]

[16] P. Curie, "Sur la formation des cristaux et sur les constantes capillaires de leurs différentes faces," *Bull. Soc. Min. de France* **8**, 145 (1885).

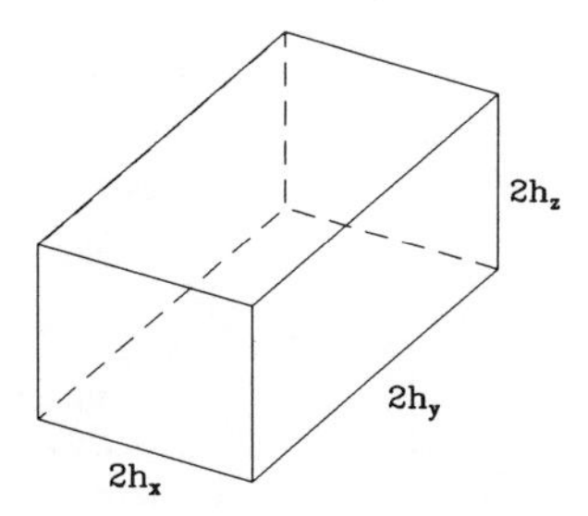

Figure 6.9: Curie's construction for a rectangular prism of fixed volume but minimum surface energy.

The generalization of this argument to all convex bodies is known as the Wulff construction[17]: the crystal shape that minimizes total surface free energy at constant volume is given by the inner envelope of "Wulff" planes perpendicular to and passing through the radius vectors of the orientation-dependent molar surface free energy $\gamma(\theta, \phi)$. This construction is illustrated in two dimensions in Figure 6.10, for a hypothetical $\gamma(\theta)$. As can be seen, this construction places low molar surface free energy orientations nearer to the center of the crystal, thereby increasing their relative surface areas, and places higher molar surface free energy orientations farther from the center of the crystal, thereby decreasing their relative surface areas. Indeed, as we shall see, orientations with very high molar surface free energies may by this construction be placed so far from the center of the crystal that their surface areas vanish entirely, and are no longer represented on the equilibrium crystal shape.

Herring's Construction

An equivalent construction, which may be called Herring's construction,[18] is illustrated in Figure 6.11. One draws spheres passing through the origin and tangent to the $\gamma(\theta, \phi)$ plot. The interior envelope of the points on the spheres diametral to the origin is the equilibrium crystal shape.

To see why, consider the three points labeled O, P, and A on the circumference of the two-dimensional projection of one such sphere. Point O

[17]G. Wulff, *Z. Kristallogr. Mineral.* **34**, 449 (1901); H. Hilton, *Mathematical Crystallography* (Oxford University Press, 1903).

[18]C. Herring, "Some theorems on the free energies of crystal surfaces," *Phys. Rev.* **82**, 87 (1951).

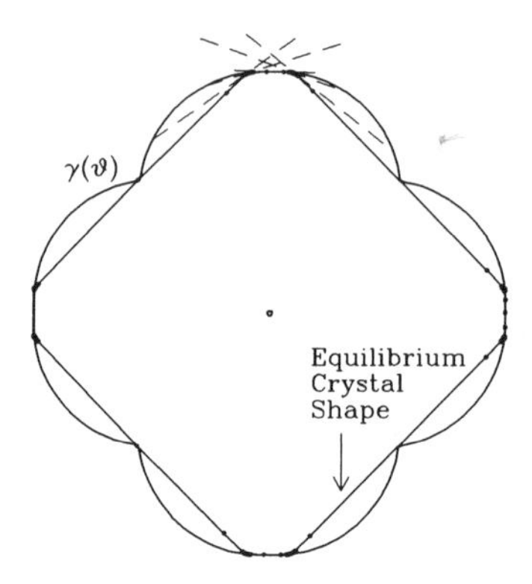

Figure 6.10: Wulff's construction for a two-dimensional crystal of fixed volume but minimum surface energy.

is the origin both of the $\gamma(\theta, \phi)$ plot as well as of the equilibrium crystal. Point A is the point *on* the $\gamma(\theta, \phi)$ plot that the sphere passes tangentially through. Point P is the point on the sphere diametral to the origin.

Because $\overline{OP}$ is a diameter of the sphere, it follows from elementary geometry that the angle $\angle OAP$ is a right angle, and that AP is a Wulff "plane" perpendicular to the $\gamma(\theta, \phi)$ plot at A. Consequently, P is a possible point bounding the equilibrium crystal. To see whether it is an actual such point, we consider two possibilities.

On the one hand, suppose, as illustrated in the left panel of Figure 6.11, that the $\gamma(\theta, \phi)$ plot passes within the tangent sphere at some other point B lying between the origin and another point C on the tangent sphere. Since, again from elementary geometry, $\angle OCP$ must be a right angle, the plane through C at right angles to $\overline{OC}$ must pass through P. Hence, the plane through B at a right angle to $\overline{OB}$ must intersect the line segment $\overline{OP}$ "interior" to the point P, precluding point P from bounding the equilibrium crystal.

On the other hand, suppose, as illustrated in the right panel of Figure 6.11, that the $\gamma(\theta, \phi)$ plot nowhere passes within the tangent sphere. Then, for every point E on the $\gamma(\theta, \phi)$ plot, there must exist some point D on the tangent sphere lying between E and the origin. Since, again, $\angle ODP$ must be a right angle, the plane through D at right angles to OD must pass through P. Hence, the plane through E at a right angle to $\overline{OE}$ must intersect the line through $\overline{OP}$ "exterior" to the point P, and cannot preclude point P from bounding the equilibrium crystal.

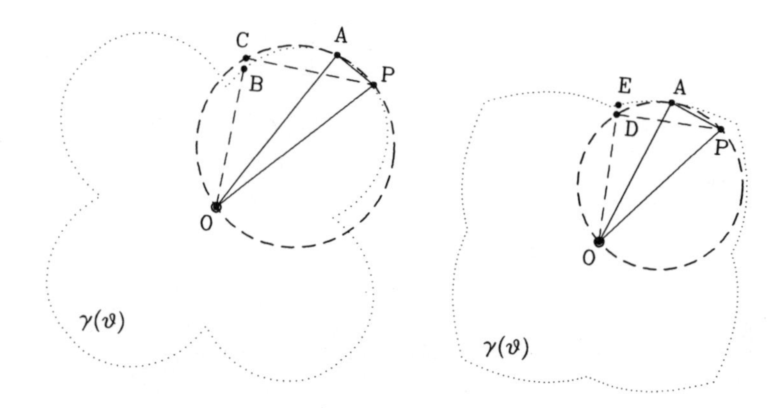

Figure 6.11: Herring's construction for deducing whether orientation A is represented on the equilibrium crystal shape at P. For the $\gamma(\theta)$ plot shown on the left, orientation A is *not* represented, because the $\gamma(\theta)$ plot at another orientation B lies within the tangent circle. For the $\gamma(\theta)$ plot shown on the right, orientation A *is* represented, because at every other orientation (e.g., E) the $\gamma(\theta)$ plot lies outside the tangent circle.

Altogether, the equilibrium crystal shape is the locus of diametral points P on all tangent spheres *not* intersected by other portions of the $\gamma(\theta, \phi)$ plot. Alternatively, one may find first the locus of diametral points P on all tangent spheres without regard to intersections with the $\gamma(\theta, \phi)$ plot, and then take the interior envelope of those points.

Andreev's Construction

In a sense, Herring's construction maps points like A in energy–orientation (θ, ϕ, γ) space onto points like P in real (x, y, z) space. In other words, it tells us where in real space a surface of a particular orientation will appear. To quantify this mapping, consider the circle shown in Figure 6.12 tangent to and passing through the $\gamma(\theta)$ plot at point A. We would like to deduce the (x, z) coordinates of the point P diametral to point O in terms of $\gamma(\theta)$ and $\gamma'(\theta)$ at point A.

First, let us deduce the x coordinates of point P. Denote the lengths of the line segments $\overline{AJ}$ and $\overline{AP}$ by l_1 and l_2, respectively. Then, the x-coordinate of the point P is

$$x = (l_1 + l_2)\cos\theta. \tag{6.28}$$

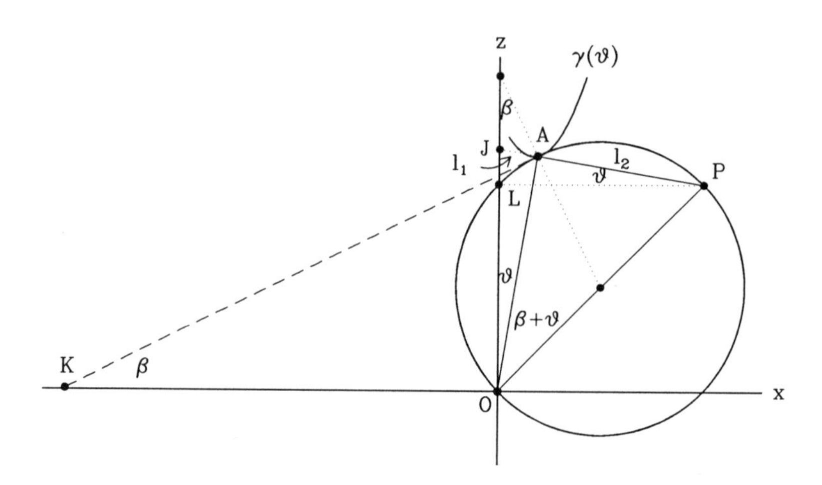

Figure 6.12: Andreev's construction for deducing the mapping between point A in energy–orientation space onto point P on the equilibrium crystal in real space.

Now we note that l_1 is $\gamma \tan\theta$, and if we let β be the angle $\angle AKO$, where the line through $\overline{AK}$ is tangent to $\gamma(\theta)$ at A, then l_2 is $\gamma\tan(\beta+\theta) = \gamma(\tan\beta+\tan\theta)/(1-\tan\beta\tan\theta)$. Hence,

$$x = \gamma\cos\theta\left(\tan\theta + \frac{\tan\beta+\tan\theta}{1-\tan\beta\tan\theta}\right). \tag{6.29}$$

Finally, since $\tan\beta$ is the slope of the $\gamma(\theta)$ plot, we can rewrite it as

$$\tan\beta = \frac{\partial(\gamma\cos\theta)}{\partial(\gamma\sin\theta)} = \frac{\partial(\gamma\cos\theta)/\partial\theta}{\partial(\gamma\sin\theta)/\partial\theta} = \frac{-\gamma\sin\theta+\gamma'\cos\theta}{\gamma\cos\theta+\gamma'\sin\theta}. \tag{6.30}$$

Inserting this expression for $\tan\beta$ into Equation 6.29 then gives, after some algebra,

$$x = \gamma'\cos\theta + \gamma\tan\theta\cos\theta = \frac{\partial(\gamma/\cos\theta)}{\partial(\tan\theta)}. \tag{6.31}$$

Second, let us deduce the z-coordinate of point P. By inspection of Figure 6.12, the z-coordinate is the difference between the lengths of the line segments $\overline{OJ}$ and $\overline{LJ}$. Since $\overline{OJ}$ is $\gamma/\cos\theta$ and $\overline{LJ}$ is $x\tan\theta$, we then have

$$z = \frac{\gamma}{\cos\theta} - x\tan\theta = \frac{\gamma}{\cos\theta} - \tan\theta\frac{\partial(\gamma/\cos\theta)}{\partial(\tan\theta)}. \tag{6.32}$$

Equations 6.31 and 6.32 are explicit algebraic expressions for the Herring construction. Note that both are expressed in terms of $f \equiv \gamma/\cos\theta$ and

$s \equiv \tan\theta$. The first is the surface free energy per unit area projected onto a reference surface of orientation $\theta = 0$ and the second is the slope of the misorientation from $\theta = 0$. In terms of f and s, Equations 6.31 and 6.32 can then be rewritten more conveniently as

$$\begin{aligned} x &= \frac{\partial f}{\partial s} \\ z &= f - s\frac{\partial f}{\partial s}. \end{aligned} \tag{6.33}$$

As illustrated in Figures 6.13 and 6.14, the x-coordinate of the surface of the equilibrium crystal having orientation $\theta = \tan^{-1} s$ is the slope $\partial f/\partial s$, and the z-coordinate is the intercept of the tangent to $f(s)$ with the $s = 0$ axis. This simple and elegant mapping, originally derived by Andreev,[19] may be called Andreev's construction.

Note that this mapping of $f(s)$ onto $z(x)$ is essentially a Legendre transformation analogous to those that map energies onto free energies.[20] For example, recall from Chapter 1 that temperature-dependent Helmholtz free energies can be written as $F(T) = U - S(\partial U/\partial S)$, where $T = \partial U/\partial S$. Hence, the equilibrium crystal shape may be regarded as a kind of free energy in which the "extensive" quantity, s, has been replaced by a conjugate "intensive" quantity, $\partial f/\partial s$.

Note also that for vicinal surfaces characterized by a terrace and step structure, s can be regarded as a step density, and $\partial f/\partial s$ can be regarded as a kind of chemical potential for steps. Viewed in this way, crystals evolve toward their equilibrium shape because their surfaces represent "open" systems with respect to interchange of "steps."

To illustrate the use of this powerful and convenient mapping, consider the $f(s)$ and corresponding $z(x)$ plots shown in Figures 6.13 and 6.14. In Figure 6.13, $f(s)$ near $s = 0$ has been assumed to take the cubic form derived in Section 6.1.3,

$$f(s) = a + bs + ds^3. \tag{6.34}$$

Then, the shape of the equilibrium crystal is given by

$$z(x) = f - s\frac{\partial f}{\partial s} = a - 2ds^3 = a - 2d\left(\frac{x-b}{3d}\right)^{3/2}, \tag{6.35}$$

where we have used the mapping $x = \partial f/\partial s = b + 3ds^2$. Hence, the rounded region of the equilibrium crystal joins the $s = 0$ facet at $x =$

[19] A.F. Andreev, "Faceting phase transitions of crystals," *Sov. Phys. JETP* **53**, 1063 (1982).

[20] C. Rottman and M. Wortis, "Statistical mechanics of equilibrium crystal shapes: interfacial phase diagrams and phase transitions," *Phys. Rep.* **103**, 59 (1984).

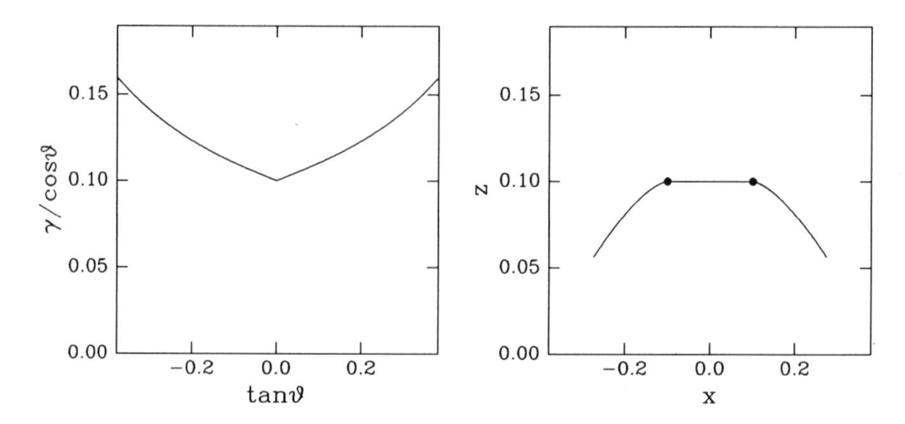

Figure 6.13: Andreev's construction near an orientation around which the projected surface free energy per unit area (left) is concave up in $\tan\theta$, and for which the $\tan\theta = 0$ facet joins the rest of the crystal (right) with a continuous first derivative.

b with a "critical exponent" of 3/2. Because the misorientation of the rounded region goes to zero continuously as the facet is approached [$z'(x)$ is continuous but $z''(x)$ is not], the junction can be thought of as a second-order phase transition.[21]

In Figure 6.14, $f(s)$ has been assumed to be concave down except for cusps at $s = 0$ and $s = \pm 1$. Then, the $z(x)$ mapping becomes "reentrant," and the $s = 0$ and $s = 1$ facets join directly. Because the orientation of the crystal changes discontinuously [$z'(x)$ is discontinuous], the junction can be thought of as a first-order phase transition.

6.2.2 Shapes of Surfaces: Facetting

In Subsection 6.2.1, we discussed various constructions and mappings for deducing the equilibrium shapes of crystals subject to the constraint of constant volume. Often, however, a different constraint is imposed, that of constant average surface orientation. In this subsection, we ask: under what conditions will such a surface be stable, and under what other conditions will it tend to "facet" into combinations of other orientations? An example of such facetting is shown in Figure 6.15.

To answer this question it will be convenient to use the quantities introduced in Subsection 6.2.1. These are the surface free energies per unit

[21] V.L. Pokrovsky and A.L. Talapov, "Ground state, spectrum, and phase diagram of two-dimensional incommensurate crystals," *Phys. Rev. Lett.* **42**, 65 (1979).

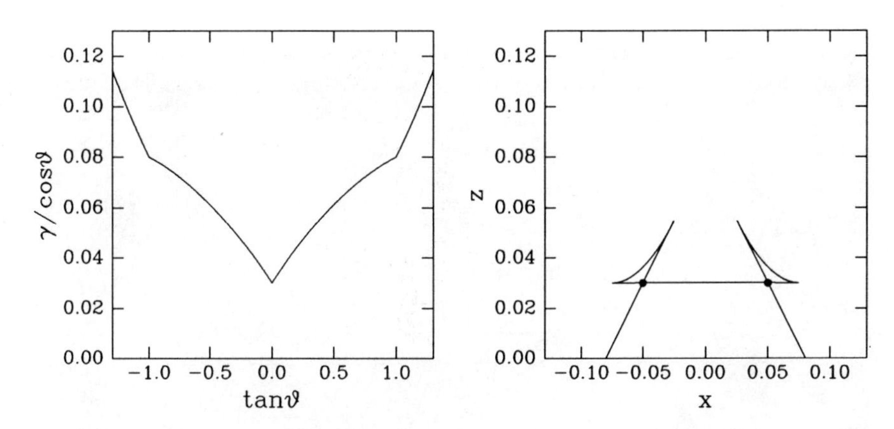

Figure 6.14: Andreev's construction for a crystal with $\tan\theta = 0$ and $\tan\theta = \pm 1$ facets, and between which the projected surface free energy per unit area (left) is concave down in $\tan\theta$. Then, the $\tan\theta = 0$ and $\tan\theta = \pm 1$ facets join (right) with discontinuous first derivatives.

area projected onto a reference surface of orientation $\theta = 0$, $f = \gamma/\cos\theta$; and the slope of the misorientation from $\theta = 0$, $s = \tan\theta$. In terms of these quantities, we will discuss, in turn, the following questions. First, under what conditions will a surface be stable against facetting? Second, if a surface is unstable against facetting, what will the misorientations of the new facetted surfaces be? Third, what is the analogous stability criterion in the more conventional $\gamma(\theta)$ representation? Fourth and finally, how can these ideas be used to generate phase diagrams on which coexistence of surfaces of differing orientations may be represented?

A Stability Criterion

Let us begin by deriving a criterion for the stability of a surface against facetting. Consider the surface depicted by the dotted lines in Figure 6.16, oriented at some angle θ with respect to the reference surface depicted by the dashed lines. Suppose that surface breaks up into the hill and valley structure depicted by the solid lines. If the two new orientations make angles θ_1 and θ_2 with respect to the reference surface, and have areas projected onto the reference surface of x_1 and x_2, then their projected vertical heights are $h_1 = x_1 \tan\theta_1$ and $h_2 = x_2 \tan\theta_2$, respectively.

Under what conditions will the original surface be stable against formation of this hill and valley structure? Since the projected free energy of the original surface is $f(\theta)(x_1 + x_2)$, and that of the two new surfaces is $f(\theta_1)x_1 + f(\theta_2)x_2$, the criterion is that there not exist straddling orienta-

Figure 6.15: Scanning tunneling micrograph of a Si surface cut 6° from (111) along an azimuth rotated about 10° away from the high symmetry $[\overline{1}\overline{1}2]$ direction. The scan area is approximately 200 × 300 Å. The surface has phase separated into two facets of different orientations, one containing 7 × 7 reconstructed terraces separted by straight $[\overline{1}\overline{1}2]$ steps and one rotated azimuthally by approximately 40° from the $[\overline{1}\overline{1}2]$ direction.[a]

[a] J. Wei, X-S Wang, N.C. Bartelt, E.D. Williams, and R.T. Tung, "The precipitation of kinks on stepped Si(111) surfaces," *J. Chem. Phys.* **94**, 8384 (1991).

tions θ_1 and θ_2 such that

$$f(\theta) > f(\theta_1)\frac{x_1}{x_1 + x_2} + f(\theta_2)\frac{x_2}{x_1 + x_2}. \tag{6.36}$$

Note that from the relations $x_1 = h_1/\tan\theta_1$, $x_2 = h_2/\tan\theta_2$, and $x_1 + x_2 = (h_1 + h_2)/\tan\theta$, the fractions of the reference surface that have the two orientations can be deduced, after some algebra, to be

$$\begin{aligned} \frac{x_2}{x_1 + x_2} &= \frac{\tan\theta - \tan\theta_1}{\tan\theta_2 - \tan\theta_1} \\ \frac{x_1}{x_1 + x_2} &= 1 - \frac{\tan\theta - \tan\theta_1}{\tan\theta_2 - \tan\theta_1}. \end{aligned} \tag{6.37}$$

These equations are equivalent to a lever rule that determines, given an average orientation θ, the amounts of two other orientations required for a

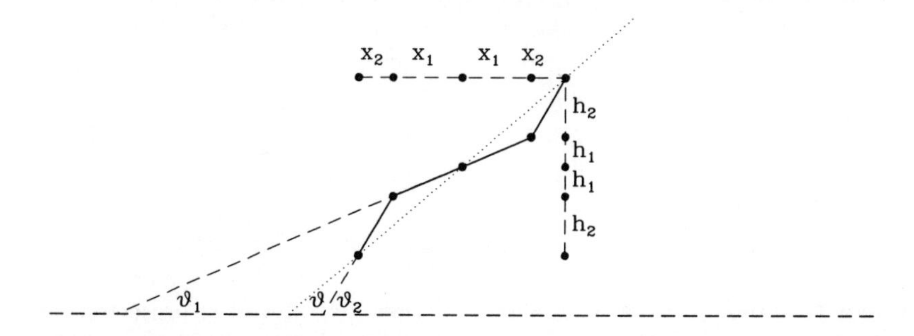

Figure 6.16: Geometry of a surface of average orientation θ that has facetted into two surfaces of orientations θ_1 and θ_2 relative to a reference surface. The projected horizontal widths and vertical heights of the two surfaces are $2x_1$, $2x_2$, $2h_1$, and $2h_2$, respectively.

continuous joining of surfaces. Altogether, the stability criterion is then

$$f(\tan(\theta) < f(\tan\theta_1)\left(1 - \frac{\tan\theta - \tan\theta_1}{\tan\theta_2 - \tan\theta_1}\right) + f(\tan\theta_2)\left(\frac{\tan\theta - \tan\theta_1}{\tan\theta_2 - \tan\theta_1}\right), \tag{6.38}$$

where we consider f to be a function of $\tan\theta$ rather than θ. In other words, a surface of orientation θ is stable if on an f vs. $\tan\theta$ plot, $f(\tan\theta)$ is less than all lever-rule-weighted sums of $f(\tan\theta_1)$ and $f(\tan\theta_2)$.

The Common Tangent Criterion

Now that we have derived a criterion for stability of a surface against facetting, let us ask the opposite question. Suppose that the original surface *is* unstable with respect to breakup? What will be the two straddling orientations, which can be considered two "phases," that will coexist stably in its stead? To answer this question, we imagine making two distinct concerted variations in the geometries of the two surfaces, and require the total free energy change to vanish.

First, we imagine varying the projected vertical height of surface 1 by dh_1, while at the same time varying the projected vertical height of surface 2 by $dh_2 = -dh_1$, so that the two surfaces continue to join perfectly. If each surface is thought of loosely as made up of steps and "missing steps," then this variation can be thought of as moving steps from surface 2 to surface 1, and at the same time moving missing steps from surface 1 to surface 2.

Since x_1 and x_2 are unchanged during this variation, the free energy changes associated with surfaces 1 and 2 are

$$\frac{\partial(f_1 x_1)}{\partial h_1} = x_1 \frac{\partial f_1}{\partial(x_1 \tan\theta_1)} = \frac{\partial f_1}{\partial \tan\theta_1} \tag{6.39}$$

and

$$-\frac{\partial(f_2 x_2)}{\partial h_2} = -x_2 \frac{\partial f_2}{\partial(x_2 \tan\theta_2)} = -\frac{\partial f_2}{\partial \tan\theta_2}, \tag{6.40}$$

where f_1 and f_2 are the surface free energies per unit area projected onto the reference surface. If the sum of these changes is to vanish, then we must have

$$\frac{\partial f_1}{\partial \tan\theta_1} = \frac{\partial f_2}{\partial \tan\theta_2}. \tag{6.41}$$

In other words, the slopes of the $f(\tan\theta)$ plot at the two orientations θ_1 and θ_2 must be equal.

Second, we imagine varying the projected area of surface 1 by dx_1, while at the same time varying the projected area of surface 2 by $dx_2 = -dx_1$, again so that the two surfaces continue to join perfectly. If each surface is thought of loosely as composed of steps and "missing steps," then this variation can be thought of as moving missing steps from surface 1 to surface 2.

Since h_1 and h_2 are unchanged during this variation, the free energy changes associated with surfaces 1 and 2 are

$$\begin{aligned}
\frac{\partial(f_1 x_1)}{\partial x_1} &= f_1 + \left(\frac{h_1}{\tan\theta_1}\right) \frac{\partial f_1}{\partial(h_1/\tan\theta_1)} \\
&= f_1 - \tan\theta_1 \left(\frac{\partial f_1}{\partial \tan\theta_1}\right) \\
-\frac{\partial(f_2 x_2)}{\partial x_2} &= -f_2 - \left(\frac{h_2}{\tan\theta_2}\right) \frac{\partial f_2}{\partial(h_2/\tan\theta_2)} \\
&= -f_2 + \tan\theta_2 \left(\frac{\partial f_2}{\partial \tan\theta_2}\right).
\end{aligned} \tag{6.42}$$

If the sum of these changes is to vanish, then we must have

$$f_1 - \tan\theta_1 \left(\frac{\partial f_1}{\partial \tan\theta_1}\right) = f_2 - \tan\theta_2 \left(\frac{\partial f_2}{\partial \tan\theta_2}\right). \tag{6.43}$$

In other words, the $\tan\theta = 0$ intercepts of the tangents to the $f(\tan\theta)$ plot at the two orientations θ_1 and θ_2 must be equal.

Altogether, Equations 6.41 and 6.43 combined tell us that both the slopes and intercepts of the two tangents must be equal, and so the tangents themselves must coincide. Therefore, the condition for coexistence of two surfaces of different orientation is that their $f(\tan\theta)$ plots *share a common tangent.*[22] Another way of viewing the origin of this construction is to think of the steps as particles. Then, equilibrium between surfaces of different orientation is analogous to equilibrium with respect to interchange of particles, hence equality of chemical potentials.[23] In a sense, $s \equiv \tan\theta$ is an extensive, rather than an intensive, variable, and can vary inhomogeneously within an equilibrium system.[24]

Herring's Criterion

The common tangent criterion for orientational stability just derived is a powerful and useful one. It implies that conditions for stability and coexistence of surface orientations are formally equivalent to the conditions for stability and coexistence of binary alloy phases. Hence, the arguments and insights derived from Chapter 3 apply directly.

For example, if the $f(\tan\theta)$ plot is concave up as in the top of Figure 6.17, then all orientations are stable. If it is concave down, as in the bottom of Figure 6.17, then only the $\tan\theta = 0$ and $\tan\theta = \pm 1$ facets are stable; all other orientations decompose into a phase mixture of those facets, in proportions given by the lever rule.

This common tangent criterion in the $f(\tan\theta)$ representation can also be understood using the more conventional $\gamma(\theta)$ representation. To see how, note that the critical shape for the $f(\tan\theta)$ plot dividing these two extremes of behavior is a straight line:

$$f(\tan\theta) = A + B\tan\theta. \tag{6.44}$$

Note that on a $\gamma(\theta)$ plot, such straight lines become circles passing through the origin,

$$\gamma(\theta) = (\cos\theta)f(\tan\theta) = A\cos\theta + B\sin\theta, \tag{6.45}$$

with origin at $(A/2, B/2)$ and radius $(A/2)^2 + (B/2)^2$. Hence, $f(\tan\theta)$ plots that are concave up correspond to $\gamma(\theta)$ plots that "bulge" out between facets less than would a sphere passing through the origin, as in the top of Figure 6.17, and $f(\tan\theta)$ plots that are concave down correspond to

[22] A.A. Chernov, "The spiral growth of crystals," *Sov. Phys.–Usp.* **4**, 116 (1961); and N. Cabrera, "The equilibrium of crystal surfaces," *Surf. Sci.* **2**, 320 (1964).

[23] P. Nozières, "Surface melting and crystal shape," *J. Phys.* **50**, 2541 (1989).

[24] N.C. Bartelt, T.L. Einstein, and C. Rottman, "First-order transitions between surface phases with different step structures," *Phys. Rev. Lett.* **66**, 961 (1991).

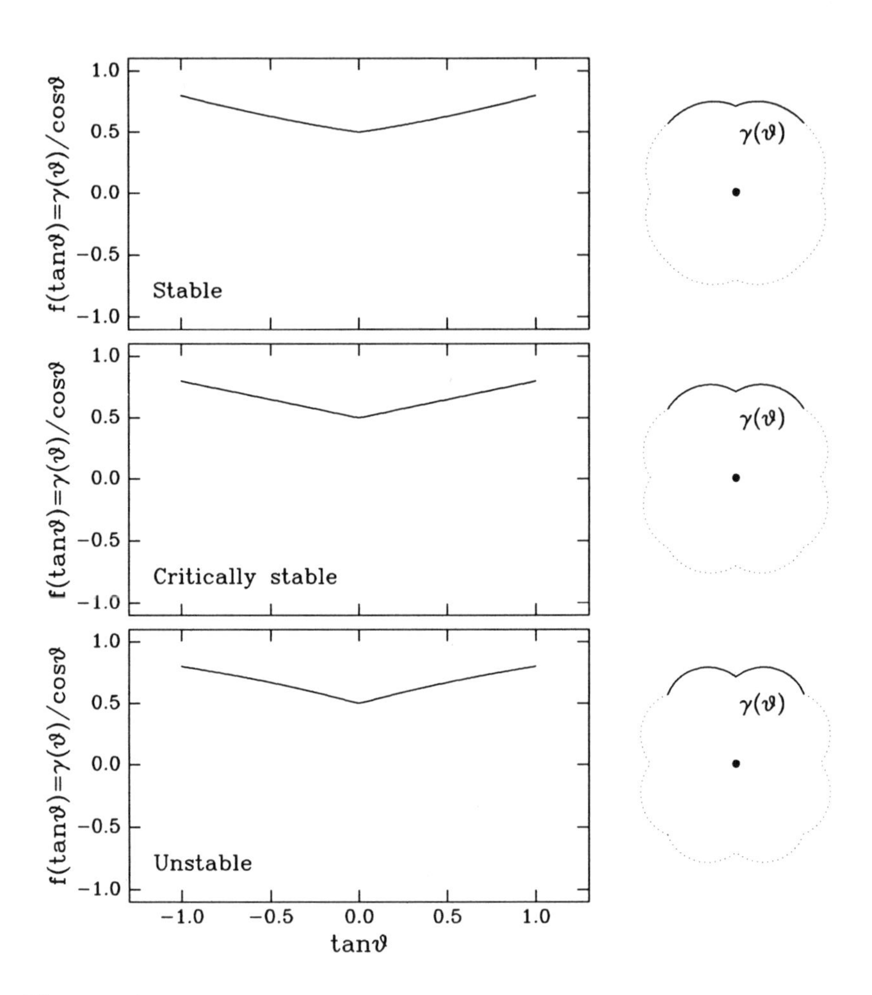

Figure 6.17: Common tangent criteria for orientational stability. The $f(\tan\theta) = \gamma/\cos\theta$ plot at top is concave up between $\tan\theta = 0$ and $\tan\theta = \pm 1$, hence surfaces whose orientations lie between those angles are stable against facetting into an inhomogeneous mix of $\theta = 0$ and $\theta = \pm\pi/4$ surfaces. The $f(\tan\theta) = \gamma/\cos\theta$ plot at bottom is concave down between $\tan\theta = 0$ and $\tan\theta = \pm 1$, hence surfaces whose orientations lie between those angles are unstable against facetting into an inhomogeneous mix of $\theta = 0$ and $\theta = \pm\pi/4$ surfaces. The $f(\tan\theta) = \gamma/\cos\theta$ plot in the middle are straight lines between $\tan\theta = 0$ and $\tan\theta = \pm 1$, hence surfaces whose orientations lie between those angles are critically stable against facetting into an inhomogeneous mix of $\theta = 0$ and $\theta = \pm\pi/4$ surfaces.

$\gamma(\theta)$ plots that bulge out between facets more than would a sphere passing through the origin, as in the bottom of Figure 6.17.[25]

Now, tangent spheres at orientations that bulge less than spherically must lie inside the $\gamma(\theta)$ plot, and hence lie on the equilibrium crystal shape, while tangent spheres at orientations that bulge more than spherically must lie outside the $\gamma(\theta)$ plot, and hence be absent from the equilibrium crystal shape. As a consequence, we also have Herring's criterion, originally proved in a different maner: those orientations are stable that are represented on the equilibrium crystal shape, and those orientations are unstable that are not represented on the equilibrium crystal shape.

Temperature-Dependent Phase Equilibria

Let us now illustrate the stability criterion and common tangent constructions just derived with a concrete example. Consider the cubic 2D crystal shown in Figure 6.18, whose lowest free energy surfaces are (11) and (01) facets. At low temperatures, we expect the $\gamma(\theta)$ plot to be deeply cusped at those orientations, leading to an equilibrium crystal bounded solely by these facets. As temperature increases, the $\gamma(\theta)$ plot becomes less and less cusped. In this case, the (01) facets are shown to roughen first, leading to an equilibrium crystal bounded by continuously curved surfaces joined to (11) facets. Then, the (11) facets roughen, leading to an equilibrium crystal bounded everywhere by continuously curved surfaces.

Another way of looking at the temperature evolution of this system is to plot, as illustrated in Figure 6.19, f vs $\tan\theta$ and $z = f - s(\partial f/\partial s)$ vs $x = \partial f/\partial s$ diagrams. At low temperatures, the $f(\tan\theta)$ plot is deeply cusped at $\tan\theta = 0$ and $\tan\theta = \pm 1$. Application of the common tangent construction then leads to the orientational gap shown in the bottom left of Figure 6.19, and to the first-order facet-facet joining shown in the bottom right of Figure 6.19. At higher temperatures, the $\gamma(\theta)$ plot becomes less and less cusped. As this happens, the orientational gap vanishes, and all orientations become stable. At the same time, the first-order facet-facet joining evolves to a second-order joining, and ultimately disappears entirely.

Finally, it is often convenient to plot these orientational gaps (in $\tan\theta$) and facet-facet phase transition positions (in $x = \partial f/\partial s$) as temperature-dependent phase diagrams. The resulting $T(\tan\theta)$ phase diagram is shown

[25] $\gamma(\theta)$ plots composed of *exactly* spherical bulges between facets, as in the middle of Figure 6.17, are also known as "raspberry" figures; see F.C. Frank, "The geometrical thermodynamics of surfaces," in W.D. Robertson and N.A. Gjostein, Eds., *Metal Surfaces: Structure, Energetics and Kinetics*, Proceedings of a joint seminar of the American Society for Metals and the Metallurgical Society of AIME, October 27-28, 1962 (American Society for Metals, Metals Park, Ohio, 1963), Chap. 1.

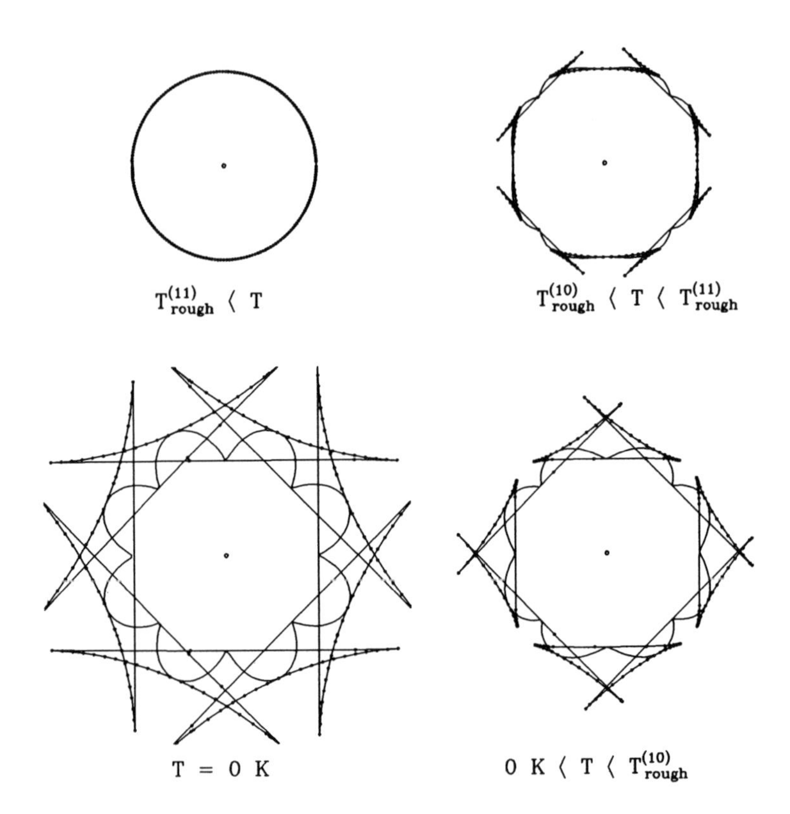

Figure 6.18: Equilibrium shapes for a hypothetical 2D crystal with both (11) and (01) facets. As temperature increases (counterclockwise from lower left), the $\gamma(\theta)$ plot becomes less and less cusped, and the equilibrium shape becomes less and less faceted.

in the middle left of Figure 6.19. It maps out the critical values of $\tan\theta$ for which surfaces of a specified average orientation will decompose into mixtures of orientations. At temperatures below 380 K, only (11) and (01) facets are stable; all other orientations decompose into lever-rule mixtures of those orientations. At temperatures above 380 K, orientations near (01) become stable; all other orientations now decompose into lever-rule mixtures of (11) facets and nonsingular orientations near (01). With increasing temperature above 380 K, orientations farther and farther from (01) become stable, until at 900 K, even orientations near (11) are stable. Above 900 K, surfaces of any average orientation will be stable against decomposition into inhomogeneous mixtures of surfaces of differing orientations.

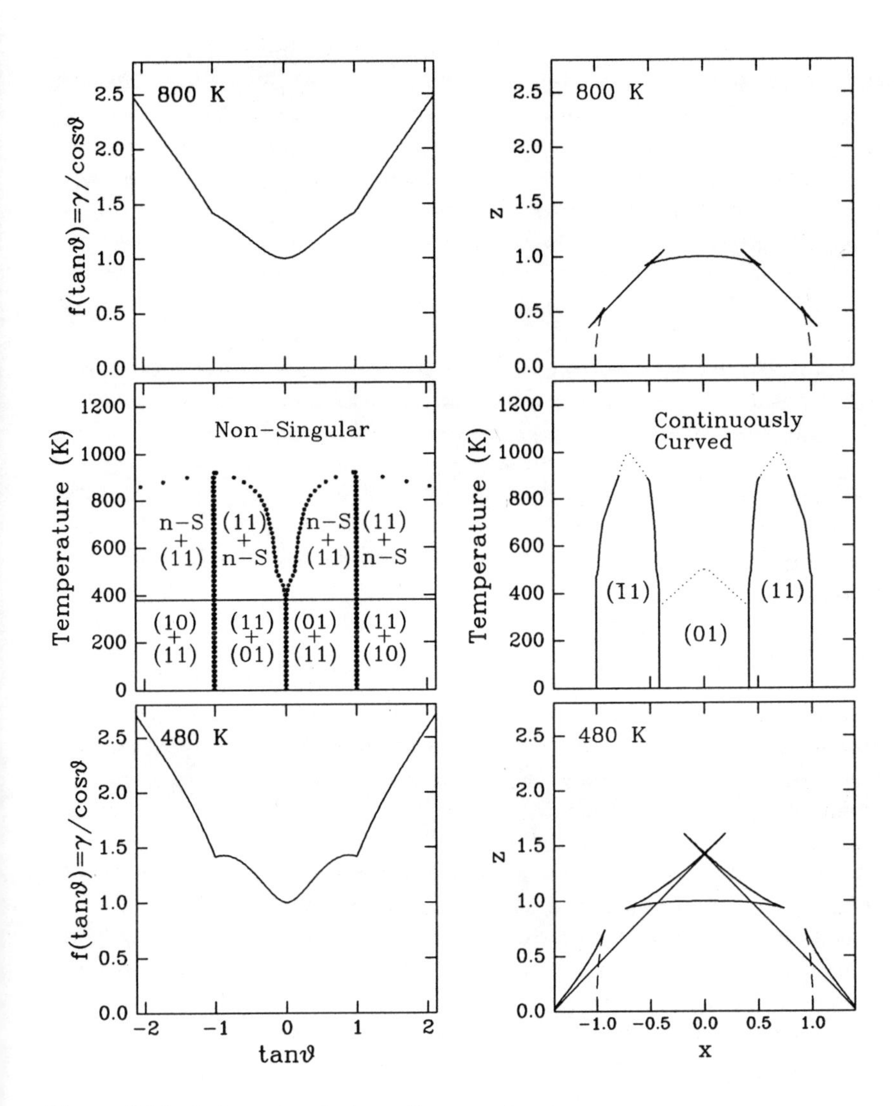

Figure 6.19: Orientational phase diagrams for the hypothetical 2D crystal illustrated in Figure 6.18. The left middle diagram shows the mixtures of orientations that a surface of a specific average orientation will decompose into. The right middle diagram shows the horizontal positions at which different facets join on the equilibrium crystal; solid and dotted lines indicate first- and second-order phase transitions, respectively. Above and below the diagrams are examples of the $\gamma/\cos\theta$ vs $\tan\theta$ and $z = f - s(\partial f/\partial s)$ vs $x = \partial f/\partial s$ plots from which these diagrams were derived.

The corresponding $T(x)$ phase diagram is shown in the middle right of Figure 6.19. It maps out the critical horizontal positions at which different facets join on the equilibrium crystal. At temperatures below 380 K, (11) facets join (01) facets in first-order phase transitions. At temperatures above 380 K, orientations near (01) begin to appear. As a consequence, (11) facets join continuously curved orientations near (01) in first-order phase transitions, while the continuously curved orientations near (01) join (01) facets in second-order phase transitions. With increasing temperature above 380 K, these alternative orientations near (01) become increasingly stable, until at 500 K the (01) facets "roughen" and disappear entirely.

Above 500 K, (01) facets are absent from the equilibrium crystal, but (11) facets are present, and continue to join continuously curved orientations near (01) in first-order phase transitions. With increasing temperature above 500 K, though, these continuously curved orientations approach more and more closely (11) orientations, until at 900 K the (11) facets begin to join these continuously curved orientations in second-order phase transitions. Finally, at 1000 K, the (11) facet itself "roughens" and disappears entirely.

6.2.3 Shapes of Thin Films: Growth Modes

Thus far, in Subsections 6.2.1 and 6.2.2, we have been concerned with single-material systems, e.g., homoepitaxial films of one material on substrates of the same material. Then, the surface free energy and, in particular, its orientation dependence, plays the most important role in determining the equilibrium morphology. However, for two-material systems, e.g., heteroepitaxial films of one material on substrates of another material, interface and volume free energies also play important roles.[26]

In this subsection, we discuss how these energies determine the equilibrium morphology, or "growth mode," of the film. We discuss two approaches in turn. The first approach considers the shape of the thickness-dependent total free energy. The second approach considers the contact angles that the film islands make with the substrate, as determined by the surface and interface energies.

Free Energies

Consider the thickness-dependent total free energy curves shown in Figure 6.20. Note that these are the total free energies of the system relative

[26] E.G. Bauer, "Phänomenologische theorie der kristallabscheidung an oberflächen. I & II," *Z. Kristallogr.* **110**, 372, 395 (1958), *NASA Technical Translations* **TT F-11**, 888 and 889 (NASA, Washington, D.C., August, 1968).

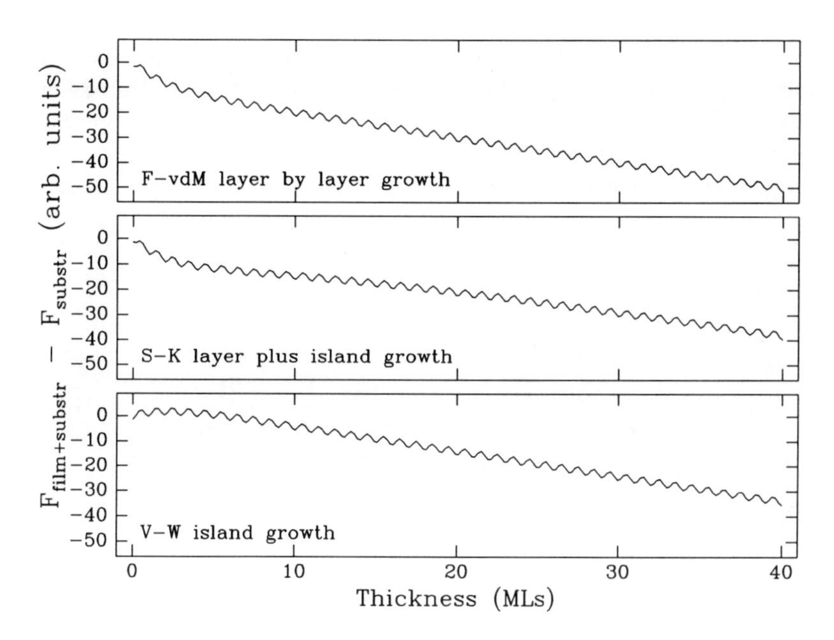

Figure 6.20: Thickness dependences of total free energies for the three classical heteroepitaxial growth modes.

to that of the bare substrate, and include both volume and surface contributions. We assume that there is a nonzero driving force for epitaxy, and so in each panel the overall trend is for the free energy to decrease with increasing thickness of the heteroepitaxial layer. We also assume that fully completed layers, with fully laterally coordinated atoms, have slightly lower energies than partially completed layers, and so in each panel the free energies are shown corrugated slightly with a monolayer periodicity.

There are three cases of interest.[27] In the top panel of Figure 6.20, the sum of the free energies associated with the free surface of the epitaxial film, $\gamma_{e/v}$, and with the interface between the substrate and the epitaxial film, $\gamma_{s/e}$, is less than or equal to that associated with the original substrate surface, $\gamma_{s/v}$:

$$\gamma_{e/v} + \gamma_{s/e} < \gamma_{s/v}. \tag{6.46}$$

Then, the overall free energy decreases faster over the first layer (or two), before settling down to a steady state slope for thicker films. The overall shape of the thickness-dependent free energy is then concave up. Therefore,

[27] M.H. Grabow and G.H. Gilmer, "Thin film growth modes, wetting and cluster nucleation," *Surf. Sci.* **194**, 333 (1988).

for every integral-monolayer thickness, the system is thermodynamically stable against breakup into inhomogeneous regions, some thicker and some thinner. This leads to what is known as the ideal Frank-van-der-Merwe layer-by-layer growth mode.[28]

In the bottom panel, the sum of $\gamma_{e/v}$ and $\gamma_{s/e}$ is greater than $\gamma_{s/v}$:

$$\gamma_{e/v} + \gamma_{s/e} > \gamma_{s/v}. \tag{6.47}$$

Then, the overall free energy *increases* at first as the first layer (or two) is deposited, before turning around and decreasing for thicker films. The overall shape of the thickness-dependent free energy is then concave down. Systems of uniform thickness are therefore thermodynamically unstable against breakup into inhomogeneous regions, some very thick and some completely uncovered. This leads to what is known as the Volmer-Weber island growth mode.[29] It is often observed in "dirty" systems in which impurities lower the free energy of the starting surface, but are buried shortly after heteroepitaxy begins.[30]

In the middle panel, the sum of $\gamma_{e/v}$ and $\gamma_{s/e}$ is, just as in the top panel, less than $\gamma_{s/v}$:

$$\gamma_{e/v} + \gamma_{s/e} > \gamma_{s/v}. \tag{6.48}$$

Therefore, the surface free energy decreases faster as the first layer (or two) is deposited. However, because of some constraint that the substrate imposes on the epilayer, the energy decreases less steeply as subsequent layers are deposited. Only for very thick films, when the epilayer decouples from the substrate, does the energy decrease as steeply as expected for a given driving force for homoepitaxy. The overall shape of the thickness-dependent free energy is therefore initially concave up, but then subsequently concave down. Films thicker than a few layers are therefore unstable to breakup into inhomogeneous regions, some very thick and some having only one (or two) layers. This leads to what is known as the Stranski-Krastanov layer plus island growth mode.[31]

---

[28]F.C. Frank and J.H. van der Merwe, "One-dimensional dislocations. I. Static theory," *Proc. R. Soc. London* **A198**, 205 (1949); F.C. Frank and J.H. van der Merwe, "One-dimensional dislocations. II. Misfitting monolayers and oriented overgrowth," *Proc. R. Soc. London* **A198**, 216 (1949); F.C. Frank and J.H. van der Merwe, "One-dimensional dislocations. III. Influence of the second harmonic term in the potential representation, on the properties of the model," *Proc. R. Soc. London* **A200**, 125 (1950); and F.C. Frank and J.H. van der Merwe, "One-dimensional dislocations. IV. Dynamics," *Proc. R. Soc. London* **A201**, 261 (1950).

[29]M. Volmer and A. Weber, "Keimbildung in übersättigten gebilden," *Z. Phys. Chem.* **119**, 277 (1926).

[30]B.A. Joyce, "The growth and structure of semiconducting thin films," *Rep. Prog. Phys.* **37**, 363 (1974).

[31]I.N. Stranski and L. Krastanow, "Zur theorie der orientierten ausscheidung von

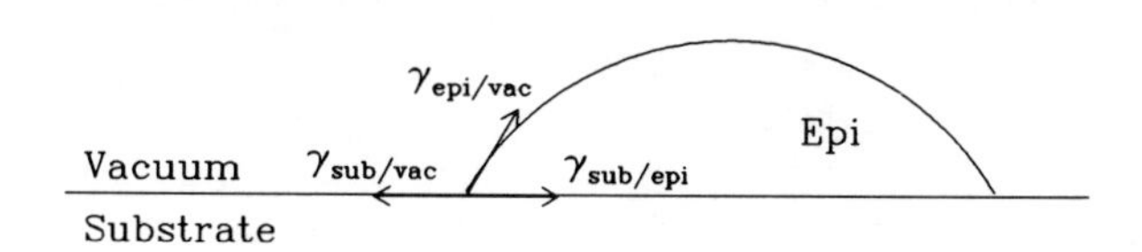

Figure 6.21: Surface tension forces acting on a heteroepitaxial nucleus on a substrate.

In practice, the growth mode that often applies to strained-layer heteroepitaxy is the Stranski-Krastanov growth mode. Films thinner than the critical thickness for strain relaxation are strained, and their free energies do not decrease with increasing thickness as steeply as the free energies of unstrained films. Films above the critical layer thickness, however, relax, and their free energies decrease at a rate approaching that for unstrained homoepitaxy.

Contact Angles

Another way of looking at these three classic growth modes is to consider the contact angle of a spherical heteroepitaxial cap on the surface.[32] If, as shown in Figure 6.21, the free energies of each interface are considered vector forces lying within their respective interfaces, then lateral force balance requires that

$$\gamma_{s/v} = \gamma_{s/e} + \gamma_{e/v} \cos \beta. \tag{6.49}$$

The contact angle will therefore be given by

$$\cos \beta = \frac{\gamma_{s/v} - \gamma_{s/e}}{\gamma_{e/v}}. \tag{6.50}$$

If $(\gamma_{s/v} - \gamma_{s/e})/\gamma_{e/v} > 1$, then there is no contact angle satisfying Equation 6.50, the cap is unstable, the heteroepitaxial layer wets the substrate, and Frank-van-der-Merwe layer-by-layer growth occurs. If $(\gamma_{s/v} - \gamma_{s/e})/\gamma_{e/v} < 1$ then there is a finite contact angle satisfying Equation 6.50, caps having that contact angle are stable, the heteroepitaxial layer does not wet the substrate, and Volmer-Weber island growth occurs.

---

ionenkristallen aufeinander," *Ber. Akademie der Wissenschaften und der Literatur, Mainz. Mathematisch-Naturwissenschaftliche Klasse*, **146**, 797 (1939).

[32]R. Kern, G. Le Lay and J.J. Metois, "Basic mechanisms in the early stages of epitaxy," in *Current Topics in Materials Science*, Vol. 3, E. Kaldis, Ed. (North-Holland, Amsterdam, 1979).

If $(\gamma_{s/v} - \gamma_{s/e})/\gamma_{e/v}$ depends on thickness, then it is possible for the first few layers to wet the substrate, for subsequent layers to island, and for the growth mode to be Stanski-Krastanov layer-plus-island. For example, for an island on a bare substrate, $(\gamma_{s/v} - \gamma_{s/e})/\gamma_{e/v}$ may be greater than unity, so that a wetting epilayer forms. Then, for an island on a wetting epilayer, $(\gamma_{w/v} - \gamma_{w/e})/\gamma_{e/v}$, where $\gamma_{w/v}$ is the surface free energy of the wetting layer and $\gamma_{w/e}$ is the interface free energy between the wetting layer and the epilayer, may be less than unity, so that islands form on the wetting epilayer.

This can come about in strained heteroepitaxy if the strain and dislocation energies of the epitaxial film are thought of as an effective interface free energy that is included as part of $\gamma_{w/e}$. Then, for very thin (monolayer) films, $\gamma_{s/v}$ may be so large that $(\gamma_{s/v} - \gamma_{s/e})/\gamma_{e/v}$ is greater than unity and the first epilayer wets the substrate. For very thick, unstrained films, the free energy of the interface between the first wetting epilayer and subsequent epilayers ($\gamma_{w/e}$) would normally vanish, and the surface free energies of the wetting epilayer ($\gamma_{w/v}$) and subsequent epilayers ($\gamma_{e/v}$) would be equal. Hence $(\gamma_{w/v} - \gamma_{w/e})/\gamma_{e/v} \rightarrow 1$, and islands would be unstable. For intermediate thickness strained films, however, with a finite $\gamma_{w/e}$, $(\gamma_{w/v} - \gamma_{w/e})/\gamma_{e/v} < 1$, and islands will be stable.

6.3 Nonequilibrium Morphology

In Section 6.2, we discussed equilibrium shapes of crystals and crystal surfaces in the *absence* of growth. In this section, we discuss nonequilibrium shapes in the *presence* of growth. We restrict our attention to homoepitaxy on simple starting surfaces at or near singular orientations, and composed of noninteracting arrays of steps. We do not treat the interesting but exceedingly complex cases of epitaxy on starting surfaces well away from singular orientations or of growth on inhomogeneous, "patterned" starting surfaces composed of multiple orientations.[33] We also do not treat the important but complex case of heteroepitaxy, in which surface morphology is often tightly coupled to the transition between coherency and semicoherency (see, e.g., Exercise 2 in Chapter 5), and in which Stranski-Krastonov layer-plus-island and Volmer-Weber island growth modes are often observed. Finally, we also neglect effects due to surface reconstructions in covalently bonded

[33] W.W. Mullins, "Flattening of a nearly plane solid surface due to capillarity," *J. Appl. Phys.* **30**, 77 (1959); W.T. Tsang and A.Y. Cho, "Growth of GaAs–$Ga_{1-x}Al_xAs$ over preferentially etched channels by molecular beam epitaxy: a technique for two-dimensional thin-film definition," *Appl. Phys. Lett.* **30**, 293 (1977); and E. Kapon, M.C. Tamargo, and D.M. Hwang, "Molecular beam epitaxy of GaAs/AlGaAs superlattice heterostructures on nonplanar substrates," *Appl. Phys. Lett.* **50**, 347 (1987).

semiconductors,[34] or due to the interplay between morphology and composition that can occur on the surfaces of binary alloys.[35]

In modeling nonequilibrium surface morphology, it is useful to distinguish between two approaches. In the first approach, surface morphology is modeled directly. At one extreme, molecular dynamics simulations track the exact positions $\{\vec{r_1}, \vec{r_2}, \ldots\}$ of all atoms as they move in response to forces between them.[36] At the other extreme, continuum models track the height h of a coarse-grained surface position (x, y). The time evolution of $h(x, y)$ is determined by various driving (e.g., growth with stochastic noise) and relaxation (e.g., diffusional) terms.[37] In between these two extremes, Monte Carlo simulations track the column heights n of discrete surface lattice sites (i, j). The time evolution of the $n(i, j)$ is determined by the probabilities of surmounting assumed energy barriers separating various configurations.[38]

In the second approach, surface morphology is not modeled directly. Instead, the surface is decomposed into defects of various kinds, such as steps, 2D islands, and adatoms. The time evolution of surface morphology is then determined by the dynamics of the motion and interactions of these defects.

In this section, we will take the second approach. Its disadvantage is

[34] See, e.g., S.A. Barnett and A. Rockett, "Monte Carlo simulations of Si(001) growth and reconstruction during molecular beam epitaxy," *Surf. Sci.* **198**, 133 (1988); and H.-J. Gossman and L.C. Feldman, "Initial stages of silicon molecular-beam epitaxy: effects of surface reconstruction," *Phys. Rev.* **B32**, 6 (1985).

[35] A. Madhukar and S.V. Ghaisas, "The nature of molecular beam epitaxial growth examined via computer simulations," *CRC Critical Reviews in Solid State and Materials Sciences* **14**, 1 (1988).

[36] M. Schneider, A. Rahman, and I.K. Schuller, "Role of relaxation in epitaxial growth: a molecular-dynamics study," *Phys. Rev. Lett.* **55**, 604 (1985); E.T. Gawlinski and J.D. Gunton, "Molecular-dynamics simulation of molecular-beam epitaxial growth of the silicon (100) surface," *Phys. Rev.* **B36**, 4774 (1987); S. Das Sarma, S.M. Paik, K.E. Khor, and A. Kobayashi, "Atomistic numerical simulation of epitaxial crystal growth," *J. Vac. Sci. Technol.* **B5**, 1179 (1987); and D. Srivastava and B.J. Garrison, "Growth mechanisms of Si and Ge epitaxial films on the dimer reconstructed Si (100) surface via molecular dynamics," *J. Vac. Sci. Technol.* **A8**, 3506 (1990).

[37] M. Kardar, G. Parisi, and Y-C Zhang, "Dynamic scaling of growing interfaces," *Phys. Rev. Lett.* **56**, 889 (1986); D.E. Wolf, "Kinetic roughening of vicinal surfaces," *Phys. Rev. Lett.* **67**, 1783 (1991); Z.-W. Lai and S. Das Sarma, "Kinetic growth with surface relaxation: continuum versus atomistic models," *Phys. Rev. Lett.* **66**, 2348 (1991).

[38] F.F. Abraham and G.H. White, "Computer simulation of vapor deposition on two-dimensional lattices," *J. Appl. Phys.* **41**, 1841 (1970); G.H. Gilmer and P. Bennema, "Simulation of crystal growth with surface diffusion," *J. Appl. Phys.* **43**, 1347 (1972); S. Clarke and D.D. Vvedensky, "Origin of reflection high-energy electron-diffraction intensity oscillations during molecular-beam epitaxy: a computational modeling approach," *Phys. Rev. Lett.* **58**, 2235 (1987); and P.A. Maksym, "Fast Monte Carlo simulation of MBE growth," *Semicond. Sci. Technol.* **3**, 594 (1988).

that it requires *a priori* knowledge of the important defect types and the ways in which they interact, knowledge that is currently far from complete. Its advantage, though, is that it simplifies and brings deeper physical understanding to a rich statistical behavior. The evolution of surface morphology is complex and highly nonlinear, often even oscillatory upon initiation of growth. Indeed, such oscillations, illustrated in Figure 3.14 have been observed by reflection high-energy electron diffraction (RHEED) and other *in situ* measurements in a variety of materials, including III-V,[39] IV-IV,[40] II-VI[41] and I-VII[42] compounds, as well as metals[43] and high-T_c superconductors.[44] Similar oscillations have also been observed during

[39] J.J. Harris, B.A. Joyce, and P.J. Dobson, "Oscillations in the surface structure of Sn-doped GaAs during growth by MBE," *Surf. Sci.* **103**, L90 (1981); C.E.C. Wood, "RED intensity oscillations during MBE of GaAs," *Surf. Sci.* **108**, L441 (1981); J.N. Eckstein, C. Webb, S.-L. Weng, and K.A. Bertness, "Photoemission oscillations during epitaxial growth," *Appl. Phys. Lett.* **51**, 1833 (1987); L.P. Erickson, M.D. Longerbone, R.C. Youngman, and B.E. Dies, "The observation of oscillations in secondary electron emission during the growth of GaAs by MBE," *J. Crystal Growth* **81**, 55 (1987); J.P. Harbison, D.E. Aspnes, A.A. Studna, L.T. Florez, and M.K. Kelly, "Oscillations in the optical response of (001) GaAs and AlGaAs surfaces during crystal growth by molecular beam epitaxy," *Appl. Phys. Lett.* **52**, 2046 (1988); and J.Y. Tsao, T.M. Brennan, and B.E. Hammons, "Oscillatory As_4 surface reaction rates during molecular beam epitaxy of AlAs, GaAs and InAs," *J. Crystal Growth* **111**, 125 (1991).

[40] T. Sakamoto, N.J. Kawai, T. Nakagawa, K. Ohta, and T. Kojima, "Intensity oscillations of reflection high-energy electron diffraction during silicon molecular beam epitaxial growth," *Appl. Phys. Lett.* **47**, 617 (1985).

[41] L.A. Kolodziejski, R.L. Gunshor, N. Otsuka, B.P. Gu, Y. Hefetz, and A.V. Nurmikko, "Use of RHEED oscillations for the growth of 2D magnetic semiconductor superlattices (MnSe/ZnSe)," *J. Cryst. Growth* **81**, 491 (1987).

[42] H. Dabringhaus and H.J. Meyer, "Untersuchung der kondensation und verdampfung von alkalihalogenid-kristallen mit molekularstrahlmethoden. II. Relaxationseffekte auf der (100)-oberfläche von KCl," *J. Cryst. Growth* **16**, 31 (1972); and H.J. Meyer and H. Dabringhaus, "Molecular processes of condensation and evaporation of alkali halides," in *Current Topics in Materials Science* Vol. 1, E. Kaldis, Ed. (North-Holland, Amsterdam, 1978), Chap. 2.

[43] Y. Namba, R.W. Vook, and S.S. Chao, "Thickness periodicity in the Auger line shape from epitaxial (111) Cu films," *Surf. Sci.* **109**, 320 (1981); T. Kaneko, M. Imafuku, C. Kokubu, R. Yamamoto, and M. Doyama, "The first observation of RHEED intensity oscillation during the growth of Cu/Mo multi-layered films," *J. Phys. Soc. Jpn.* **55**, 2903 (1986); S.T. Purcell, B. Heinrich, and A.S. Arrott, "Intensity oscillations for electron beams reflected during epitaxial growth of metals," *Phys. Rev.* **B35**, 6458 (1987); C. Koziol, G. Lilienkamp, and E. Bauer, "Intensity oscillations in reflection high-energy electron diffraction during molecular beam epitaxy of Ni on W (110)," *Appl. Phys. Lett.* **51**, 901 (1987); and D.A. Steigerwald and W.F. Egelhoff, Jr., "Observation of intensity oscillations in RHEED during the epitaxial growth of Cu and fcc Fe on Cu (100)," *Surf. Sci.* **192**, L887 (1987).

[44] T. Terashima, Y. Bando, K. Iijima, K. Yamamoto, K. Hirata, K. Hayashi, K. Kamigaki, and H. Terauchi, "Reflection high-energy electron diffraction oscillations during epitaxial growth of high-temperature superconducting oxides," *Phys. Rev. Lett.* **65**, 2684 (1990).

Péclet number	Growth Regime
$L^2j/D \ll 1$	"Diffusional" Step Flow
$L^2j/D \approx 1$	"Convective" Step Flow
$L^2j/D > 1$	2D Nucleation and Growth
$L^2j/D \gg 1$	Statistical Growth

Table 6.1: Magnitudes of Péclet numbers and the corresponding type of growth.

other kinds of crystal growth, such as electrocrystallization[45] and gas source or chemical beam epitaxy.[46]

To organize our treatment, we consider in the following Subsections the four regimes of behavior on vicinal (stepped) surfaces indicated in Table 6.1. These regimes are distinguished by the ratio between the velocity at which the steps move as they consume adatoms and the velocity at which adatoms diffuse to the steps. If j is the deposition rate in monolayers per second, and if L is the average spacing between the steps, then the velocity at which the steps move is roughly $v_{\text{step}} = jL$. If D is the adatom diffusivity, then the velocity of adatom diffusion to the steps is roughly $v_{\text{adat}} \approx D/L$. The ratio between the velocities is therefore L^2j/D. This ratio is a kind of Péclet number, in that it is a dimensionless measure of the relative importance of convective over diffusional mass flow. Low Péclet numbers imply high temperatures and a dominance of diffusional mass flow; high Péclet numbers imply low temperatures and a dominance of convective mass flow. Another way of understanding the Péclet number is to note that it is also the ratio between the diffusion time across the terraces, L^2/D, and the adatom arrival time, $\tau_{\text{ML}} = 1/j$. Low ratios imply either low growth rates or high adatom diffusivities; high ratios imply either high growth rates or low adatom diffusivities.

6.3.1 Fast Adatoms and "Diffusive" Step Flow

In this subsection, we discuss how surface morphology evolves if Péclet numbers are much less than unity, so that adatom diffusion to nearby steps is fast relative both to step flow and to the rate at which adatoms arrive from the vapor. Then, adatom coverages will be low, adatom-adatom interactions can be neglected, and growth will proceed exclusively by the flow

[45] V. Bostanov, R. Roussinova, and E. Budevski, "Multinuclear growth of dislocation-free planes in electrocrystallization," *J. Electrochem. Soc.* **119**, 1346 (1972).

[46] W.T. Tsang, T.H. Chiu, J.E. Cunningham, and A. Robertson, "Observations on intensity oscillations in reflection high-energy electron diffraction during chemical beam epitaxy," *Appl. Phys. Lett.* **50**, 1376 (1987).

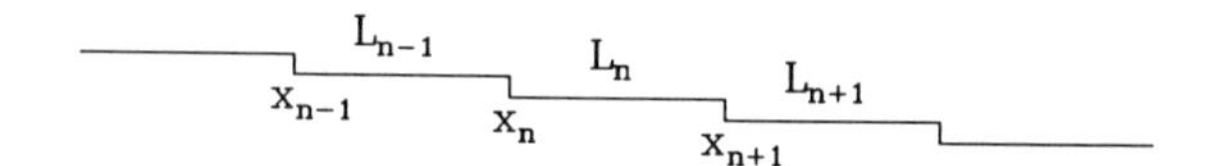

Figure 6.22: An array of steps at horizontal positions $\{x_n\}$ separated by terraces of widths $\{L_n\}$.

of steps across the surface. In other words, adatoms are fast, and it is their diffusion that mediates step flow.

To understand how the morphology of a surface evolves in this step flow regime, consider the array of steps illustrated in Figure 6.22. Suppose adatoms on terrace n have probability p^+ of attaching at the "up" step n on the left, and probability $p^- = 1 - p^+$ of attaching at the "down" step $n+1$ on the right. If L_n is the width of terrace n in monolayers, then jL_n adatoms land on that terrace each second, of which p^+ attach at step n and p^- attach at step $n+1$. Terrace n makes an "up" contribution to the velocity of step n of jL_np^+ and a "down" contribution to the velocity of step $n+1$ of jL_np^-. Alternatively, the velocity of step n can be viewed as containing an "up" contribution from terrace n of jL_np^+, and a "down" contribution from step $n-1$ of $jL_{n-1}p^-$. In other words,

$$\dot{x}_n = v_n = jL_n\left(\frac{p^+}{p^+ + p^-}\right) + jL_{n-1}\left(\frac{p^-}{p^+ + p^-}\right). \tag{6.51}$$

Since the width of the nth terrace is $L_n = x_{n+1} - x_n$, we can also write

$$\dot{x}_n = j(x_{n+1} - x_n)\left(\frac{p^+}{p^+ + p^-}\right) + j(x_n - x_{n-1})\left(\frac{p^-}{p^+ + p^-}\right), \tag{6.52}$$

which is a set of difference equations for the time evolution of the positions of the steps in the array.

If the incorporation probabilities are rewritten as

$$\begin{aligned} p^+ &= \frac{(p^+ + p^-)}{2} + \frac{(p^+ - p^-)}{2} \\ p^- &= \frac{(p^+ + p^-)}{2} - \frac{(p^+ - p^-)}{2}, \end{aligned} \tag{6.53}$$

then Equation 6.52 can be recast, after some algebra, into the form

$$\dot{x}_n = j\left(\frac{x_{n+1} - x_{n-1}}{2}\right) + j\left(\frac{p^+ - p^-}{p^+ + p^-}\right)\left(\frac{x_{n+1} - x_n}{2} - \frac{x_n - x_{n-1}}{2}\right). \tag{6.54}$$

The first term on the right-hand side of Equation 6.54 is a simple difference between step positions, while the second term is a difference between differences. Hence, the continuum equivalent of Equation 6.54 is

$$\dot{x}(n,t) = j\frac{\partial x}{\partial n} + j\left(\frac{p^+ - p^-}{p^+ + p^-}\right)\frac{\partial^2 x}{\partial n^2}, \tag{6.55}$$

which is a single differential equation for the evolution of the step positions. An identical equation may be derived for the evolution of the terrace widths by inserting Equation 6.51 into $\dot{L}_n = \dot{x}_{n+1} - \dot{x}_n$:

$$\dot{L}(n,t) = j\frac{\partial L}{\partial n} + j\left(\frac{p^+ - p^-}{p^+ + p^-}\right)\frac{\partial^2 L}{\partial n^2}. \tag{6.56}$$

The first derivative terms in both of these equations give rise to wave behavior such that, for a fixed horizontal coordinate x, the step index n decreases as time increases. In particular, as steps move to the right during growth, the indices of the steps seen by a stationary observer decrease as $\partial n/\partial t = -j$.

The second derivative terms in both of these equations are dispersion terms that tend to either damp or amplify fluctuations. Suppose, e.g., a surface at time $t = 0$ is composed of terraces having average widths of L_{avg}, but with an additional small sinusoidal variation of amplitude ΔL over step index changes of n_λ, i.e., $L(n) = L_{\text{avg}} + \Delta L \sin(n/n_\lambda)$. Then, its time evolution can be shown (see Exercise 8) to be given by

$$L(n,t) = L_{\text{avg}} + \Delta L \sin 2\pi\left(\frac{n + jt}{n_\lambda}\right)e^{-t/\tau_D}, \tag{6.57}$$

where the rate at which the sinusoidal variation decays is[47]

$$\frac{1}{\tau_D} = j\left(\frac{2\pi}{n_\lambda}\right)^2(p^+ - p^-). \tag{6.58}$$

The decay rate depends inversely on the square of the wavelength of the perturbation. As a consequence, growth will tend to smoothen short-wavelength perturbations sooner than long-wavelength ones, and very long-wavelength perturbations will tend to smoothen exceedingly slowly.[48]

[47] R.L. Schwoebel, "Step motion on crystal surfaces. II," *J. Appl. Phys.* **40**, 614 (1969); and T. Fukui, H. Saito, and Y. Tokura, "Superlattice structure observation for $(AlAs)_{1/2}(GaAs)_{1/2}$ grown on (001) vicinal GaAs substrates," *Japan. J. Appl. Phys.* **27**, L1320 (1988).

[48] H.-J. Gossman, F.W. Sinden, and L.C. Feldman, "Evolution of terrace size distributions during thin-film growth by step-mediated epitaxy," *J. Appl. Phys.* **67**, 745 (1990).

Note that it is the *anisotropy* between the up and down step incorporation probabilities that determines whether the perturbation will grow or shrink. If $p^+ > p^-$, then the perturbation decays; if $p^+ < p^-$, then the perturbation grows. This can be understood by inspection of Figure 6.22. If L_n is at some instant wider than its neighbors, then if adatoms on that terrace preferentially attach at the "up" step, L_n will decrease and the perturbation will decay, while if they preferentially attach at the "down" step, L_n will increase and the perturbation will grow.

Note that although Equation 6.56 describes a wave moving backward in step index with increasing time, the horizontal position $x \approx L_{\text{avg}}(n + jt)$ of a given step index itself moves forward with time as steps flow to the right. Hence, Equation 6.57 can be rewritten approximately as

$$L(x,t) = L_{\text{avg}} + \Delta L \sin 2\pi \left(\frac{x}{L_{\text{avg}} n_\lambda} \right) e^{-t/\tau_D}. \tag{6.59}$$

In real space, terrace width perturbations propagate nearly vertically, even though the steps themselves propagate horizontally to the right. This behavior is illustrated in Figure 6.23, which shows the evolution of an array of steps having an initial Gaussian perturbation centered at $x_n = 80$.

Finally, we note that, in deriving Equation 6.56, adatoms were assumed to attach only at adjacent steps. If, instead, adatoms cross adjacent steps and ultimately attach at more distant steps, then higher order derivatives appear in Equations 6.55 and 6.56 that can cause perturbations to propagate to the right.[49]

6.3.2 Slow Adatoms and "Convective" Step Flow

In Subsection 6.3.1, we discussed how surface morphology evolves if Péclet numbers are much less than unity, so that adatom diffusion to nearby steps is fast. In this subsection, we discuss how surface morphology evolves if Péclet numbers are on the order of unity, so that adatom diffusion to nearby steps is comparable to the step flow velocity. Then, as we shall see, adatom annihilation occurs not only by adatom diffusion to steps, but by step flow over adatoms. As a consequence, there can arise an oscillatory interplay between accumulation of adatoms between the steps, and sweeping of adatoms by step flow.

To quantify this, consider the equi-spaced array of steps illustrated in Figure 6.24, with a space and time-dependent adatom coverage $\theta(x,t)$. On the terrace bounded by steps at x_{L} and x_{R}, the coverage increases with time

[49] S.A. Chalmers, J.Y. Tsao, and A.C. Gossard, "Lateral motion of terrace width distributions during step-flow growth," *Appl. Phys. Lett.* **61**, 645 (1992).

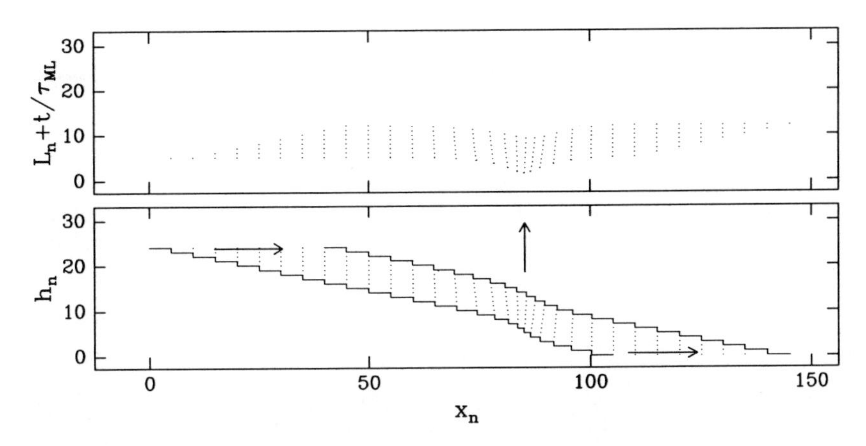

Figure 6.23: Snapshots in time of the terrace widths, $L_n(t)$, heights, $h_n(t)$ and lateral positions, $x_n(t)$, of an array of steps after successive monolayers have been deposited. The terraces are on average five lattice parameters wide, but with an initial Gaussian bunching centered at $x_n \approx 80$. Bottom: The heights h_n of the steps are constant, but their lateral positions x_n increase with time as the steps flow to the right. Top: Even though the steps flow to the right, the perturbation in the terrace widths propagates vertically up. For clarity, the terrace widths L_n are shown offset by successive monolayers $t/\tau_{\rm ML}$.

due to deposition at a rate j, and the spatial distribution of the coverage broadens in time due to diffusion at a rate $D\partial^2\theta/\partial x^2$. If evaporation back into the vapor is negligible, then the coverage evolves according to[50]

$$\dot{\theta}(x,t) = j + D\frac{\partial^2\theta}{\partial x^2}. \tag{6.60}$$

At the left step edge, the rate at which adatoms attach will be proportional to the adatom coverage, $k^+_{\rm att}\theta(x_{\rm L},t)$, where $k^+_{\rm att}$ is a kinetic rate constant for successful adatom attachment at an up step. If adatoms can also detach from steps, then there will be a competing rate, $k^+_{\rm det}$, where $k^+_{\rm det}$ is a kinetic rate constant for successful adatom detachment from an up step.

The difference between these two rates must be exactly balanced by the diffusive flow of adatoms into the step, $D[\partial\theta/\partial x]_{x_{\rm L}}$. Hence, we have at the left step

$$k^+_{\rm att}\theta(x_{\rm L},t) - k^+_{\rm det} = D\left[\frac{\partial\theta}{\partial x}\right]_{x_{\rm L}}. \tag{6.61}$$

[50] W.K. Burton, N. Cabrera, and F.C. Frank, "The growth of crystals and the equilibrium structure of their surfaces," *Philos. Trans. R. Soc. London Ser.* **A243**, 299 (1951).

Figure 6.24: Steady-state adatom coverages on an equi-spaced array of "fast" steps.

Using similar reasoning, we also have at the right step

$$k_{\rm att}^{-}\theta(x_{\rm R},t) - k_{\rm det}^{-} = -D\left[\frac{\partial\theta}{\partial x}\right]_{x_{\rm R}}, \tag{6.62}$$

where $k_{\rm att}^{-}$ and $k_{\rm det}^{-}$ are kinetic rate constants for successful adatom attachment and detachment from the down step. These two boundary conditions determine, along with Equation 6.60, the time evolution of the adatom coverage.[51]

Note, however, that these boundary conditions are complicated by the fact that, as adatoms attach at the steps, the steps themselves move, so that the positions in space at which the boundary conditions must be applied also move. Since the velocities at which the steps move is determined by the sum of the attachment rates of adatoms coming from the left and the right of each step, we have

$$\begin{aligned} v(t) &= [k_{\rm att}^{+}\theta(x_{\rm L},t) - k_{\rm det}^{+}] + [k_{\rm att}^{-}\theta(x_{\rm R},t) - k_{\rm det}^{-}] \\ &= D\left[\frac{\partial\theta}{\partial x}\right]_{x_{\rm L}} - D\left[\frac{\partial\theta}{\partial x}\right]_{x_{\rm R}}. \end{aligned} \tag{6.63}$$

To remove this complication, it is convenient to transform into a coordinate system, $x' = x + \int v dt$, that itself moves with the steps. Then, the boundary conditions given by Equations 6.61 and 6.62 may be applied at fixed $x'_{\rm L}$ and $x'_{\rm R}$, but the differential Equation 6.60 becomes

$$\dot{\theta}(x',t) = j + D\frac{\partial^2\theta}{\partial x'^2} + v\frac{\partial\theta}{\partial x'}. \tag{6.64}$$

The equation now contains both a "diffusive" term, $D\partial^2\theta/\partial x'^2$, as well as a "convective" term, $v\partial\theta/\partial x'$, due to the motion of the step.[52]

[51] R. Ghez and S.S. Iyer, "The kinetics of fast steps on crystal surfaces and its application to the molecular beam epitaxy of silicon," *IBM J. Res. Develop.* **32**, 804 (1988).

[52] K. Voigtlander, H. Risken, and E. Kasper, "Modified growth theory for high supersaturation," *Appl. Phys.* **A39**, 31 (1986); and V. Fuenzalida and I. Eisele, "High supersaturation layer-by-layer growth: application to Si MBE," *J. Crystal Growth* **74**, 597 (1986).

For simplicity, let us now assume that adatom detachment from steps is negligible, so that $k_{\text{det}}^{+} = k_{\text{det}}^{-} = 0$. Let us also assume that the local attachment rates are extremely fast, so that $k_{\text{att}}^{+} \to \infty$ and $k_{\text{att}}^{-} \to \infty$. Then, the boundary conditions given by Equations 6.61 and 6.62 simplify to

$$\theta(x'_{\text{L}}) = \theta(x'_{\text{R}}) = 0, \tag{6.65}$$

and the step velocity becomes

$$v(t) = D\left[\frac{\partial\theta}{\partial x'}\right]_{x'_{\text{L}}} - D\left[\frac{\partial\theta}{\partial x'}\right]_{x'_{\text{R}}}. \tag{6.66}$$

Equations 6.64, 6.65 and 6.66 together form a simplified set of equations for the time evolution of the adatom coverage in a reference frame moving at velocity $v(t)$.

The behavior of this set of equations is illustrated in Figure 6.25, which shows numerical simulations of the adatom coverage and step velocity at various times after the onset of growth. It can be seen that the step velocity *oscillates* in time during growth. The reason is that the adatom coverage initially builds up preferentially in the middle of the terrace, so the step moves slowly. As the step approaches the high-coverage region of the terrace, it accelerates and consumes the adatoms. Then, after most of the adatoms have been consumed, the step slows and the cycle continues.[53]

Also shown in Figure 6.25 is the time evolution of a simple measure of the smoothness of the terrace, $I = (1-2\theta_{\text{avg}})^2$, where $\theta_{\text{avg}} \equiv \int_{x'_{\text{L}}}^{x'_{\text{R}}} \theta(x',t)dx'/L$. This quantity is that which would be measured in a kinematic surface diffraction experiment under conditions for which diffraction from the uncovered terrace $(1-\theta_{\text{avg}})$ is out of phase with that from the adatoms (θ_{avg}):

$$I = [(1-\theta_{\text{avg}}) - (\theta_{\text{avg}})]^2 = (1-2\theta_{\text{avg}})^2. \tag{6.67}$$

The terrace smoothness also oscillates in time during growth, as the steps alternately accelerate and decelerate through high and low adatom coverage regions.

Ultimately, the oscillations damp out, and the adatom coverage approaches a steady-state distribution given by

$$\theta(x', t \to \infty) = \frac{1 - e^{-x'Lj/D}}{1 - e^{-L^2 j/D}} - \frac{x'}{L}. \tag{6.68}$$

This distribution is illustrated in the left half of Figure 6.26 for various values of the Péclet number, L^2j/D.

[53] G.S. Petrich, P.R. Pukite, A.M. Wowchak, G.J. Whaley, P.I. Cohen, and A.S. Arrott, "On the origin of RHEED intensity oscillations," *J. Cryst. Growth* **95**, 23 (1989).

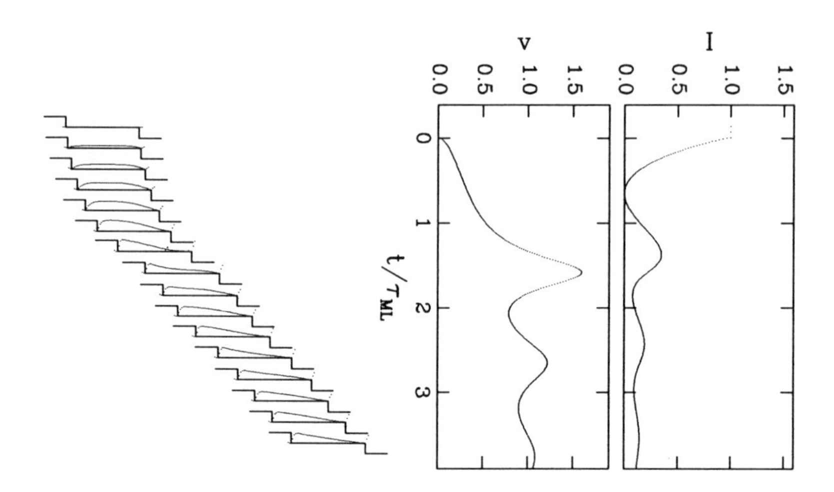

Figure 6.25: Non-steady-state adatom coverage (left), step velocity (middle), and terrace smoothness (right) during oscillatory flow of an equi-spaced array of "fast" steps. The adatom coverages are shown as snapshots taken every 0.25 monolayer.

For Péclet numbers less than unity, the step velocity is low relative to the adatom diffusive velocity. The adatom distribution becomes nearly symmetric, and approaches

$$\theta(x', t \to \infty) = \left[1 - \frac{(2x' - L)^2}{L^2}\right] \frac{jL^2}{8D}. \tag{6.69}$$

However, as the Péclet number increases beyond unity, the step velocity increases relative to the adatom diffusive velocity. The adatom distribution becomes more and more skewed, due to "pile-up" in front of the moving step.

The steady-state average adatom coverage on each terrace is

$$\begin{aligned}
\theta_{\text{avg}}(t \to \infty) &= \int_{x'_{\text{L}}}^{x'_{\text{R}}} \frac{\theta(x')}{L} dx' \\
&= \frac{1 + \left(e^{-L^2 j/D} - 1\right) D/(L^2 g)}{1 - e^{-L^2 j/D}} - \frac{1}{2} \\
&= \left(\frac{1}{2}\right) \frac{1 + e^{-L^2 j/D}}{1 - e^{-L^2 j/D}} - \frac{D}{jL^2}.
\end{aligned} \tag{6.70}$$

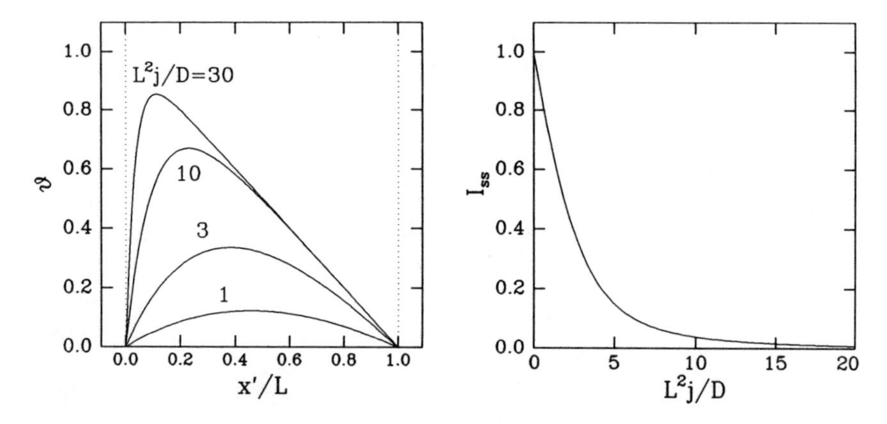

Figure 6.26: Left: Steady-state adatom coverages on an array of "fast" steps for various values of L^2j/D, the Péclet number. Right: Dependence of steady-state average terrace smoothness on the Péclet number.

The steady-state kinematic surface diffraction intensity corresponding to this average coverage is shown in the right half of Figure 6.26 as a function of the Péclet number. As can be seen, it decreases quickly as the step velocity increases relative to the adatom diffusivity, and hence as the average adatom coverage builds up on each terrace.[54]

6.3.3 2D Cluster Nucleation, Growth and Coalescence

In Subsection 6.3.2, we discussed how surface morphology evolves if Péclet numbers are on the order of unity, so that adatom diffusion to nearby steps is comparable to the step flow velocity. In this subsection, we discuss how surface morphology evolves if Péclet numbers are greater than unity, so that adatom diffusion to nearby steps is slow relative both to step flow and to the rate at which adatoms arrive from the vapor. Then, adatoms accumulate and interact on the terraces between the steps, and ultimately form 2D clusters.

If the clusters are transient, in that they break apart faster than they grow, then their main consequence will be to impede adatom diffusion. Adatoms diffusing toward steps will occasionally meet and merge with a

[54] J.H. Neave, P.J. Dobson, B.A. Joyce, and J. Zhang, "Reflection high-energy electron diffraction oscillations from vicinal surface – a new approach to surface diffusion measurements," *Appl. Phys. Lett.* **47**, 100 (1985); and T. Nishinaga and K-I Cho, "Theoretical study of mode transition between 2d-nucleation and step flow in MBE growth of GaAs," *Japan. J. Appl. Phys.* **27**, L12 (1988).

cluster or another adatom. Assuming the cluster itself is relatively immobile, the adatom will be unable to continue its journey until it breaks free from the cluster. Then, the effective adatom diffusivity decreases with increasing adatom coverage,[55] so that step flow becomes less and less "diffusive" and more and more "convective."

If the clusters are permanent, in that they form stable growing nuclei, then the kinetics of growth are altered drastically. In a sense, the clusters take on a life of their own. Their boundaries represent "extrinsic" steps that compete for adatoms with the intrinsic steps always present on a vicinal surface. As a consequence, the clusters grow and ultimately coalesce, often in a complex, oscillatory way.

To understand why, consider epitaxy on a singular surface, or on a vicinal surface whose terraces are very wide compared to the spacing of the clusters. Enumerate the layers by $n = 0, 1, 2, \ldots$, where $n = 0$ is the initially completely occupied substrate surface layer, $n = 1$ is the first, initially completely unoccupied epilayer, and so on. Associate with each of these layers three coverages: α_n, the total coverage of mobile adatoms created by impingement from the vapor; η_n, the total coverage of nuclei centers created by interaction between mobile adatoms; and θ_n, the total coverage of immobile atoms permanently incorporated into clusters. These coverages are represented in Figure 6.27 by the open circles, filled squares, and open squares, respectively.

Mobile Adatoms

For simplicity, assume that mobile adatoms are created exclusively by impingement from the vapor (rather than by detachment from clusters). Then, the rate at which the mobile adatom coverage in layer n increases is equal to the flux times the exposed coverage of layer $n-1$, or $(\theta_{n-1} - \theta_n)/\tau_{\mathrm{ML}}$.

Once mobile adatoms in layer n are created, they may diffuse to and attach at the edges of both layer $n-1$ and layer n clusters. The rates at which they do so will be proportional to the product of the mobile adatom coverage (α_n), the coverage of layer $n-1$ and layer n nuclei centers (η_{n-1} and η_n), and the capture numbers, or efficiencies, associated with those nuclei. These capture numbers are essentially the geometric cross sections that the clusters present to diffusing adatoms, and have been the subject of considerable study.[56] Here, we take them to be constant. The rate at which

---

[55] A.K. Myers-Beaghton and D.D. Vvedensky, "Nonlinear equation for diffusion and adatom interactions during epitaxial growth on vicinal surfaces," *Phys. Rev.* **B42**, 5544 (1990).

[56] G. Zinsmeister, "Theory of thin film condensation. Part D: Influence of a variable collision factor," *Thin Solid Films* **7**, 51 (1971); J.A. Venables, "Rate equation ap-

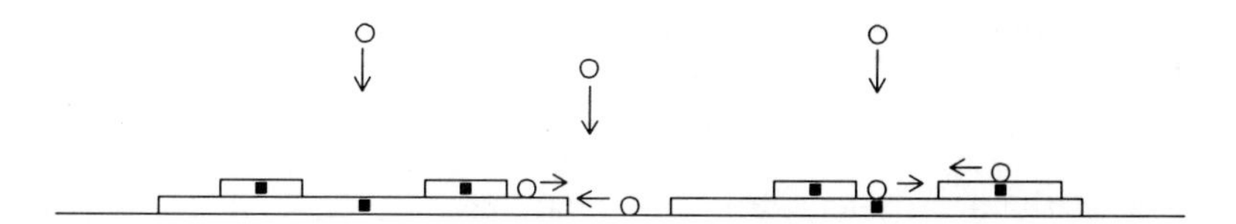

Figure 6.27: Adatom arrival, diffusion and attachment at step edges, accompanied by cluster nucleation, growth, and coalescence. The open circles on the surface represent mobile adatoms; the filled squares represent nuclei centers; and the open squares represent immobile atoms permanently incorporated into clusters.

adatoms in level n are captured by clusters in level $n-1$ is then $k_{\rm att}^{-}\alpha_n\eta_{n-1}$, where $k_{\rm att}^{-}$ is the kinetic rate constant for attachment at "down" steps; and the rate at which they are captured by clusters in level n is $k_{\rm att}^{+}\alpha_n\eta_n$, where $k_{\rm att}^{+}$ is the kinetic rate constant for attachment at "up" steps.

At the same time, mobile adatoms in layer n may also hop over steps into layers $n-1$ and $n+1$, while mobile adatoms in layers $n-1$ and $n+1$ may hop over steps into layer n. Here, we assume these adatom exchange rates to be proportional to the mobile adatom coverage in the layer the adatoms are jumping from and the exposed coverage of the layer the adatoms are jumping to. Hence, the exchange rate out of layer n is $k_{\rm exch}^{+}\alpha_n(\theta_n-\theta_{n+1})+k_{\rm exch}^{-}\alpha_n(\theta_{n-2}-\theta_{n-1})$, and the exchange rate into layer n is $k_{\rm exch}^{+}\alpha_{n-1}(\theta_{n-1}-\theta_n)+k_{\rm exch}^{-}\alpha_{n+1}(\theta_{n-1}-\theta_n)$, where $k_{\rm exch}^{+}$ and $k_{\rm exch}^{-}$ are the rates of hopping over "up" and "down" steps, respectively.

Altogether, the coverage of mobile adatoms evolves in time approximately as

$$\begin{aligned}\dot{\alpha}_n &= \frac{\theta_{n-1}-\theta_n}{\tau_{\rm ML}} - \alpha_n\left(k_{\rm att}^{-}\eta_{n-1}+k_{\rm att}^{+}\eta_n\right)\\ &\quad - k_{\rm exch}^{+}\alpha_n(\theta_n-\theta_{n+1}) - k_{\rm exch}^{-}\alpha_n(\theta_{n-2}-\theta_{n-1})\\ &\quad + k_{\rm exch}^{+}\alpha_{n-1}(\theta_{n-1}-\theta_n) + k_{\rm exch}^{-}\alpha_{n+1}(\theta_{n-1}-\theta_n). \qquad (6.71)\end{aligned}$$

It increases due to deposition and to exchange from adjacent layers, but decreases due to incorporation into growing clusters and to exchange into adjacent layers.

proaches to thin film nucleation kinetics," *Phil. Mag.* **27**, 697 (1973); B. Lewis and G.J. Rees, "Adatom migration, capture and decay among competing nuclei on a substrate," *Phil. Mag.* **29**, 1253 (1974); and R. Kariotis and M.G. Lagally, "Rate equation modeling of epitaxial growth," *Surf. Sci.* **216**, 557 (1989).

Immobile Adatoms

As mobile adatoms attach at steps, the coverage of immobile atoms permanently incorporated into clusters must increase correspondingly. Since, as illustrated in Figure 6.27, the coverage of immobile atoms in layer n depends on the attachment of mobile adatoms in layers n and $n+1$, we can write

$$\dot{\theta}_n = k_{\mathrm{att}}^{+} \alpha_n \eta_n + k_{\mathrm{att}}^{-} \alpha_{n+1} \eta_n. \tag{6.72}$$

As in Equation 6.71, k_{att}^{+} and k_{att}^{-} are kinetic rate constants for attachment of mobile adatoms at up and down steps, respectively.

Note that in this simple treatment we neglected possible anisotropies in the shapes of the clusters. Such anisotropies can arise from anisotropic attachment or diffusion rates, and have been observed during growth of semiconductors having strong and anisotropic surface reconstructions.[57]

Nuclei Centers

Finally, the coverage of nuclei centers itself increases, as mobile adatoms collide to form 2D clusters, and then decreases as the clusters grow, impinge on each other, and ultimately coalesce. In general, nucleation is a complex process by which a distribution of clusters of various sizes evolves in time in response to kinetic adatom attachment and detachment rates and to highly nonlinear size and shape dependencies to cluster energetics.[58] Nucleation may also be "heterogeneous," in the sense of being catalyzed by defects on the surface.[59] In this simple treatment, we assume that two adatoms are sufficient to form a stable cluster, and that the nucleation rate is proportional to the collision rate between adatoms, $k_{\mathrm{nuc}} \alpha_n^2$.

Coalescence of clusters is also a complex process that depends on the distribution of clusters in both size and space. At one extreme, if the nuclei centers are distributed randomly in space, then their initial coalescence rate can be shown to be proportional to both the coverage of cluster centers

[57] R.J. Hamers, "Nucleation and growth of epitaxial layers on Si(001) and Si(111) surfaces by scanning tunneling microscopy," *Ultramicroscopy* **31**, 10 (1989); J.Y. Tsao, E. Chason, U. Koehler, and R. Hamers, "Dimer strings, anistropic growth, and persistent layer-by-layer epitaxy," *Phys. Rev.* **B40**, 11951 (1989); and Y.-W. Mo, B.S. Swartzentruber, R. Kariotis, M.B. Webb, and M.G. Lagally, "Growth and equilibrium structures in the epitaxy of Si on Si (001)," *Phys. Rev. Lett.* **63**, 2393 (1989).

[58] See, e.g., D. Walton, "Nucleation of vapor deposits," *J. Chem. Phys.* **37**, 2182 (1962); K.F. Kelton, A.L. Greer, and C.V. Thompson, "Transient nucleation in condensed systems," *J. Chem. Phys.* **79**, 6261 (1983).

[59] Anti-phase boundaries between equivalent reconstruction domains on the surface are an example. See, e.g., R.J. Hamers, "Nucleation and growth of epitaxial silicon on Si(001) and Si(111) surfaces studied by scanning tunneling microscopy," *Ultramicroscopy* **31**, 10 (1989).

and the rate of change of the coverage of immobile adatoms incorporated into the clusters, or $2\eta_n\dot{\theta}_n$.[60] At the other extreme, if their centers are distributed equally in space, then the initial coalescence rate will be zero, increasing sharply when the clusters just begin to impinge on each other.[61] Here, we assume a coalescence rate between these two extremes: $\eta_n\dot{\theta}_n/(1-\theta_n)$. This form of the coalescence rate guarantees that the coverage of nuclei centers decreases smoothly to zero as the coverage of immobile adatoms incorporated into the clusters approaches unity, or that $\eta_n \to 0$ as $\theta_n \to 1$.

Altogether, the coverage of nuclei centers evolves in time approximately as

$$\dot{\eta}_n = k_{\text{nuc}}\alpha_n^2 - \left(\frac{\eta_n}{1-\theta_n}\right)\dot{\theta}_n. \tag{6.73}$$

Note that in deriving Equation 6.73, we have neglected, for simplicity, elimination of nuclei centers in the absence of growth. More comprehensive treatments must allow for such effects, which are due to surface tension. Small clusters, because of their large perimeter length to cluster area ratio, are thermodynamically less stable than, and will ultimately "ripen" into, increasingly larger clusters.[62]

Numerical Solutions

Equations 6.71, 6.72 and 6.73 form a set of coupled rate equations, three for each layer, describing the evolution of the coverages of mobile adatoms, immobile adatoms, and nuclei centers. They may be solved analytically in some simple limiting cases,[63] but in general require numerical integra-

---

[60] R. Vincent, "A theoretical analysis and computer simulation of the growth of epitaxial films," *Proc. Roy. Soc. Lond.* **A321**, 53 (1971); and M.J. Stowell, "Thin film nucleation kinetics," *Phil. Mag.* **26**, 361 (1972).

[61] J.A. Venables, "Rate equation approaches to thin film nucleation kinetics," *Phil. Mag.* **27**, 697 (1973).

[62] See, e.g., I.M. Lifschitz and V.V. Slyozov, "The kinetics of precipitation from supersaturated solid solutions," *J. Phys. Chem. Solids* **19**, 35 (1961); C. Wagner, "Theorie der alterung von niederschlägen durch umlösen," *Z. Electrochem.* **65**, 581 (1961); P.W. Voorhees and M.E. Glicksman, "Solution to the multi-particle diffusion problem with applications to Ostwald ripening – I. Theory," *Acta Met.* **32**, 2001 (1984); C.V. Thompson, "Coarsening of particles on a planar substrate: interface energy anisotropy and application to grain growth in thin films," *Acta Met.* **36**, 2929 (1988); and H.A. Atwater and C.M. Yang, "Island growth and coarsening in thin films – conservative and nonconservative systems," *J. Appl. Phys.* **67**, 6202 (1990).

[63] See, e.g., A.N. Kolmogoroff, *Bull. Acad. Sci. URSS* (Cl. Sci. Math. Nat.) **3**, 355 (1937); M. Avrami, "Kinetics of phase change I. General theory," *J. Chem. Phys.* **7**, 1103 (1939); M. Avrami, "Kinetics of phase change II. Transformation-time relations for random distribution of nuclei," *J. Chem. Phys.* **8**, 212 (1940); M. Avrami, "Kinetics of phase change III. Granulation, phase change and microstructure," *J. Chem. Phys.* **9**, 177 (1941); W.B. Hillig, "A derivation of classical two-dimensional nucleation kinetics

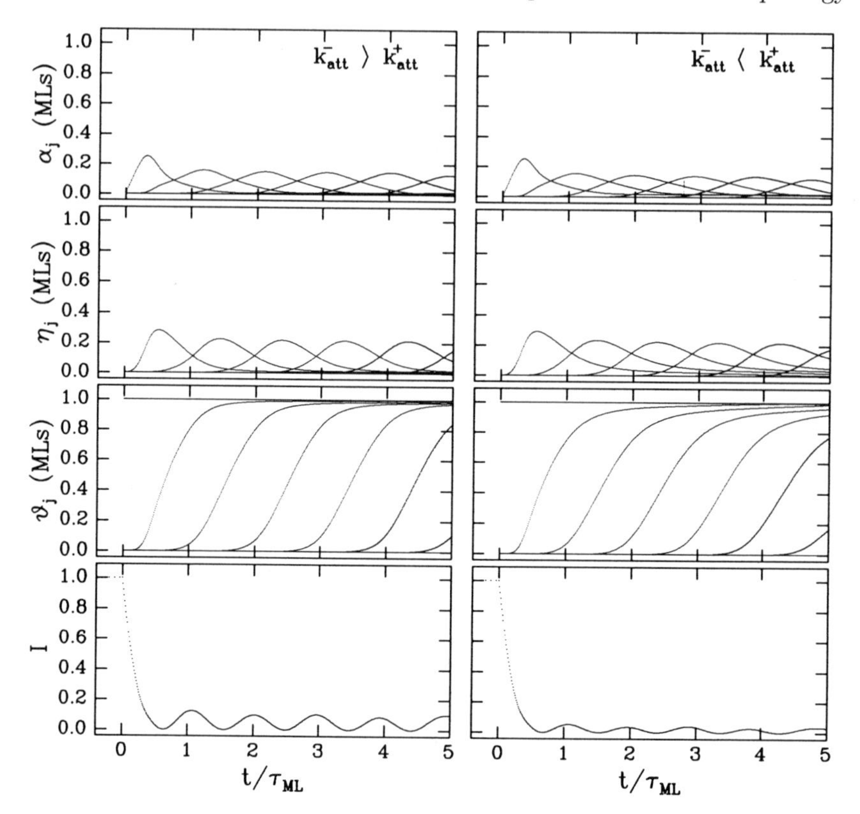

Figure 6.28: Time evolution, from top to bottom, of mobile adatom coverages (α_n), nuclei center coverages (η_n), immobile adatom coverages coverages (θ_n), and overall surface smoothness. The kinetic parameters were all taken to be $20/\tau_{\text{ML}}$ except k_{att}^{-}, which was slightly higher ($30/\tau_{\text{ML}}$) in the left panels, and slightly lower ($15/\tau_{\text{ML}}$) in the right panels.

tion. Two examples of time evolutions deduced by numerical integration are shown in Figure 6.28.

As growth commences, at $t/\tau_{\text{ML}} = 0$, the mobile adatom coverage in layer 1 increases from zero at a rate $1/\tau_{\text{ML}}$. At a critical coverage, clusters in layer 1 begin to nucleate and grow, and as they do so, the mobile adatom coverage in layer 1 begins to decrease while the immobile atom coverage in layer 1 begins to increase. Finally, the clusters begin to coalesce, the

and the associated crystal growth laws," *Acta Met.* **14**, 1868 (1966); and D. Kaschiev, "Growth kinetics of dislocation-free interfaces and growth mode of thin films," *J. Crystal Growth* **40**, 29 (1977).

nuclei center coverage decreases, and the rate at which mobile adatoms incorporate into permanent clusters also decreases.

In the meantime, as clusters in layer 1 form, mobile adatoms begin to be created in layer 2. In this way, successive layers are born by a burst of nucleation and growth of clusters, only to die by being covered by a burst of nucleation and growth of higher level clusters.[64] If these bursts are well separated in time, then growth is smooth, and successive layers are born only after previous layers have died. If the bursts overlap in time, then growth is rough, and successive layers are born even before previous layers have died.

Also shown is the time evolution of a generalization of Equation 6.67 for the smoothness of the surface,

$$I = \left\{ \sum_{n=0}^{\infty} (-1)^{n+1} [(\alpha_n + \theta_n) - (\alpha_{n+1} + \theta_{n+1})] \right\}^2 . \tag{6.74}$$

Just as that defined by Equation 6.67, this quantity is that which would be measured in a kinematic surface diffraction experiment under conditions for which diffraction from adjacent exposed surface layers is out of phase.

In both cases shown in Figure 6.28, the smoothness of the surface oscillates in time with a monolayer periodicity. The strength of the oscillations is, however, very sensitive to the values of the kinetic parameters. For example, they are stronger when adatom attachment is faster at down steps than at up steps (left side of Figure 6.28), rather than vice-versa (right side of Figure 6.28). The reason is that if adatoms preferentially attach at down steps, then the mobile adatom coverage in higher layers will be lower, and cluster nucleation in these higher layers will tend to be suppressed until the lower layers are fully complete.

Note that the oscillations predicted by Equations 6.71, 6.72 and 6.73, even when weak, are relatively persistent. In practice, faster decays are nearly always observed, and are thought to be due to effects such as a small amount of step flow (see Figure 6.25) or slight nonuniformities in growth fluxes arriving at the surface (see Exercise 10).

As a final comment, note that this treatment neglected evaporation of mobile adatoms back into the vapor. At the low to medium temperatures typical of most MBE growth, this assumption is reasonable. At high temperatures, however, evaporation can become significant. Then, the os-

[64] S. Stoyanov, "Layer growth of epitaxial films and superlattices," *Surf. Sci.* **199**, 226 (1988); and P.I. Cohen, G.S. Petrich, P.R. Pukite, G.J. Whaley, and A.S. Arrott, "Birth-death models of epitaxy I. Diffraction oscillations from low index surfaces," *Surf. Sci.* **216**, 222 (1989).

cillations in surface smoothness and in mobile adatom coverages can also manifest themselves as oscillations in the growth rate itself.[65]

6.3.4 Statistical Roughening

In Subsection 6.3.3, we discussed how surface morphology evolves if Péclet numbers are greater than unity, so that adatom diffusion to nearby steps is slow relative to the step flow velocity. In this subsection, we discuss how surface morphology evolves if Péclet numbers are *much* greater than unity, so that the rate at which adatoms diffuse, even to adjacent lattice sites, becomes slower than the rate at which they arrive from the vapor. In other words, suppose adatoms "stick" wherever they happen to land. If they arrive randomly, then they will be uncorrelated in space, and it is sufficient to know the probability p that any particular column on the surface will have a height n. If they arrive randomly in time according to Poisson statistics, then this probability will be

$$p(n) = \frac{\theta_{\text{tot}}^n}{n!} e^{-\theta_{\text{tot}}}. \tag{6.75}$$

In this equation, θ_{tot} is the total coverage of deposited atoms, so that $\sum_{n=0}^{\infty} p(n) = 1$ and $\sum_{n=0}^{\infty} np(n) = \theta_{\text{tot}}$. As illustrated in the left half of Figure 6.29, the column height probabilities are roughly centered at $n = \theta_{\text{tot}}$, but become more and more dispersed as θ_{tot} increases. Ultimately, for large θ_{tot}, the asymmetric Poissonian distribution approaches a symmetric Gaussian distribution.[66]

If we again generalize Equation 6.67 to calculate the smoothness of the surface, then we can write

$$I = \left[\sum_{n=0}^{\infty}(-1)^n p(n)\right]^2 = \left[\sum_{n=0}^{\infty}\frac{(-\theta_{\text{tot}})^n}{n!}e^{-\theta_{\text{tot}}}\right]^2 = e^{-4\theta_{\text{tot}}}. \tag{6.76}$$

As illustrated in the right half of Figure 6.29, the surface smoothness decreases exponentially with increasing total coverage, at a rate four times faster than the simple deposition rate.

Suggested Reading

1. A.A. Chernov, *Modern Crystallography III. Crystal Growth* (Springer-Verlag, Berlin, 1984).

[65] G.H. Gilmer, "Transients in the rate of crystal growth," *J. Cryst. Growth* **49**, 465 (1980).

[66] E. Chason and J.Y. Tsao, "Adatoms, strings and epitaxy on singular surfaces," *Surf. Sci.* **234**, 361 (1990).

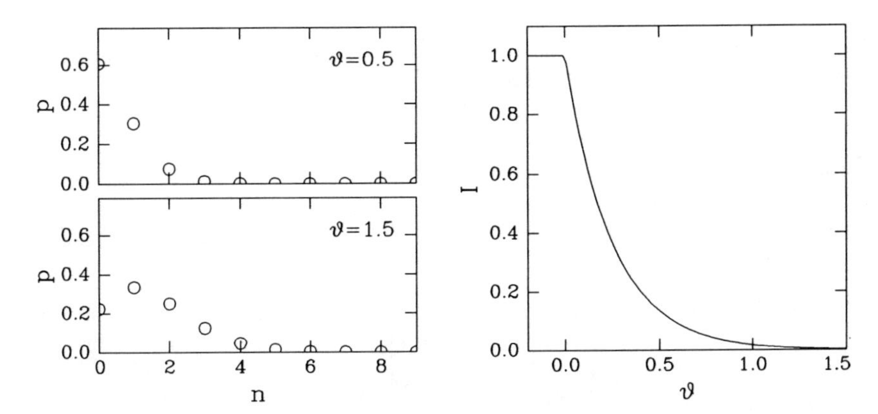

Figure 6.29: Probability p that a particular column will have height n (left) and decay of kinematic out-of-phase diffraction intensity (right) after statistical deposition of various average numbers of monolayers $\theta_{\rm tot}$.

2. R. Kern, G. Le Lay, and J.J. Metois, "Basic Mechanisms in the Early Stages of Epitaxy," in *Current Topics in Materials Science*, Vol. 3, E. Kaldis, Ed. (North-Holland, Amsterdam, 1979).

3. P.K. Larsen and P.J. Dobson, Eds., *Reflection High-Energy Electron Diffraction and Reflection Electron Imaging of Surfaces* (Plenum Press, New York, 1988).

4. B. Lewis and J.C. Anderson, *Nucleation and Growth of Thin Films* (Academic Press, New York, 1978).

5. J.W. Matthews, Ed., *Epitaxial Growth* Parts A and B, (Academic Press, New York, 1975).

6. W.D. Robertson and N.A. Gjostein, Eds., *Metal Surfaces: Structure, Energetics and Kinetics*, Proceedings of a joint seminar of the American Society for Metals and the Metallurgical Society of AIME, October 27-28, 1962 (American Society for Metals, Metals Park, Ohio, 1963).

7. J.A. Venables, G.D.T. Spiller, and M. Hanbucken, "Nucleation and growth of thin films," *Rep. Prog. Phys.* **47**, 399 (1984).

8. E.D. Williams and N.C. Bartelt, "Thermodynamics of surface morphology," *Science* **251**, 393 (1991).

9. A. Zangwill, *Physics at Surfaces* (Cambridge University Press, Cambridge, 1988).

Exercises

1. Derive Equation 6.12, the equilibrium probabilities of plus, minus and missing kinks.

2. Show that angles such as $\angle OAP$ in Figure 6.11, with origin on the circumference of a circle and with legs passing through the ends of a diameter of the circle, are right angles.

3. Consider the facetted 2D crystal illustrated in Figure 6.30 bounded by four faces of surface energy γ_o oriented perpendicular to rays along $\theta = 0, \pi/2, \pi, 3\pi/2$, and four faces of surface energy γ_1, oriented perpendicular to rays along $\theta = \pi/4, 3\pi/4, 5\pi/4, 7\pi/4$. Show that the pyramids that make up this polyhedra obey the "common vertex" relations

$$\begin{aligned} h_o &= b_o/2 + b_1/\sqrt{2} \\ h_1 &= b_o/\sqrt{2} + b_1/2. \end{aligned} \tag{6.77}$$

   Using these relations, show that the pyramidal heights of the polyhedron with minimum surface energy, $E = 4(\gamma_o b_o + \gamma_1 b_1)$, at constant area, $A = 4(h_o b_o/2 + h_1 b_1/2)$, are proportional to the surface energies of the bases, $\gamma_o/h_o = \gamma_1/h_1$, in agreement with the Wulff construction.

4. Derive Equations 6.31 for the relationship between the fractional surface areas, $x_1/(x_1 + x_2)$ and $x_2/(x_1 + x_2)$, and the tangents of the orientation angles of those surfaces, $\tan\theta_1$ and $\tan\theta_2$.

5. Suppose $f(s)$ in Equation 6.34 were quadratic rather than cubic. What would be the shape of the equilibrium crystal near the $s = 0$ facet?

6. What is the functional form of $f(s)$, where $f \equiv \gamma/\cos\theta$ and $s \equiv \tan\theta$, for an orientation-independent molar surface free energy $\gamma(\theta) =$ constant? Is it concave up or down?

7. Is there an equilbrium island size for Volmer-Weber island growth, or will larger islands continuously grow in time at the expense of smaller islands?

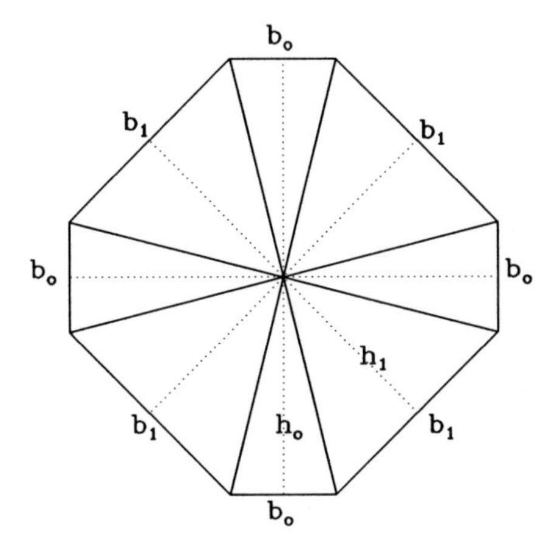

Figure 6.30: Decomposition of a facetted 2D crystal into pyramids.

8. Show that $1/\tau_D$ in Equation 6.57 is given by Equation 6.58.

9. Consider a surface whose layer coverages obey a power law, $\theta_n = [\theta_{\rm tot}/(1+\theta_{\rm tot})]^n$. Show that the total coverage of deposited atoms is $\theta_{\rm tot} = \sum_{n=1}^{\infty} \theta_n$. Justify the equation

$$I = \left[\sum_{n=0}^{\infty} (-1)^n (\theta_{n-1} - \theta_n) \right]^2 \tag{6.78}$$

for the kinematic surface diffraction intensity in an out-of-phase condition, and use it to calculate the smoothness of this surface. Does it decrease more or less quickly with $\theta_{\rm tot}$ than if the layer distribution were distributed according to Poissonian statistics? What if the layer coverages were distributed according to Gaussian statistics?

10. Suppose adatoms arrive at a surface with a nonuniformity of 10%. How might this cause an "artificial" decay in the amplitude of observed growth oscillations and what would be the decay rate?

Chapter 7

Surface Composition

In Chapter 6, we discussed the equilibrium and nonequilibrium morphology of a surface assuming that the *composition* of the surface was unimportant. In this chapter, we discuss the equilibrium and nonequilibrium composition of a surface assuming that the *morphology* of the surface is unimportant. In both of these chapters, therefore, we neglect possible interdependencies between morphology and composition, interdependencies that are clearly present but thus far poorly understood. For example, we do not discuss, except casually, the various reconstructions of the surfaces of III/V semiconductors, and how they might depend on the ratio between the column III and column V atom coverages on the surface.[1] Instead, we discuss those interesting and important aspects of surface composition that are to first order independent of surface morphology.

We begin, in Section 7.1, by describing a thermodynamic framework

[1] A.Y. Cho, "GaAs epitaxy by a molecular beam method: observations of surface structure on the (001) face," *J. Appl. Phys.* **42**, 2074 (1971); J.R. Arthur, "Surface stoichiometry and structure of GaAs," *Surf. Sci.* **43**, 449 (1974); M.D. Pashley, K.W. Haberern, W. Friday, J.M. Woodall, and P.D. Kirchner, "Structure of GaAs (001) (2x4)–c(2x8) determined by scanning tunneling microscopy," *Phys. Rev. Lett.* **60**, 2176, (1988); D.K. Biegelsen, R.D. Bringans, J.E. Northrup, and L.-E. Swartz, "Surface reconstructions of GaAs (100) observed by scanning tunneling microscopy," *Phys. Rev.* **B41**, 5701 (1990); C. Deparis and J. Massies, "Surface stoichiometry variation associated with GaAs (001) reconstruction transitions," *J. Cryst. Growth* **108**, 157 (1991); R. Ludeke, R.M. King, and E.H.C. Parker, "MBE surface and interface studies," in E.H.C. Parker, ed., *The Technology and Physics of Molecular Beam Epitaxy* (Plenum Press, New York, 1985), pp. 555-628; H.H. Farrell and C.J. Palmstrøm, "Reflection high energy electron diffraction characteristic absences in GaAs (100) (2x4)–As: a tool for determining surface stoichiometry," *J. Vac. Sci. Technol.* **B8**, 903 (1990); and J.Y. Tsao, T.M. Brennan, J.F. Klem, and B.E. Hammons, "Surface-stoichiometry dependence of As_2 desorption and As_4 'reflection' from GaAs (001)," *J. Vac. Sci. Techn.* **A7**, 2138 (1989).

for understanding surface alloys. For simplicity, we develop the framework within the approximation that the surface is exactly one monolayer thick. As a consequence, the framework, like those of other monolayer models, cannot be used to understand phenomena that depend on surface effects greater than one monolayer deep.[2] Nevertheless, the framework is intuitive, leads to a deep physical understanding of the relationship between bulk and surface alloy phases, and can be easily used in semi-empirical modeling.

Then, in Section 7.2, we apply the framework to equilibria and nonequilibria between vapor and monolayer adsorbate phases, treating the adsorbate phase as a surface alloy of adsorbates and "missing" adsorbates. In doing so, we will derive familiar equilibrium constructs, such as adsorption isotherms and adsorption isobars, as well as discuss less familiar nonequilibrium phenomena, such as transient and coverage-dependent adsorption and desorption.

Finally, in Section 7.3, we will apply the framework to the technologically important phenomena of segregation and trapping of dopants or other impurities at surfaces during MBE. This phenomenon is especially complex, in that it involves equilibria and nonequilibria between vapor, surface, *and* bulk crystalline phases.

7.1 Monolayer Thermodynamics

In this section, we discuss the equilibrium thermodynamics of the surface of a bulk alloy. We begin, in Subsection 7.1.1, by establishing a nomenclature consistent with that introduced in Chapter 3. We then ask, in the first half of Subsection 7.1.2: given a composition of the bulk alloy, what is the composition of the surface alloy that is in equilibrium with that bulk alloy? In general, the surface and bulk compositions will not be the same in equilibrium, in that one component of the alloy will tend to segregate to the surface, displacing the other component back into the bulk. We finally ask, in the second half of Subsection 7.1.2: given the compositions of the surface and bulk alloys, what is the free energy required to create new surface at that composition? This free energy is the surface work (also often called the surface tension), and is minimum if the surface composition is such that the surface alloy phase is in equilibrium with the bulk alloy phase.

---

[2] See, e.g., J.K. Strohl and T.S. King, "A multicomponent, multilayer model of surface segregation in alloy catalysts," *J. Catal.* **118**, 53 (1989).

7.1.1 Surface Free Energies and Chemical Potentials

Let us begin, in this subsection, by establishing our nomenclature. Consider a binary crystalline alloy phase, β, containing $N_{\rm a}$ moles (or atoms) of component a and $N_{\rm b}$ moles (or atoms) of component b. As in Chapter 3, we write the molar Gibbs free energy of this bulk phase as

$$g^{\beta} \equiv \frac{G^{\beta}}{N_{\rm a} + N_{\rm b}}, \tag{7.1}$$

where G^{β} is the total Gibbs free energy. Again, as in Chapter 3, the chemical potentials of the two components a and b in β are the intercepts with the $x = 0$ and $x = 1$ axes of the tangents to g^{β}:

$$\begin{aligned} \mu_{\rm a}^{\beta} &= g^{\beta} - x^{\beta}\frac{\partial g^{\beta}}{\partial x^{\beta}} \\ \mu_{\rm b}^{\beta} &= g^{\beta} + (1 - x^{\beta})\frac{\partial g^{\beta}}{\partial x^{\beta}}, \end{aligned} \tag{7.2}$$

where $x^{\beta} \equiv N_{\rm b}/(N_{\rm a} + N_{\rm b})$ is the composition of β.

Consider second a surface of the bulk crystal, σ, characterized by $N_{\rm a}^{\sigma}$ exposed atoms of component a and $N_{\rm b}^{\sigma}$ exposed atoms of component b. Associate with the exposed atoms on this surface a Gibbs free energy equal to the difference between the total Gibbs free energy and the Gibbs free energy of the *nonsurface* atoms still in the bulk crystal:

$$G^{\sigma}(N_{\rm a}^{\sigma}, N_{\rm b}^{\sigma}) = G^{\rm tot}(N_{\rm a}^{\sigma}, N_{\rm b}^{\sigma}, N_{\rm a}^{\beta}, N_{\rm b}^{\beta}) - G^{\beta}(N_{\rm a}^{\beta}, N_{\rm b}^{\beta}). \tag{7.3}$$

In general, G^{σ} depends not only on $N_{\rm a}^{\sigma}$ and $N_{\rm b}^{\sigma}$, but on $N_{\rm a}^{\beta}$ and $N_{\rm b}^{\beta}$ as well. Here, we neglect this dependence, and note that such a dependence is nontrivial to include in a way that self-consistently treats bonding within the surface layer and bonding between the surface layer and the bulk layers below.[3]

Let us therefore consider this surface to be a 2D monolayer phase having its own thermodynamic properties apart from those of the bulk. In this way, we can adopt the nomenclature and definitions developed originally for bulk phases. For example, by analogy to Equation 7.1, the molar Gibbs free energy of the exposed surface atoms can be defined as

$$g^{\sigma} \equiv \frac{G^{\sigma}}{N_{\rm a}^{\sigma} + N_{\rm b}^{\sigma}}, \tag{7.4}$$

[3] J.W. Belton and M.G. Evans, "Studies in the molecular forces involved in surface formation. II. The surface free energies of simple liquid mixtures," *Trans. Faraday Soc.* **41**, 1 (1945); A. Schuchowitzky, *Acta Physicochim. URSS* **19** (2-3), 176 (1944); R. Defay and I. Prigogine, "Surface tension of regular solutions," *Trans. Faraday Soc.* **46**, 199 (1950); and S. Ono and S. Kondo, *Handb. Physik* **10**, 134 (1960).

and, by analogy to Equations 7.2, the chemical potentials of a and b in σ are the intercepts with the $x=0$ and $x=1$ axes of the tangents to g^σ:

$$\begin{aligned}\mu_{\rm a}^\sigma &= g^\sigma - x^\sigma \frac{\partial g^\sigma}{\partial x^\sigma} \\ \mu_{\rm b}^\sigma &= g^\sigma + (1-x^\sigma)\frac{\partial g^\sigma}{\partial x^\sigma}.\end{aligned} \tag{7.5}$$

where $x^\sigma \equiv N_{\rm b}^\sigma/(N_{\rm a}^\sigma + N_{\rm b}^\sigma)$ is the composition of σ.

Furthermore, the composition dependence of the molar Gibbs free energy of surface phases may be semi-empirically modeled in the same way that the molar Gibbs free energy of bulk phases is often modeled, as ideal solutions, or as one of a heirarchy of regular solutions (see Table 3.1). Examples of such composition-dependent surface and bulk molar Gibbs free energies for the Ag–Au system are shown in Figure 7.1. In this system, the molar Gibbs free energies are thought to be characterized by the sub-regular forms

$$\begin{aligned}g^{\langle \mathrm{Ag}_{1-x^\beta}\mathrm{Au}_{x^\beta}\rangle} &= (1-x^\beta)g^{\langle \mathrm{Ag}\rangle} + x^\beta g^{\langle \mathrm{Au}\rangle} - s_{\rm mix,ideal}T + \Omega^\beta(1-x^\beta)x^\beta \\ g^{\rangle \mathrm{Ag}_{1-x^\sigma}\mathrm{Au}_{x^\sigma}(} &= (1-x^\sigma)g^{\rangle \mathrm{Ag}(} + x^\sigma g^{\rangle \mathrm{Au}(} - s_{\rm mix,ideal}T + \Omega^\sigma(1-x^\sigma)x^\sigma,\end{aligned} \tag{7.6}$$

and are linear interpolations between the molar Gibbs free energies of the pure-component phases, plus entropic and enthalpic "mixing" terms.

Note that in writing these equations, we have extended the notation of Section 2.4 so that interface phases are represented by *mismatched* pairs of brackets, braces, and parentheses to denote the bulk phases the interface is sandwiched between. In this notation, the two phases of interest, the crystalline bulk and surface phases, are denoted $\langle \mathrm{Ag}_{1-x^\beta}\mathrm{Au}_{x^\beta}\rangle$ and $\rangle \mathrm{Ag}_{1-x^\sigma}\mathrm{Au}_{x^\sigma}($, and their compositions are denoted x^β and x^σ.

For the crystalline solid, $g^{\langle \mathrm{Ag}\rangle}$ and $g^{\langle \mathrm{Au}\rangle}$ are the known molar Gibbs free energies of the pure-component phases Ag and Au,[4] and

$$\Omega^{\langle \mathrm{Ag}_{1-x^\beta}\mathrm{Au}_{x^\beta}\rangle} = A + Bx^\beta + CT \tag{7.7}$$

is a known composition and temperature-dependent interaction parameter.[5]

[4]The molar Gibbs free energies of the pure crystals were calculated according to the prescription described in Chapter 2, using the heat capacity expression $c_p = (c_o + c_1 T)T^2/(T^2 + \Theta_T^2)$. The heat capacity parameters for $\langle$Ag$\rangle$ were c_o = 0.253 meV/(atomK), c_1 = 0.0553 μeV/atom/(atomK^2), Θ_T = 55.4 K; the parameters for $\langle$Au$\rangle$ were c_o = 0.248 meV/(atomK), c_1 = 0.563 μeV/atom/(atomK^2) and Θ_T = 43.5 K.

[5]Following J.L. White, R.L. Orr, and R. Hultgren, "The thermodynamic properties

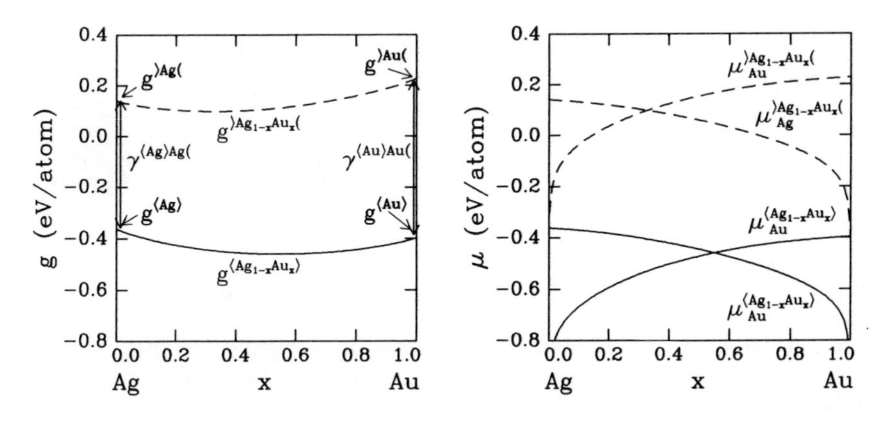

Figure 7.1: Left: Molar Gibbs free energies of the crystalline bulk and surface phases of the Ag–Au system at 700 K. Right: Chemical potentials of Ag and Au in these phases. As in Figure 3.1, the intercepts of the tangents to the molar Gibbs free energies with the $x = 0$ and $x = 1$ axes are the chemical potentials. As the tangents sweep around the arc of the molar Gibbs free energy curves, the intercepts of those tangents trace out the chemical potentials at the various compositions.

For the crystalline surface *at* the endpoint compositions, the molar Gibbs free energies are those for pure crystalline $\langle\text{Ag}\rangle$ and $\langle\text{Au}\rangle$, but offset upward by their known surface tensions. In other words,

$$\begin{aligned} g^{\rangle\text{Ag}(} &= g^{\langle\text{Ag}\rangle} + \gamma^{\langle\text{Ag}\rangle\text{Ag}(} \\ g^{\rangle\text{Au}(} &= g^{\langle\text{Au}\rangle} + \gamma^{\langle\text{Au}\rangle\text{Au}(}, \end{aligned} \tag{7.8}$$

where the experimentally measured values[6] for $\gamma^{\langle\text{Ag}\rangle}$ and $\gamma^{\langle\text{Au}\rangle}$, the work per unit area required to form new surfaces of pure crystalline Ag and Au, have been normalized by the number of atoms per unit area on close-packed (111) planes.

For the crystalline surface *away* from the endpoint compositions, the molar Gibbs free energy has been found to be consistent with a sub-regular solution behavior that mimics that of the crystalline bulk phase.[7] In other

of silver-gold alloys," *Acta Metall.* **5**, 747 (1957) and H. Okamoto and T.B. Massalski, in *Phase Diagrams of Binary Gold Alloys*, H. Okamoto and T.B. Massalski, Eds. (ASM International, Metals Park, Ohio, 1987), pp. 4-12, the sub-regular solution parameters were taken to be $A = -0.210$ eV/atom, $B = 0.0347$eV/atom and $C = 0.0000596$ eV/(atomK).

[6] We use the values $\gamma^{\langle\text{Ag}\rangle\text{Ag}(} = 0.50$ eV/atom and $\gamma^{\langle\text{Au}\rangle\text{Au}(} = 0.63$ eV/atom [H. Jones, "The surface energy of solid metals," *Met. Sci. J.* **5**, 15 (1971)].

[7] J.Y. Tsao, "Graphical representation of Ag–Au surface segregation," *Surf. Sci.* **262**, 382 (1992).

words,

$$\Omega^{\sigma} = \Omega^{\beta}. \tag{7.9}$$

As drawn in the left panel of Figure 7.1, the shape of the molar Gibbs free energy of the surface alloy is the same as that of the bulk alloy, but is offset upward by amounts that vary linearly from $\gamma^{\langle \mathrm{Ag}\rangle \mathrm{Ag}(}$ on one end to $\gamma^{\langle \mathrm{Au}\rangle \mathrm{Au}(}$ on the other end.

7.1.2 Atom Transfers between Surface and Bulk

Having defined, in Subsection 7.1.1, the thermodynamic functions for the crystalline bulk and surface phases, let us consider, in this subsection, *transferring* atoms between the two phases. Such transfers can take place in two ways, and are discussed separately in the following two subsubsections.

Parallel Tangents and Equilibrium Segregation

In the first way of transferring atoms between the two phases, the overall number of surface sites is preserved. Then, if we move, e.g., a Au atom from the bulk to the surface, we must at the same time move a Ag atom from the surface to the bulk: atom transfers between bulk and surface must be atom *exchanges*. Hence, they are accompanied by free energy changes equal to the *difference* between (a) the "excess" chemical potentials required to move a Au atom from the bulk to the surface, or

$$\mu_{\mathrm{Au}}^{\mathrm{exc}} \equiv \mu_{\mathrm{Au}}^{\rangle \mathrm{Ag}_{1-x^{\sigma}}\mathrm{Au}_{x^{\sigma}}(} - \mu_{\mathrm{Au}}^{\langle \mathrm{Ag}_{1-x^{\beta}}\mathrm{Au}_{x^{\beta}}\rangle}, \tag{7.10}$$

and (b) the "excess" chemical potentials required to move a Ag atom from the bulk to the surface, or

$$\mu_{\mathrm{Ag}}^{\mathrm{exc}} \equiv \mu_{\mathrm{Ag}}^{\rangle \mathrm{Ag}_{1-x^{\sigma}}\mathrm{Au}_{x^{\sigma}}(} - \mu_{\mathrm{Ag}}^{\langle \mathrm{Ag}_{1-x^{\beta}}\mathrm{Au}_{x^{\beta}}\rangle}. \tag{7.11}$$

In other words, they are accompanied by a free energy change of

$$\begin{aligned} \mu_{\mathrm{Au}}^{\mathrm{exc}} - \mu_{\mathrm{Ag}}^{\mathrm{exc}} \quad \equiv \quad & \left(\mu_{\mathrm{Au}}^{\rangle \mathrm{Ag}_{1-x^{\sigma}}\mathrm{Au}_{x^{\sigma}}(} - \mu_{\mathrm{Au}}^{\langle \mathrm{Ag}_{1-x^{\beta}}\mathrm{Au}_{x^{\beta}}\rangle}\right) \\ & - \left(\mu_{\mathrm{Ag}}^{\rangle \mathrm{Ag}_{1-x^{\sigma}}\mathrm{Au}_{x^{\sigma}}(} - \mu_{\mathrm{Ag}}^{\langle \mathrm{Ag}_{1-x^{\beta}}\mathrm{Au}_{x^{\beta}}\rangle}\right). \end{aligned} \tag{7.12}$$

This free energy change can be rewritten as

$$\begin{aligned} \mu_{\mathrm{Au}}^{\mathrm{exc}} - \mu_{\mathrm{Ag}}^{\mathrm{exc}} \quad \equiv \quad & \left(\mu_{\mathrm{Au}}^{\rangle \mathrm{Ag}_{1-x^{\sigma}}\mathrm{Au}_{x^{\sigma}}(} - \mu_{\mathrm{Ag}}^{\rangle \mathrm{Ag}_{1-x^{\sigma}}\mathrm{Au}_{x^{\sigma}}(}\right) \\ & - \left(\mu_{\mathrm{Au}}^{\langle \mathrm{Ag}_{1-x^{\beta}}\mathrm{Au}_{x^{\beta}}\rangle} - \mu_{\mathrm{Ag}}^{\langle \mathrm{Ag}_{1-x^{\beta}}\mathrm{Au}_{x^{\beta}}\rangle}\right), \end{aligned} \tag{7.13}$$

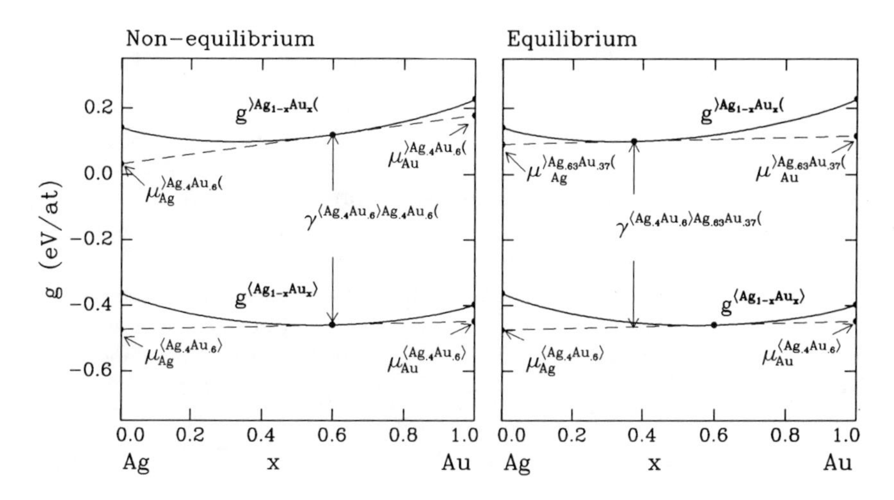

Figure 7.2: Surface and bulk phase equilibria in the Ag–Au system at 700 K. Left: The two tangents to the molar Gibbs free energies have different slopes and there is a driving force for Ag or Au atoms to segregate to the surface. Center: the two tangents are parallel and the surface is in equilibrium with the bulk.

and can be seen to be the difference between the slope of the tangent to $\mu^{\rangle Ag_{1-x^\sigma}Au_{x^\sigma}\langle}$ at x^σ and the slope of the tangent to $\mu^{\langle Ag_{1-x^\beta}Au_{x^\beta}\rangle}$ at x^β. When $\mu_{Au}^{exc} - \mu_{Ag}^{exc}$ is positive, as in the left panel of Figure 7.2, Au surface atoms will tend to exchange with Ag bulk atoms, and the surface will become enriched in Ag. When it is negative, then Ag surface atoms will tend to exchange with Au bulk atoms. When it is zero, as in the right panel of Figure 7.2, then the surface is in equilibrium with the bulk. In other words, the crystal and surface phases are in equilibrium with each other when the tangents to their molar Gibbs free energies have the same slopes, or, equivalently, when the tangents are parallel.[8]

Now, according to this parallel tangent criterion, to find the composition of a surface in equilibrium with a bulk crystal of a particular composition, we must solve $\partial g^{\rangle Ag_{1-x^\sigma}Au_{x^\sigma}\langle}/\partial x^\sigma = \partial g^{\langle Ag_{1-x^\beta}Au_{x^\beta}\rangle}/\partial x^\beta$ by varying x^σ for fixed x^β. Equivalently, and sometimes more conveniently, one can (see

[8] M. Hillert, "The role of interfaces in phase transformations," in *The Mechanism of Phase Transformations in Crystalline Solids*, Monograph and Report Series No. 33 (The Institute of Metals, London, 1969), pp. 231-247; and M. Guttmann, "Grain boundary segregation, two dimensional compound formation, and precipitation," *Met. Trans.* **8A**, 1383 (1977).

Exercise 1) minimize the function

$$\eta(x^\sigma, x^\beta) \equiv g^{\rangle \mathrm{Ag}_{1-x^\sigma}\mathrm{Au}_{x^\sigma}(} - x^\sigma \frac{\partial g^{\langle \mathrm{Ag}_{1-x^\beta}\mathrm{Au}_{x^\beta}\rangle}}{\partial x^\beta}, \tag{7.14}$$

by varying x^σ for fixed x^β. Both numerical prescriptions are general, and can be used even if the molar Gibbs free energies of the surface and bulk are represented by very complicated semi-empirical forms.

For example, consider the relationship between the equilibrium surface and bulk compositions of strictly regular bulk and surface phases β and σ. In the limit of small x^β and x^σ, this relationship can be shown (see Exercise 4) to be given by

$$\kappa_{\mathrm{equ}} \equiv \frac{x^\beta}{x^\sigma} = e^{-[\gamma^{\langle \mathrm{b}\rangle \mathrm{b}(} - \gamma^{\langle \mathrm{a}\rangle \mathrm{a}(} + \Omega^\sigma - \Omega^\beta]/kT}, \tag{7.15}$$

where $\gamma^{\langle \mathrm{a}\rangle \mathrm{a}(}$ and $\gamma^{\langle \mathrm{b}\rangle \mathrm{b}(}$ are the surface tensions of the pure a and pure b phases. The quantity κ_{equ}, the ratio between the equilibrium bulk and surface compositions, can be thought of as an equilibrium "partition" coefficient, in that it describes the physical partitioning of a dilute impurity between two adjacent phases.

More generally, Equation 7.14 must be solved numerically. For the Ag–Au system, the resulting dependence of the surface composition on bulk composition is shown as the segregation isotherm in Figure 7.3. Note that at all compositions, the surface tends to be enriched in Ag relative to the bulk. The reason is that, even though $g^{\rangle \mathrm{Ag}_{1-x^\sigma}\mathrm{Au}_{x^\sigma}(}$ has the same *shape* as $g^{\langle \mathrm{Ag}_{1-x^\beta}\mathrm{Au}_{x^\beta}\rangle}$, its *offset* relative to $g^{\langle \mathrm{Ag}_{1-x^\beta}\mathrm{Au}_{x^\beta}\rangle}$ increases linearly with composition because pure Au has a higher surface tension than does pure Ag. As a consequence, at the same composition, the slope of the tangent to $g^{\rangle \mathrm{Ag}_{1-x^\sigma}\mathrm{Au}_{x^\sigma}(}$ will be greater than the slope of the tangent to $g^{\langle \mathrm{Ag}_{1-x^\beta}\mathrm{Au}_{x^\beta}\rangle}$, and is compensated for by a decrease in the composition of $g^{\rangle \mathrm{Ag}_{1-x^\sigma}\mathrm{Au}_{x^\sigma}(}$.

Surface Work

In the second way of transferring atoms between the crystalline bulk and surface phases, the overall number of surface sites is not preserved. Instead, as Ag or Au are transferred from the bulk to the surface, new surface sites are created to accommodate them. The work per atom required to create new surface of composition x^σ from bulk crystal of composition x^β is now the *sum* of the changes in the chemical potentials of the two components, weighted by their mole fractions on the surface:

$$\gamma^{\langle \mathrm{Ag}_{1-x^\beta}\mathrm{Au}_{x^\beta}\rangle \mathrm{Ag}_{1-x^\sigma}\mathrm{Au}_{x^\sigma}(} \equiv (1 - x^\sigma)\mu_{\mathrm{Ag}}^{\mathrm{exc}} + x^\sigma \mu_{\mathrm{Au}}^{\mathrm{exc}}. \tag{7.16}$$

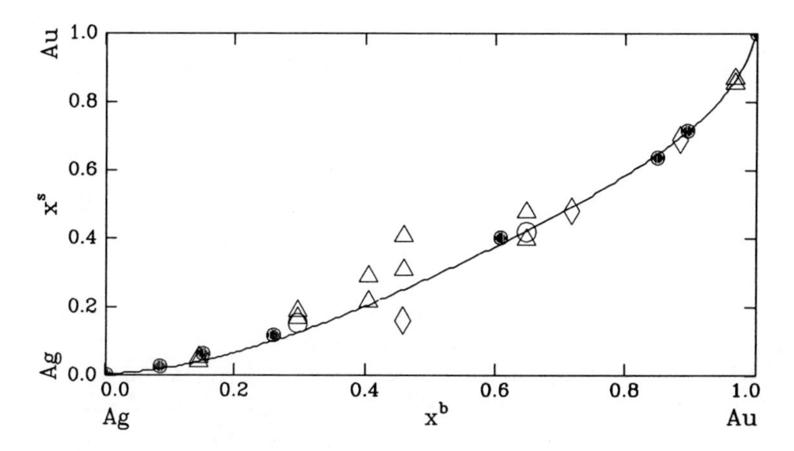

Figure 7.3: Surface segregation isotherm for the Ag–Au system at 700 K, deduced from the parallel tangent construction illustrated in Figure 7.2. Measured data are represented by open circles,[a] triangles,[b] diamonds,[c] and filled circles.[d]

[a]G.C. Nelson, "Determination of the surface versus bulk composition of silver–gold alloys by low energy ion scattering spectroscopy," *Surf. Sci.* **59**, 310 (1976).

[b]S.H. Overbury and G.A. Somorjai, "The surface composition of the silver–gold system by Auger electron spectroscopy," *Surf. Sci.* **55**, 209 (1976).

[c]M.J. Kelley, D.G. Swartzfager, and V.S. Sundaram, "Surface segregation in the Ag–Au and Pt–Cu systems," *J. Vac. Sci. Technol.* **16**, 664 (1979).

[d]K. Meinel, M. Klaua, and H. Bethge, "Segregation and sputter effects on perfectly smooth (111) and (100) surfaces of Au–Ag alloys studied by AES," *Phys. Stat. Sol.* **A106**, 133 (1988).

This equation defines the surface work, $\gamma^{\langle \mathrm{Ag}_{1-x^\beta}\mathrm{Au}_{x^\beta}\rangle \mathrm{Ag}_{1-x^\sigma}\mathrm{Au}_{x^\sigma}(}$, in terms of the surface and bulk compositions and chemical potentials.

Let us now expand $\mu_{\mathrm{Au}}^{\mathrm{exc}}$ and $\mu_{\mathrm{Ag}}^{\mathrm{exc}}$ using Equations 7.10 and 7.11, and apply the identities

$$\begin{aligned} g^{\langle \mathrm{Ag}_{1-x^\beta}\mathrm{Au}_{x^\beta}\rangle} &= (1-x^\beta)\mu_{\mathrm{Ag}}^{\langle \mathrm{Ag}_{1-x^\beta}\mathrm{Au}_{x^\beta}\rangle} + x^\beta \mu_{\mathrm{Au}}^{\langle \mathrm{Ag}_{1-x^\beta}\mathrm{Au}_{x^\beta}\rangle} \\ g^{\rangle \mathrm{Ag}_{1-x^\sigma}\mathrm{Au}_{x^\sigma}(} &= (1-x^\sigma)\mu_{\mathrm{Ag}}^{\rangle \mathrm{Ag}_{1-x^\sigma}\mathrm{Au}_{x^\sigma}(} + x^\sigma \mu_{\mathrm{Au}}^{\rangle \mathrm{Ag}_{1-x^\sigma}\mathrm{Au}_{x^\sigma}(}, \end{aligned} \quad (7.17)$$

which can be derived from Equations 7.2 and 7.5. Then, the surface work can be written as

$$\begin{aligned} &\gamma^{\langle \mathrm{Ag}_{1-x^\beta}\mathrm{Au}_{x^\beta}\rangle \mathrm{Ag}_{1-x^\sigma}\mathrm{Au}_{x^\sigma}(} = \\ &\quad \left(g^{\rangle \mathrm{Ag}_{1-x^\sigma}\mathrm{Au}_{x^\sigma}(} - g^{\langle \mathrm{Ag}_{1-x^\beta}\mathrm{Au}_{x^\beta}\rangle}\right) \\ &\quad - (x^\sigma - x^\beta)\left(\mu_{\mathrm{Au}}^{\langle \mathrm{Ag}_{1-x^\beta}\mathrm{Au}_{x^\beta}\rangle} - \mu_{\mathrm{Ag}}^{\langle \mathrm{Ag}_{1-x^\beta}\mathrm{Au}_{x^\beta}\rangle}\right). \end{aligned} \quad (7.18)$$

If we define the "excess" molar Gibbs free energy to be $g^{\rm exc} \equiv g^{\rangle {\rm Ag}_{1-x^\sigma}{\rm Au}_{x^\sigma}\langle} - g^{\langle {\rm Ag}_{1-x^\beta}{\rm Au}_{x^\beta}\rangle}$, then we also have

$$\begin{aligned} g^{\rm exc} &= \gamma^{\langle {\rm Ag}_{1-x^\beta}{\rm Au}_{x^\beta}\rangle {\rm Ag}_{1-x^\sigma}{\rm Au}_{x^\sigma}\langle} \\ &\quad + (x^\sigma - x^\beta)\left(\mu_{\rm Au}^{\langle {\rm Ag}_{1-x^\beta}{\rm Au}_{x^\beta}\rangle} - \mu_{\rm Ag}^{\langle {\rm Ag}_{1-x^\beta}{\rm Au}_{x^\beta}\rangle}\right). \end{aligned} \tag{7.19}$$

This last expression for the relationship between the excess molar Gibbs free energy and the surface work can be understood graphically by inspection of Figure 7.2. If the surface and bulk compositions are the same, as in the left panel, then $x^\sigma = x^\beta$ and $g^{\rm exc} = \gamma^{\langle {\rm Ag}_{1-x^\beta}{\rm Au}_{x^\beta}\rangle {\rm Ag}_{1-x^\sigma}{\rm Au}_{x^\sigma}\langle}$ itself. Otherwise, as in the right panel, we must add a correction term equal to the slope of the tangent to $g^{\langle {\rm Ag}_{1-x^\beta}{\rm Au}_{x^\beta}\rangle}$ times the difference between the surface and bulk compositions. The surface work can thus be seen to be the vertical distance between the tangent to the molar Gibbs free energy of the surface at composition x^σ, evaluated at x^σ, and the tangent to the molar Gibbs free energy of the bulk at composition x^β, also evaluated at x^σ. Importantly, this graphical interpretation of the surface work holds whether or not the tangents are parallel, hence *whether or not the surface and bulk are in equilibrium with each other.*

To make contact with standard treatments of surface thermodynamics, note that Equation 7.19 can be rewritten in yet another equivalent form:

$$g^{\rm exc} = \gamma^{\langle {\rm Ag}_{1-x^\beta}{\rm Au}_{x^\beta}\rangle {\rm Ag}_{1-x^\sigma}{\rm Au}_{x^\sigma}\langle} + x_{\rm Ag}^{\rm exc}\mu_{\rm Ag}^{\langle {\rm Ag}_{1-x^\beta}{\rm Au}_{x^\beta}\rangle} + x_{\rm Au}^{\rm exc}\mu_{\rm Au}^{\langle {\rm Ag}_{1-x^\beta}{\rm Au}_{x^\beta}\rangle}. \tag{7.20}$$

This equation reproduces the well-established relation[9] (at constant temperature) between the excess molar Gibbs free energy of the surface, the surface work, and the excess Ag and Au at the surface, $x_{\rm Ag}^{\rm exc} \equiv x^\beta - x^\sigma$ and $x_{\rm Au}^{\rm exc} \equiv x^\sigma - x^\beta$.

Finally, let us return to Equation 7.16, to understand more clearly the difference between the work required to transfer atoms to the surface in the two different ways. In the first way, we form new surface area at fixed composition. The work required is then $\gamma^{\langle {\rm Ag}_{1-x^\beta}{\rm Au}_{x^\beta}\rangle {\rm Ag}_{1-x^\sigma}{\rm Au}_{x^\sigma}\langle}$. In the second way, we change the composition of the surface at fixed surface area. The work required is then $\partial\gamma^{\langle {\rm Ag}_{1-x^\beta}{\rm Au}_{x^\beta}\rangle {\rm Ag}_{1-x^\sigma}{\rm Au}_{x^\sigma}\langle}/\partial x^\sigma = \mu_{\rm Au}^{\rm exc} - \mu_{\rm Ag}^{\rm exc}$ (see Exercise 3). In equilibrium, that work must be zero, as in the discussion following Equations 7.12 and 7.13.

[9] A.W. Adamson, *Physical Chemistry of Surfaces*, 4th Ed. (John Wiley and Sons, New York, 1982).

7.2 Adsorption and Desorption

In Section 7.1, we outlined a simple semi-empirical framework for understanding surface thermodynamics. The framework hinged on approximating the outermost exposed atomic monolayer as a phase whose composition and thermodynamic properties are distinct from those of the bulk. In fact, this approximation is most inaccurate for the surfaces of condensed alloy phases, whose composition and thermodynamic properties vary gradually over more than one atomic layer into the bulk.

In this section, we apply the framework to monolayer adsorbate phases on one-component bulk solids. We assume, as is often the case, that the adsorbate component does not indiffuse into the bulk and hence remains on the surface. Then, the composition of the system does change abruptly between the outermost surface monolayer and the bulk, and our approximate treatment is much more realistic. We will begin, in Subsection 7.2.1, by deriving two important equilibrium constructs: adsorption isotherms and adsorption isobars. Then, in Subsection 7.2.2, we discuss nonequilibrium adsorption and desorption.

7.2.1 Adsorption Isotherms and Isobars

Let us start, in this subsection, by deriving the equilibrium adsorbate coverages associated with an ambient vapor at a particular pressure and temperature. Consider a low-vapor-pressure bulk crystal composed of a single component, "m," bathed in a vapor composed of a single component, "a." As indicated in the left panel of Figure 7.4, the molar Gibbs free energy of the crystal is denoted $g^{\langle \mathrm{m} \rangle}$, and the molar Gibbs free energy of the vapor is denoted $g^{(\mathrm{a})}$.

In the absence of atoms of component a on the surface, the molar Gibbs free energy of the surface, $g^{\rangle \mathrm{m} \langle}$, is just offset upward from $g^{\langle \mathrm{m} \rangle}$ by the surface tension, $\gamma^{\langle \mathrm{m} \rangle \mathrm{m} \langle}$. In the presence of a full monolayer of atoms of component a on the surface, the molar Gibbs free energy of the surface is denoted $g^{\rangle \mathrm{a} \langle}$.

At intermediate compositions, as discussed in the previous section, the molar Gibbs free energy is a linearly weighted interpolation between $g^{\rangle \mathrm{m} \langle}$ and $g^{\rangle \mathrm{a} \langle}$, plus entropy and enthalpy of mixing terms. For example, a strictly regular solution would be written

$$\begin{aligned} g^{\rangle \mathrm{m}_{1-\theta} \mathrm{a}_\theta \langle} &= (1-\theta) g^{\rangle \mathrm{m} \langle} + \theta g^{\rangle \mathrm{a} \langle} + kT[\theta \ln \theta + (1-\theta) \ln(1-\theta)] \\ &\quad + \Omega \theta (1-\theta), \end{aligned} \tag{7.21}$$

where θ is the "composition" of the surface phase. In a sense, the surface phase can be considered a mixture of surface sites covered by adatoms and

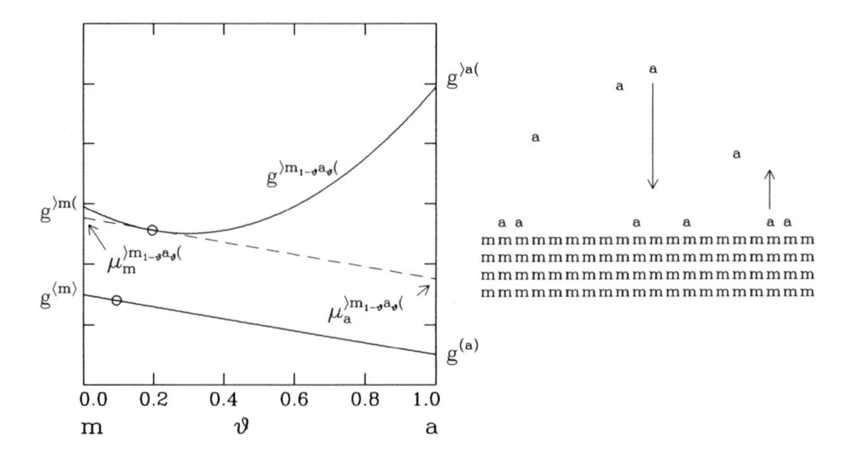

Figure 7.4: Vapor and surface adsorbate phase equilibria. Left: The two tangents to the molar Gibbs free energies are parallel, hence there is no driving force for the coverage of a atoms on the surface to change. Right: Schematic of adsorption and desorption of a atoms from the vapor onto a surface and from the surface back into the vapor.

surface sites not covered by adatoms, so that θ is also the average adatom coverage on the surface.

To find, given these molar Gibbs free energies, the equilibrium adatom coverage, we can apply the same arguments we applied in Section 7.1. Suppose, as illustrated in the left half of Figure 7.4, the surface coverage is θ_o, so that the chemical potential of atoms a is the intercept of the tangent to $g^{\rangle \mathrm{m}_{1-\theta}\mathrm{a}_\theta \langle}(\theta_o)$ with the $\theta = 1$ axis,

$$\mu_{\mathrm{a}}^{\rangle \mathrm{m}_{1-\theta}\mathrm{a}_\theta \langle} = g^{\rangle \mathrm{m}_{1-\theta}\mathrm{a}_\theta \langle} + (1-\theta)\frac{\partial g^{\rangle \mathrm{m}_{1-\theta}\mathrm{a}_\theta \langle}}{\partial \theta}. \tag{7.22}$$

and the chemical potential of atoms m is the intercept with the $\theta = 0$ axis,

$$\mu_{\mathrm{m}}^{\rangle \mathrm{m}_{1-\theta}\mathrm{a}_\theta \langle} = g^{\rangle \mathrm{m}_{1-\theta}\mathrm{a}_\theta \langle} - \theta\frac{\partial g^{\rangle \mathrm{m}_{1-\theta}\mathrm{a}_\theta \langle}}{\partial \theta}, \tag{7.23}$$

If we adsorb an atom a from the vapor, then the free energy of the system increases by $\mu_{\mathrm{a}}^{\rangle \mathrm{m}_{1-\theta}\mathrm{a}_\theta \langle} - g^{(\mathrm{a})}$ due to the movement of atom a from the vapor to the surface phase, but it decreases by $g^{\langle \mathrm{m} \rangle} - \mu_{\mathrm{m}}^{\rangle \mathrm{m}_{1-\theta}\mathrm{a}_\theta \langle}$ because the m atom that was covered has moved from the surface phase into the bulk. The equilibrium condition is therefore

$$\mu_{\mathrm{a}}^{\rangle \mathrm{m}_{1-\theta}\mathrm{a}_\theta \langle} - \mu_{\mathrm{m}}^{\rangle \mathrm{m}_{1-\theta}\mathrm{a}_\theta \langle} = g^{(\mathrm{a})} - g^{\langle \mathrm{m} \rangle}, \tag{7.24}$$

which is equivalent to the parallel tangent construction derived in Section 7.1.

Now, recall from Equation 2.47 that the molar Gibbs free energy of an elemental vapor is

$$g^{(\mathrm{a})}(p,T) = g^{(\mathrm{a})}(p_o,T) + kT \ln\left(\frac{p}{p_o}\right), \tag{7.25}$$

where p_o is a reference pressure. Hence, as the pressure of (a) increases, the molar Gibbs free energy of (a) also increases. As a consequence, the slope of $g^{(\mathrm{a})} - g^{\rangle\mathrm{m}(}$ increases, causing the parallel tangent to pivot around the $g^{\rangle\mathrm{m}_{1-\theta}\mathrm{a}_\theta(}$ curve, and ultimately causing the equilibrium coverage θ_{equ} itself to increase. For a strictly regular adsorbate phase, this parallel tangent condition is expressed by combining Equations 7.21, 7.22, 7.23, 7.24 and 7.25, giving

$$\frac{\theta_{\mathrm{equ}}}{1-\theta_{\mathrm{equ}}} e^{\Omega(1-\theta_{\mathrm{equ}})/kT} = \frac{p}{p_o} e^{\Delta g_{\mathrm{des}}/kT}, \tag{7.26}$$

where $\Delta g_{\mathrm{des}} = (g^{(\mathrm{a})}(p_o,T) - g^{\rangle\mathrm{a}(}) - (g^{\langle\mathrm{m}\rangle} - g^{\rangle\mathrm{m}(})$ is the "activation" free energy of desorption at the reference pressure p_o. This equation defines the coverage of the surface phase in equilibrium with a vapor at pressure p and temperature T, and can be used to construct both adsorption isotherms (the pressure dependence of the coverage at constant temperature) and adsorption isobars (the temperature dependence of the coverage at constant pressure).

For example, if $\Omega = 0$, so that the solution is ideal, then

$$\theta_{\mathrm{equ}} = \frac{p}{p + p_o e^{-\Delta g_{\mathrm{des}}/kT}}, \tag{7.27}$$

which reproduces what is known as Langmuir's isotherm. The adatom coverage increases linearly at first with increasing pressure, then saturates beyond a critical temperature-dependent pressure, $p_o e^{-\Delta g_{\mathrm{des}}/kT}$.

If $\Omega \neq 0$, then the solution is nonideal. On the one hand, if $\Omega > 0$, then adatoms and "missing" adatoms repel each other, which is equivalent physically to adatoms attracting each other. The adatom coverage increases more rapidly at first with increasing pressure, before again saturating beyond a critical temperature-dependent pressure. On the other hand, if $\Omega < 0$, then adatoms and "missing" adatoms attract each other, which is equivalent physically to adatoms repelling each other. The adatom coverage increases less rapidly at first with increasing pressure, before again saturating beyond a critical temperature-dependent pressure.

7.2.2 Sticking Coefficients and Desorption

In Subsection 7.2.1, we discussed the composition, or coverage, of an adsorbate surface phase in equilibrium with its vapor. Physically, that equilibrium can also be viewed as the balancing of a dynamic competition between adsorption of atoms or molecules from the vapor and desorption of atoms or molecules back into the vapor. As a consequence, if we know the adsorption rate, then, at equilibrium, we know the desorption rate as well. In this subsection, we derive expressions for this desorption rate, as well as for the rates at which coverages, perturbed away from their equilibrium values, will return to those equilibrium values.

From the kinetic theory of gases, the rate at which atoms or molecules in a vapor impinge upon a surface, per lattice site, is $p\lambda^2/\sqrt{2\pi mkT}$, where p and T are the pressure and temperature of the vapor, m is the atomic or molecular mass, and λ^2 is the area per lattice site of the surface. If $s(\theta, T)$ is the coverage and temperature dependent fraction of impinging atoms or molecules that "stick" to the surface, then the adsorption rate will be

$$j_{\rm des} = \frac{p\lambda^2 s(\theta,T)}{\sqrt{2\pi mkT}}. \tag{7.28}$$

At equilibrium, atoms or molecules must, by detailed balance, desorb exactly as fast as they adsorb. Since, at equilibrium, the coverage of a strictly regular solution surface phase is related to the pressure by Equation 7.26, the equilibrium desorption rate can also be expressed in terms of coverage as

$$j_{\rm des} = \frac{p\lambda^2 s(\theta,T)}{\sqrt{2\pi mkT}} = \frac{p_o\lambda^2 s(\theta,T)}{\sqrt{2\pi mkT}}\left(\frac{\theta}{1-\theta}\right) e^{\Omega(1-\theta)/kT} e^{-\Delta j_{\rm des}/kT}. \tag{7.29}$$

If we now assume that desorption depends directly on coverage, and only indirectly on the equilibrium pressure required to achieve that coverage, then Equation 7.29 holds even away from equilibrium. Hence, the net adsorption rate for a regular solution surface phase is

$$\begin{aligned}
\dot{\theta} &= j_{\rm net} \\
&= j_{\rm ads} - j_{\rm des} \\
&= \frac{\lambda^2 s(\theta,T)}{\sqrt{2\pi mkT}}\left[p - p_o\left(\frac{\theta}{1-\theta}\right) e^{\Omega(1-\theta)/kT} e^{-\Delta g_{\rm des}/kT}\right],
\end{aligned} \tag{7.30}$$

which is a first-order differential equation for the time evolution of the coverage.

Often, the sticking coefficient decreases linearly with coverage as $s(\theta,T) = s_o(1-\theta)$. Then,

$$\dot{\theta} = j_{\text{net}} = \frac{s_o\lambda^2}{\sqrt{2\pi mkT}}\left[p(1-\theta) - p_o\theta e^{\Omega(1-\theta)/kT}e^{-\Delta g_{\text{des}}/kT}\right]. \tag{7.31}$$

The net adsorption rate can be seen to be the difference between the rate at which atoms or molecules stick on uncovered portions of the substrate, and the rate at which atoms or molecules desorb from the covered portions of the substrate.

If the surface phase is an ideal solution, then $\Omega = 0$, and Equation 7.31 simplifies to

$$\dot{\theta} = j_{\text{net}} = \frac{s_o\lambda^2}{\sqrt{2\pi mkT}}\left[p - \theta(p+p_o)e^{-\Delta g_{\text{des}}/kT}\right]. \tag{7.32}$$

A surface having initially a coverage of θ_{ini} approaches exponentially the equilibrium coverage given by Equation 7.27 with a time constant τ given by

$$\frac{1}{\tau} = \frac{s_o\lambda^2(p+p_oe^{-\Delta g_{\text{des}}/kT})}{\sqrt{2\pi mkT}} = \frac{s_o\lambda^2 p_o e^{-\Delta g_{\text{des}}/kT}}{\sqrt{2\pi mkT}(1-\theta_{\text{equ}})}. \tag{7.33}$$

In other words,

$$\theta = \theta_{\text{equ}} + (\theta_{\text{ini}} - \theta_{\text{equ}})e^{-t/\tau}. \tag{7.34}$$

Note that for small deviations from the equilibrium coverage at pressure p_{equ}, the rate at which the surface will return to its equilibrium coverage is

$$\frac{d(\Delta\theta)}{dt} = j_{\text{net}}(\theta_{\text{equ}} + \Delta\theta) = j_{\text{net}}(\theta_{\text{equ}}) + \Delta\theta\left[\frac{\partial j_{\text{net}}}{\partial\theta}\right]_{\theta_{\text{equ}}}. \tag{7.35}$$

Since $j_{\text{net}}(\theta_{\text{equ}}) = 0$ at equilibrium,

$$\frac{1}{\tau} = -\frac{d(\Delta\theta)}{\Delta\theta dt} = -\left[\frac{j_{\text{net}}}{\partial\theta}\right]_{\text{equ}} \tag{7.36}$$

is the "small signal" approach rate back toward the equilibrium coverage.

For an ideal solution surface phase, Equations 7.32 and 7.36 give

$$\frac{1}{\tau} = \frac{s_o\lambda^2(p+p_oe^{-\Delta g_{\text{des}}/kT})}{\sqrt{2\pi mkT}} = \frac{s_o\lambda^2 p_o e^{-\Delta g_{\text{des}}/kT}}{\sqrt{2\pi mkT}(1-\theta_{\text{equ}})}, \tag{7.37}$$

which reproduces Equation 7.33. For a strictly regular solution surface phase, Equations 7.31 and 7.36 give

$$\frac{1}{\tau} = \frac{s_o\lambda^2}{\sqrt{2\pi mkT}}\left(\frac{1}{\theta_{\text{equ}}} - \frac{2\Omega\theta_{\text{equ}}}{kT}\right)e^{-\Delta g_{\text{des}}/kT}, \tag{7.38}$$

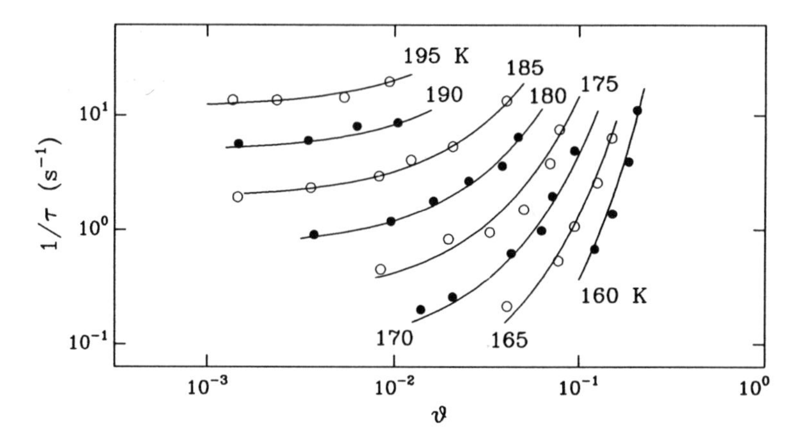

Figure 7.5: Coverage dependences of the small-signal vapor-adsorbate equilibration rate for CO on Cu(111). The open and filled circles are data measured[a] at the various indicated temperatures. The solid lines are the predictions of Equation 7.39, with a desorption molar Gibbs free energy of $\Delta g_{\text{des}} = (0.67 \text{ eV}) - (37.8kT)$, a mixing enthalpy of $\Omega_{\text{h}} = 0.107$ eV, and a mixing entropy of $\Omega_{\text{s}} = 20.7k$.

[a]B.J. Hinch and L.H. Dubois, "First-order corrections in modulated molecular beam desorption experiments," *Chem. Phys. Lett.* **171**, 131 (1990).

where θ_{equ} is given by Equation 7.26. For a regular solution with both an enthalpy and entropy of mixing, $\Omega = \Omega_{\text{h}} - T\Omega_{\text{s}}$, and

$$\frac{1}{\tau} = \frac{s_o\lambda^2}{\sqrt{2\pi mkT}} \left(\frac{1}{\theta_{\text{equ}}} - \frac{2\Omega_{\text{h}}\theta_{\text{equ}}}{kT} + \frac{2\Omega_{\text{s}}\theta_{\text{equ}}}{k} \right) e^{-\Delta g_{\text{des}}/kT}. \tag{7.39}$$

Examples of such coverage and temperature-dependent small-signal equilibration rates are illustrated in Figure 7.5 for CO on Cu(111). In this case there are both positive enthalpies and entropies of mixing. A positive enthalpy of mixing implies a repulsion between adatoms and missing adatoms, or, equivalently, an attraction between adatoms. Hence, the enthalpic barrier to desorption increases with increasing coverage. A positive entropy of mixing, however, implies an entropic barrier to desorption that decreases with increasing coverage. The two effects "compensate" each other to some extent, although, as illustrated in Figure 7.5, the balance tilts toward increasing the desorption rate with increasing coverage.

7.3 Surface Segregation and Trapping

In Section 7.1 we discussed the preferential segregation of one component from the bulk to the surface. In equilibrium, such segregation occurs when there are differences either between the surface tensions of the pure-component endpoint materials or between the free energies of mixing in the surface and bulk phases. Away from equilibrium, such segregation may or may not be significant, and will depend on the relative kinetics of crystal growth and interdiffusion between the surface and bulk phases. In this section, we discuss these dependences.

We will begin, in Subsection 7.3.1, by discussing the important simple case of segregation of a dilute solute under steady-state growth conditions.[10] This discussion will lead to an expression for the nonequilibrium partitition coefficient, κ, governing the ratio between the solute concentrations in the bulk and surface phases.

Then, in Subsection 7.3.2, we will make the assumption that, under non-steady-state conditions, this nonequilibrium partition coefficient still applies *locally* to the ratio between solute concentrations in the bulk phase just adjacent to the surface phase and in the surface phase itself. In this way, the nonequilibrium partition coefficient can be used to define a boundary condition connecting the non-steady-state evolution of solute concentrations in the bulk and surface phases.

7.3.1 Steady-State Compositional Partitioning

In this subsection, we consider steady-state segregation of a dilute solute b in a host solvent a. As illustrated in Figure 7.6, there are three phases to consider: the vapor, $(\mathrm{a}_{1-x^{\mathrm{v}}}\mathrm{b}_{x^{\mathrm{v}}})$, at composition x^{v}, the bulk solid, $\langle \mathrm{a}_{1-x^{\beta}}\mathrm{b}_{x^{\beta}} \rangle$, at composition x^{β}, and the surface monolayer dividing the two, $\rangle \mathrm{a}_{1-x^{\sigma}}\mathrm{b}_{x^{\sigma}} \langle$, at composition x^{σ}.

Between these three phases there are two basic kinetic processes that compete with each other.[11] First, vapor condenses, forming simultaneously a new surface layer (layer 1 in the right side of Figure 7.6), and transforming the previous surface layer into a new bulk solid layer (layer 2 in the right side of Figure 7.6). If condensation is "partitionless," in that the composition of the new surface layer mimics the composition of the vapor, then the system

[10]We do not treat the more complicated case of a nondilute solute; see, e.g., J.M. Moison, C. Guille, F. Houzay, F. Barthe, and M. Van Rompay, "Surface segregation of third-column atoms in group III-V arsenide compounds: ternary alloys and heterostructures," *Phys. Rev.* **B40**, 6149 (1989).

[11]J.J. Harris, D.E. Ashenford, C.T. Foxon, P.J. Dobson, and B.A. Joyce, "Kinetic limitations to surface segregation during MBE growth of III-V compounds: Sn in GaAs," *Appl. Phys.* **A33**, 87 (1984).

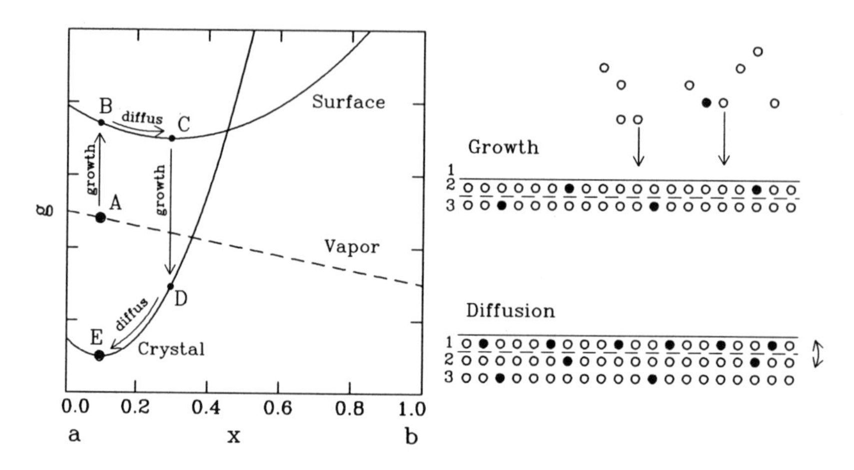

Figure 7.6: Left: Schematic molar Gibbs free energies of crystal $\langle a_{1-x^\beta} b_{x^\beta} \rangle$, vapor $(a_{1-x^v} b_{x^v})$ and surface $\rangle a_{1-x^\sigma} b_{x^\sigma} ($ phases. Right: Schematic of two competing kinetic processes: partitionless growth followed by solute partitioning via diffusion.

moves from point A to point B in the left side of Figure 7.6. At the same time, the previous surface layer, possibly enriched in solute, is transformed into a new bulk layer. Therefore, the system also moves from point C to point D in the left side of Figure 7.6.

Second, if the previous surface layer (layer 2 in the right side of Figure 7.6) were enriched in solute, then as it becomes a new bulk layer, it will also be enriched in solute. As a consequence, solute will tend to diffuse out into the new surface layer (layer 1 in the right side of Figure 7.6), moving the system from point D to point E and from B to C on the left side of Figure 7.6. In other words, partitionless condensation from vapor to surface to bulk solid is followed by partitioning by interdiffusion between the surface and the bulk solid. Note that, from start to finish, the system has moved *downward* from point A to point E in the left side of Figure 7.6, so that there is a net driving force for condensation.

Two extremes of behavior may be imagined, depending on the ratio between the rate of growth, j (in ML/s), and the rate of interdiffusion between bulk and surface layers, D_i/a^2. As in Section 6.3, this ratio, $a^2 j/D_i$, is a kind of Péclet number in that it is a dimensionless measure of the relative importance of convective over diffusional mass flow. Also as in Section 6.3, another way of understanding this Péclet number is to note that it is also the ratio between the time required for diffusion between the surface layer and its adjacent bulk layer, a^2/D_i, and the monolayer growth time, $\tau_{\mathrm{ML}} = 1/j$.

On the one hand, if $a^2 j/D_i \ll 1$, then interlayer diffusion is fast relative to growth. The surface layer will be in compositional equilibrium with its adjacent bulk layer, and the ratio between their compositions will be given by the equilibrium partition coefficient κ_{equ}. This extreme of behavior is therefore characterized by equilibrium solute segregation. On the other hand, if $a^2 j/D_i \gg 1$, then interlayer diffusion is slow relative to growth. The surface layer and its adjacent bulk layer will not have time during a monolayer growth cycle to reach composition equilibrium, and the ratio between their compositions, κ, will approach unity. This extreme of behavior is therefore characterized by nonequilibrium solute trapping.

Periodic and Aperiodic Step-Wise Growth

To quantify the dependence of κ on the Péclet number, consider a simple model in which growth proceeds by the passage of steps on a vicinal surface.[12] Suppose the composition of the surface layer just ahead of a moving step is x^σ. At time $t = 0$, just after the step has passed, that surface layer has become a bulk layer. If the new bulk layer has preserved its composition, then

$$\left[x^\beta\right]_{t=0} = x^\sigma. \tag{7.40}$$

During the subsequent time interval $\tau_{\text{ML}} = 1/j$ until yet another step passes, solute atoms in the bulk layer will diffuse to the surface layer, at a rate proportional to the deviation of the composition of the bulk layer from its equilibrium value, $\kappa_{\text{equ}} x^\sigma$. In other words,

$$\frac{\partial x^\beta}{\partial t} = \frac{-D_i}{a^2}(x^\beta - \kappa_{\text{equ}} x^\sigma). \tag{7.41}$$

From Equations 7.40 and 7.41, the solute concentration in the bulk decays exponentially with time according to

$$x^\beta = \kappa_{\text{equ}} x^\sigma + (x^\sigma - \kappa_{\text{equ}} x^\sigma) e^{-D_i t/a^2}. \tag{7.42}$$

Suppose now that once this bulk layer has been covered by yet another surface layer, further interdiffusion becomes negligible. There are two extreme possibilities for the ways in which the next layer may arrive.

On the one hand, if the steps on the surface are equispaced, then they pass over the surface *periodically*, at time intervals separated by $\tau_{\text{ML}} = 1/j$.

[12] M.J. Aziz, "Model for solute redistribution during rapid solidification," *J. Appl. Phys.* **53**, 1158 (1982).

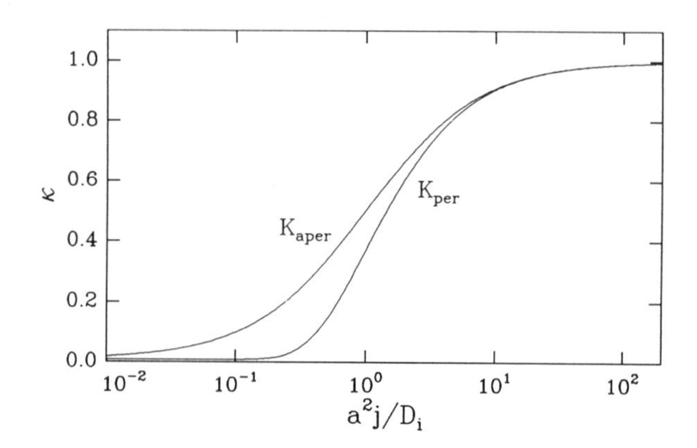

Figure 7.7: Dependence of nonequilibrium partition coefficients on the Péclet number, a^2j/D_i, for periodic and aperiodic step flow. The equilibrium partition coefficient in both cases was taken to be 10^{-2}.

Then, the steady-state composition of the bulk layer will be that which it has reached at time $\tau_{\mathrm{ML}} = 1/j$, or

$$x^\beta = \kappa_{\mathrm{equ}} x^\sigma + (x^\sigma - \kappa_{\mathrm{equ}} x^\sigma) e^{-D_i/(a^2 j)}. \tag{7.43}$$

In other words, when segregation occurs by interdiffusion of solute punctuated by the periodic passage of steps, then the steady-state ratio between bulk and surface compositions is

$$\kappa_{\mathrm{per}} = \frac{x^\beta}{x^\sigma} = \kappa_{\mathrm{equ}} + (1 - \kappa_{\mathrm{equ}}) e^{-D_i/(a^2 j)}. \tag{7.44}$$

As illustrated in Figure 7.7, κ is κ_{equ} for a^2j/D_i much less than unity, but increases to unity as a^2j/D_i approaches and exceeds unity.

On the other hand, if the steps on the surface are distributed randomly, then they pass over the surface *aperiodically*.[13] If this aperiodic passage obeys a Poisson arrival distribution, then the probability that a step will pass in an interval dt after time t will be $e^{-t/\tau_{\mathrm{ML}}} dt/\tau_{\mathrm{ML}} = ge^{-gt}dt$. Hence, the average composition of the bulk layer will be its composition after time t, weighted by this probability, or

$$x^\beta = \int_0^\infty \left[\kappa_{\mathrm{equ}} x^\sigma + (x^\sigma - \kappa_{\mathrm{equ}} x^\sigma) e^{-D_i/(a^2 j)}\right] je^{-jt} dt$$

[13]L.M. Goldman and M.J. Aziz, "Aperiodic stepwise growth model for the velocity and orientation dependence of solute trapping," *J. Mater. Res.* **2**, 524 (1987).

$$= \kappa_{\text{equ}} x^\sigma + (x^\sigma - \kappa_{\text{equ}} x^\sigma) \frac{j}{j + D_i/a^2}. \tag{7.45}$$

In other words, when segregation occurs by interdiffusion of solute punctuated by the aperiod passage of steps, then the steady-state ratio between bulk and surface composition is

$$\kappa_{\text{aper}} = \frac{x^\beta}{x^\sigma} = \frac{\kappa_{\text{equ}} + ja^2/D_i}{1 + ja^2/D_i}. \tag{7.46}$$

Again, as illustrated in Figure 7.7, κ is κ_{equ} for $a^2 j/D_i$ much less than unity, but increases to unity as $a^2 j/D_i$ approaches and exceeds unity. The increase is not as steep, however, as it is for κ_{per}.

A Segregating Dopant: Sb on Si (001)

To illustrate this behavior, consider the well-established[14] segregation of Sb impurities during MBE of Si on Si (001). Figure 7.8 shows measurements of the partition coefficient κ at various growth rates and temperatures. As temperature increases the partition coefficient initially decreases as Sb interdiffuses more and more quickly to the surface. At high temperatures, Sb diffusion is so fast that equilibrium is reached, and the partition coefficient approaches the equilibrium partition coefficient κ_{equ}. Finally, as temperature continues to increase, the surface and bulk phase compositions tend to equalize, and κ_{equ} itself approaches unity (see Equation 7.15). Therefore, as temperature continues to increase, ultimately κ begins to increase again, due to an increase in κ_{equ}.

Also shown in Figure 7.8 are the predictions of Equation 7.44 for segregation mediated by periodic step flow. As can be seen, the predictions agree reasonably well with the data, although there is some disagreement at the lower growth temperatures for the higher growth rates. The disagreement may be due to the onset of a 2D nucleation and growth mode, hence the onset of segregation mediated by aperiodic step flow.

7.3.2 Non-Steady-State Compositional Partitioning

In Subsection 7.3.1, we derived expressions for the nonequilibrium partition coefficient, κ. There, we assumed a steady-state solute concentration in the surface layer. In other words, we assumed that surface solute depletion due to incorporation into the bulk was just compensated for by adsorption from the vapor.

[14] J.C. Bean, "Arbitrary doping profiles produced by Sb-doped Si MBE," *Appl. Phys. Lett.* **33**, 654 (1978).

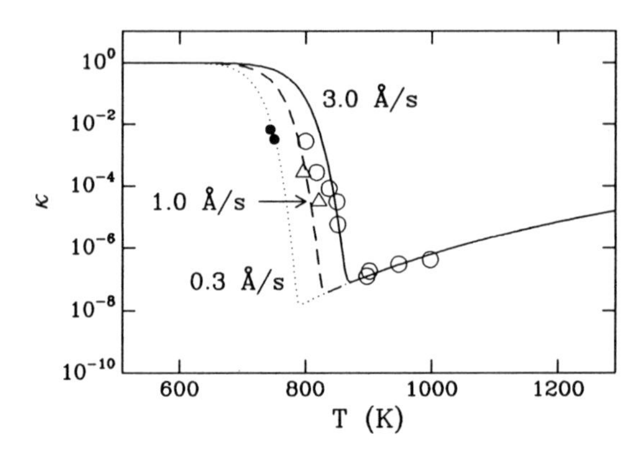

Figure 7.8: Temperature dependence of nonequilibrium partition coefficients for Sb during MBE of Si on Si (001). The data points[a] and predictions of Equation 7.44 are for growth rates of 3.0 Å/s (open circles and solid line), 1.0 Å/s (open triangles and dashed line), and 0.3 Å/s (filled circles and dotted line). The diffusivity was assumed to be Arrhenian, and given by $D_i/a^2 = 2\times10^{11}\exp(-1.67eV/kT)$, while the equilibrium partition coefficient was assumed to be $\kappa_{\text{equ}} = cxp(-1.23eV/kT)$, consistent with the form given by Equation 7.15.

[a] H. Jorke, "Surface segregation of Sb on Si (100) during molecular beam epitaxy growth," *Surf. Sci.* **193**, 569 (1988).

In this subsection, we relax this assumption, and allow the solute concentration in the surface layer to evolve. To do so, recall that the nonequilibrium partition coefficient, κ, is the ratio between solute concentrations in a bulk layer just adjacent to the surface layer and in the surface layer itself. It can therefore be thought of as the fraction of solute in the surface layer that becomes "trapped" in the adjacent bulk layer during each monolayer growth cycle. If the overall growth velocity is v, then the rate of decrease of solute in the surface layer due to trapping will be $v\kappa x^\sigma/a$, where a is a monolayer step height.

At the same time, solute may also adsorb from the vapor onto the surface, or desorb back into the vapor from the surface. If $v_{\text{ads}} = j_{\text{ads}}^{\text{solute}} a$ is the adsorption "velocity" and $v_{\text{des}} = j_{\text{des}}^{\text{solute}} x^\sigma a$ is the desorption "velocity" of solute, then the overall rate of change of solute concentration in the surface layer will be[15]

$$\dot{x}^\sigma = \frac{v_{\text{ads}}}{a} - \frac{(v_{\text{des}} + v\kappa)x^\sigma}{a}. \tag{7.47}$$

[15] C.E.C. Wood and B.A. Joyce, "Tin-doping effects in GaAs films grown by molecular beam epitaxy," *J. Appl. Phys.* **49**, 4854 (1978).

This equation describes the time evolution of the solute concentration in the surface layer during growth. It increases due to adsorption from the vapor, and decreases due to a combination of desorption back into the vapor and trapping in the bulk.

Note, though, that even after the solute has become trapped in the bulk, it may still diffuse, albeit at rates determined by the bulk diffusivities, which may be much slower than the diffusivity for exchange between the surface layer and its adjacent bulk layer. Therefore, the bulk solute concentration will evolve, after trapping, according to

$$\frac{\partial x^\beta}{\partial t} = D_\beta \frac{\partial^2 x^\beta}{\partial z^2}, \tag{7.48}$$

where z is a distance scale perpendicular to the surface in a stationary reference frame.[16]

The boundary condition on this diffusion equation is the solute concentration most recently trapped in the bulk layer just adjacent to the surface, or

$$\left[x^\beta\right]_{z=z_\sigma(t)} = \kappa x^\sigma. \tag{7.49}$$

In this equation, $z_\sigma(t) = z_{\sigma,o} - \int_0^t v dt$ is the position of the interface between the surface layer and its adjacent bulk layer. Equations 7.47, 7.48 and 7.49 together completely describe the time evolution of the overall bulk solute concentration due to nonequilibrium segregation followed by bulk diffusion. They are complicated, however, by the boundary condition in Equation 7.49, which must be applied at a moving surface. It is convenient, therefore, to transform into a reference frame, $z' = z + \int v dt'$, that moves with the surface.[17] In this reference frame, Equation 7.48 becomes

$$\frac{\partial x^\beta}{\partial t} = D_\beta \frac{\partial^2 x^\beta}{\partial z'^2} - v \frac{\partial x^\beta}{\partial z'}, \tag{7.50}$$

and Equation 7.49 becomes

$$\left[x^\beta\right]_{z'=0} = \kappa x^\sigma. \tag{7.51}$$

To illustrate the use of these equations, Figure 7.9 shows time evolutions of the spatial distributions of solute during growth of a structure

[16]We neglect electrostatic effects near the surface, which may cause solute "drift" toward or away from the surface. See, e.g., E.F. Schubert,, J.M. Kuo, R.F. Kopf, A.S. Jordan, H.S. Luftman, and L.C. Hopkins,, "Fermi-level-pinning-induced impurity redistribution in semiconductors during epitaxial growth," *Phys. Rev.* **B42**, 1364 (1990).

[17]S.A. Barnett and J.E. Greene, "Si molecular beam epitaxy: a model for temperature dependent incorporation probabilities and depth distributions of dopants exhibiting strong surface segregation," *Surf. Sci.* **151**, 67 (1985).

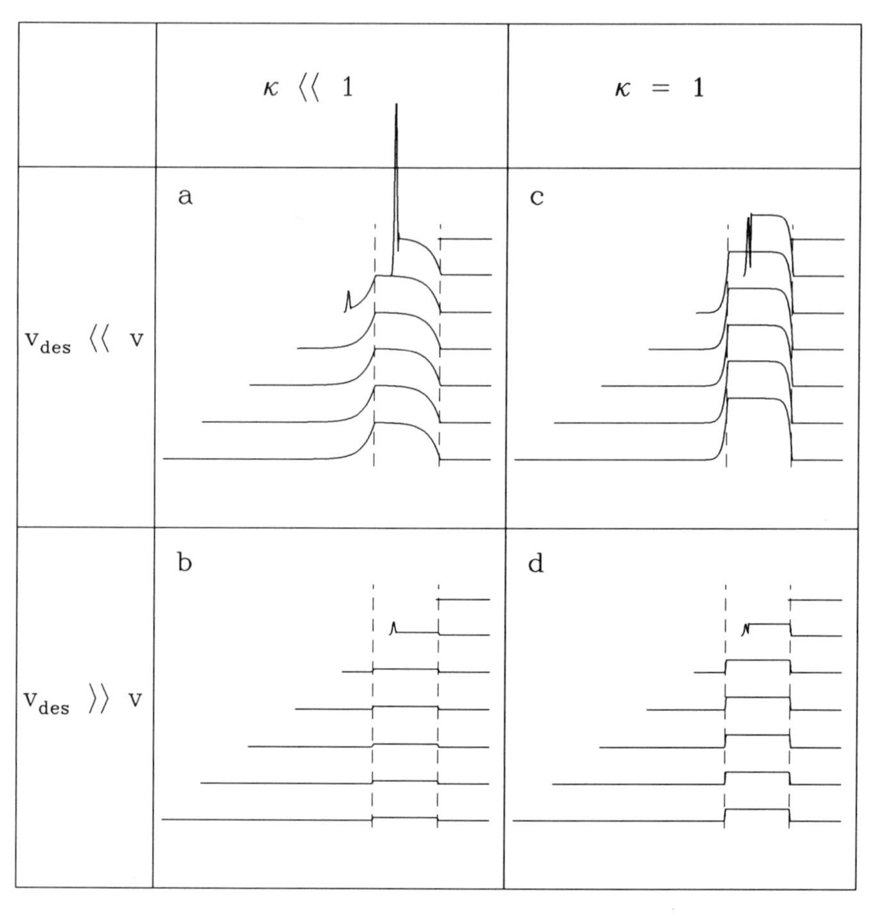

Figure 7.9: Four series of snapshots in time of solute composition profiles during MBE. All four series correspond to a square doping pulse indicated by the dashed lines, but differ according to whether v_{des} is much less than or greater than v and whether κ is much less than unity or equal to unity.

having a square pulse of solute. In the absence of bulk diffusion, there are four extremes of behavior, depending on (1) the relative rate between the growth velocity and the effective desorption velocity and (2) whether the nonequilibrium partition coefficient is near-unity or very different from unity.

Consider first the cases where the effective desorption velocity is much lower than the growth velocity. Then, a negligible fraction of the solute atoms that land on the surface leave, and virtually all ultimately incorporate into the growing crystal. On the one hand, if the partition coefficient is unity, as in panel (c), then all the solute atoms in the surface layer are incorporated into the bulk as growth proceeds. The depth profile of the final solute concentration mimics within one to two monolayers the square pulsed arrival rate of solute. On the other hand, if the partition coefficient is much less than unity, as in panel (a), then only a fraction of the solute atoms in the surface layer is incorporated into the bulk as growth proceeds. The depth profile of the final solute concentration now tails off gradually, as solute "rides" and gradually accumulates on the growing surface, and continues to be incorporated into the crystal even after the square pulse of solute has ended.

Consider second the cases where the effective desorption velocity is much higher than the growth velocity. Then, many of the solute atoms that land on the surface leave, and only a fraction ultimately incorporates into the growing crystal. That fraction is $\kappa v/(v_{\text{des}} + \kappa v) \approx \kappa v/v_{\text{des}}$, and increases linearly with the partition coefficient. The depth profile of the final solute concentration again mimics the square pulsed arrival rate of solute, because any solute in the surface layer that does not incorporate in the bulk desorbs from, rather than "rides" on, the surface.[18] Note that as growth proceeds, there is a competition between desorption and trapping of solute. On the one hand, if the partition coefficient is much less than unity, as in panel (b), then most of the solute atoms in the surface layer eventually desorb, and the absolute concentration of solute in the bulk is low. On the other hand, if the partition coefficient is unity, as in panel (d), then more of the solute atoms in the surface layer incorporate into the bulk, and the absolute concentration of solute in the bulk is higher.

Of the four extremes of behavior just discussed, only one results in a solute composition profile that is broadened beyond the square solute arrival pulse. Unfortunately, that extreme is a commonly observed one, in which appreciable solute segregates to and rides on the surface, rather than either incorporating or desorbing. It may be circumvented to some extent by reduced growth temperature, which reduces solute diffusion from the bulk to the surface.[19]

---

[18] S.S. Iyer, R.A. Metzger, and F.G. Allen, "Sharp profiles with high and low doping levels in silicon grown by molecular beam epitaxy," *J. Appl. Phys.* **52**, 5608 (1981).

[19] H.J. Gossman, E.F. Schubert, D.J. Eaglesham, and M. Cerullo, "Low-temperature Si molecular beam epitaxy: Solution to the doping problem," *Appl. Phys. Lett.* **57**, 2440 (1990).

Suggested Reading

1. A.W. Adamson, *Physical Chemistry of Surfaces*, 4th Ed. (John Wiley and Sons, New York, 1982).

2. J.M. Blakely, *Introduction to the Properties of Crystal Surfaces* (Pergamon Press, Oxford, 1973).

3. R. Defay, I. Prigogine, A. Bellemans and D.H. Everett, *Surface Tension and Adsorption* (John Wiley and Sons, New York, 1966).

Exercises

1. Verify that minimizing Equation 7.14 is equivalent to the parallel tangent construction.

2. Show that, for a given bulk composition, the surface work given by Equations 7.16 and 7.18 is minimum when the surface composition is such that the surface and bulk phases are in equilibrium with each other, i.e., when the parallel tangent construction is satisfied.

3. Show, beginning with Equation 7.16, that

$$\partial\gamma^{\langle \mathrm{Ag}_{1-x^\beta}\mathrm{Au}_{x^\beta}\rangle \mathrm{Ag}_{1-x^\sigma}\mathrm{Au}_{x^\sigma}(}/\partial x^\sigma = \mu_{\mathrm{Au}}^{\mathrm{exc}} - \mu_{\mathrm{Ag}}^{\mathrm{exc}}. \tag{7.52}$$

   Note that both $\mu_{\mathrm{Au}}^{\mathrm{exc}}$ and $\mu_{\mathrm{Ag}}^{\mathrm{exc}}$ depend on x^σ.

4. Derive Equation 7.15, the equilibrium partition coefficient between bulk and surface phases for strictly regular bulk and surface phases.

5. Derive Equation 7.26 for the dependence of the equilibrium coverage of a strictly regular adsorbate phase on pressure.

6. For a given Péclet number, the nonequilibrium partition coefficient, κ, is higher for aperiodic than for periodic passage of steps. Physically, why is this so?

7. In principle, solute segregation and trapping may occur at a number of stages in the growth cycle. Solute may ride ahead of the edges of steps sweeping laterally over terraces by horizontal diffusion following kink flow; they may also ride on the surface by vertical diffusion following step flow. Suppose the horizontal and vertical interdiffusivities at the step edges and at the surface are D_{hor} and D_{ver}, and that the average terrace width is L/a, in units of lattice spacings.

Assuming periodic *partitionless* kink flow followed by horizontal diffusive segregation, what is the nonequilibrium partition coefficient κ_{step} associated with segregation ahead of the moving step? Then, assuming periodic, *non*-partitionless step flow followed by vertical diffusive segregation, what is the nonequilibrium partition coefficient κ_{terr} associated with segregation on top of the growing terraces?

Index